Lecture Notes in Physics

The Editorial Policy for Proceedings

The series Lecture Notes in Physics reports new developments in physical research and teaching – quickly, informally, and at a high level. The proceedings to be considered for publication in this series should be limited to only a few areas of research, and these should be closely related to each other. The contributions should be of a high standard and should avoid lengthy redraftings of papers already published or about to be published elsewhere. As a whole, the proceedings should aim for a balanced presentation of the theme of the conference including a description of the techniques used and enough motivation for a broad readership. It should not be assumed that the published proceedings must reflect the conference in its entirety. (A listing or abstracts of papers presented at the meeting but not included in the proceedings could be added as an appendix.)

When applying for publication in the series Lecture Notes in Physics the volume's editor(s) should submit sufficient material to enable the series editors and their referees to make a fairly accurate evaluation (e.g. a complete list of speakers and titles of papers to be presented and abstracts). If, based on this information, the proceedings are (tentatively) accepted, the volume's editor(s), whose name(s) will appear on the title pages, should select the papers suitable for publication and have them refereed (as for a journal) when appropriate. As a rule discussions will not be accepted. The series editors and Springer-Verlag will normally not interfere with the detailed editing except in fairly obvious cases or on technical matters.

Final acceptance is expressed by the series editor in charge, in consultation with Springer-Verlag only after receiving the complete manuscript. It might help to send a copy of the authors' manuscripts in advance to the editor in charge to discuss possible revisions with him. As a general rule, the series editor will confirm his tentative acceptance if the final manuscript corresponds to the original concept discussed, if the quality of the contribution meets the requirements of the series, and if the final size of the manuscript does not greatly exceed the number of pages originally agreed upon.

The manuscript should be forwarded to Springer-Verlag shortly after the meeting. In cases of extreme delay (more than six months after the conference) the series editors will check once more the timeliness of the papers. Therefore, the volume's editor(s) should establish strict deadlines, or collect the articles during the conference and have them revised on the spot. If a delay is unavoidable, one should encourage the authors to update their contributions if appropriate. The editors of proceedings are strongly advised to inform contributors about these points at an early stage.

The final manuscript should contain a table of contents and an informative introduction accessible also to readers not particularly familiar with the topic of the conference. The contributions should be in English. The volume's editor(s) should check the contributions for the correct use of language. At Springer-Verlag only the prefaces will be checked by a copy-editor for language and style. Grave linguistic or technical shortcomings may lead to the rejection of contributions by the series editors.

A conference report should not exceed a total of 500 pages. Keeping the size within this bound should be achieved by a stricter selection of articles and not by imposing an upper limit to the length of the individual papers.

Editors receive jointly 30 complimentary copies of their book. They are entitled to purchase further copies of their book at a reduced rate. As a rule no reprints of individual contributions can be supplied. No royalty is paid on Lecture Notes in Physics volumes. Commitment to publish is made by letter of interest rather than by signing a formal contract. Springer-Verlag secures the copyright for each volume.

The Production Process

The books are hardbound, and quality paper appropriate to the needs of the authors is used. Publication time is about ten weeks. More than twenty years of experience guarantee authors the best possible service. To reach the goal of rapid publication at a low price the technique of photographic reproduction from a camera-ready manuscript was chosen. This process shifts the main responsibility for the technical quality considerably from the publisher to the authors. We therefore urge all authors and editors of proceedings to observe very carefully the essentials for the preparation of camera-ready manuscripts, which we will supply on request. This applies especially to the quality of figures and halftones submitted for publication. In addition, it might be useful to look at some of the volumes already published.

As a special service, we offer free of charge LATEX and TEX macro packages to format the text according to Springer-Verlag's quality requirements. We strongly recommend that you make use of this offer, since the result will be a book of considerably improved technical quality.

To avoid mistakes and time-consuming correspondence during the production period the conference editors should request special instructions from the publisher well before the beginning of the conference. Manuscripts not meeting the technical standard of the series will have to be returned for improvement.

For further information please contact Springer-Verlag, Physics Editorial Department V, Tiergartenstrasse 17, W-6900 Heidelberg, FRG

H. Mitter H. Gausterer (Eds.)

Recent Aspects of Quantum Fields

Proceedings of the XXX Int. Universitätswochen für Kernphysik, Schladming, Austria
February and March 1991

Springer-Verlag Berlin Heidelberg GmbH

Editors

H. Mitter
H. Gausterer
Institut für Theoretische Physik
Karl-Franzens-Universität
Universitätsplatz 5, A-8010 Graz, Austria

This book was processed by the authors using the $\TeX$ macro package from Springer-Verlag

ISBN 978-3-662-13827-4 ISBN 978-3-540-46628-4 (eBook)
DOI 10.1007/978-3-540-46628-4

Originally published by Springer-Verlag Berlin Heidelberg New York in 1991
Softcover reprint of the hardcover 1st edition 1991

58/3140-543210 - Printed on acid-free paper

Preface

This volume contains the written versions of invited lectures presented at the 30th "Internationale Universitätswochen für Kernphysik" in Schladming, in February and March 1991.

The generous support of our sponsors, the Austrian Ministry of Science and Research, the Government of Styria, and others, made it possible to invite expert lecturers. The topics of the course were centered on aspects of quantum field theory, which are relevant for the recent developments in particle physics. It is a pleasure for us to thank all the speakers for their excellent presentation and their efforts in prepearing the lecture notes.

After the school the lecture notes were revised by the authors and partly rewritten in TeX. We are also indepted to Mrs. Neuhold for careful typing of those notes which we did not receive in TeX. The lectures by J. Ellis and J. Steinberger unfortunately could not be included.

Graz, Austria
August 1991

H. Mitter
H. Gausterer

Contents

Light-Cone Quantization of Quantum Chromodynamics

Conceptual and Geometrical Problems in Quantum Gravity

Two-Dimensional Gravities and Supergravities as Integrable Systems

Recent Developments in Relativistic Thermal Field Theories

Nonexistence of Scattering Theory at Finite Temperature

Contributors

Brodsky, S.J.

Stanford Linear Accelerator Center, Stanford University,
P.O. Box 4349, Stanford, CA 94309 USA

Fritzsch, H.

Max-Planck-Institut für Physik und Astrophysik,
Werner Heisenberg Institut für Physik,
Föhringer Ring 6, W-8000 München 40, Germany

Isham, C.J.

Blackett Laboratory, Imperial College of Science
and Technology, South Kensington, London SW7 2BZ, UK

Le Bellac, M.

Institut Non Linéaire de Nice, Université de Nice Parc Valrose,
Unité Mixte de Recherche 129 du CNRS, F-06034 Nice Cedex, France

Leutwyler, H.

Institute for Theoretical Physics, University of Bern,
Sidlerstrasse 5, CH-3012 Bern, Switzerland

Narnhofer, H.

Institut für Theoretische Physik, Universität Wien,
Boltzmanngasse 5, A-1090 Wien, Austria

Nicolai, H.

II. Institut für Theoretische Physik, Universität Hamburg,
Luruper Chaussee 149, W-2000 Hamburg 50, Germany

Pauli, H.-Ch.

Max-Planck-Institut für Kernphysik, W-6900 Heidelberg 1, Germany

Chiral Effective Lagrangians

H. Leutwyler

Institute for Theoretical Physics
University of Bern
Sidlerstrasse 5, CH-3012 Bern, Switzerland

1 . Generalities

These lectures concern the properties of field theories at low energies and related issues such as the behaviour of the partition function at low temperatures. The low energy structure is particularly simple if the spectrum of the theory contains a mass gap - let us start the discussion with this case, assuming that the mass M of the lightest particle is different from zero. Consider, e.g., the elastic scattering amplitude T of this particle. Assuming, for simplicity, that the particle is spinless, the partial wave expansion takes the form

$$T(s,t) = 32\pi \sum_l (2l+1) P_l(\cos\theta) t_l(q) \tag{1.1}$$

where q and θ are the momentum and the scattering angle in the center of mass system, related to the invariants s, t by

$$s = 4(M^2+q^2); t = -4q^2 \sin^2 \frac{\theta}{2}. \tag{1.2}$$

For small momenta, the partial wave amplitudes t_l can be expanded as

$$Ret_l(q) = q^{2l}\{a_l + b_l q^2 + O(q^4)\}. \tag{1.3}$$

The first two coefficients are referred to as the scattering length a_l and the effective range b_l. If the momenta are small compared to the mass gap M of the theory, the scattering is purely elastic and is dominated by the S-wave, such that $d\sigma/d\Omega \simeq \mid a_0 \mid^2$.

Similarly, the electromagnetic form factor of a spinless particle of unit charge

$$< p' \mid j_\mu \mid p >= (p'_\mu + p_\mu) F(t) \tag{1.4}$$

can be expanded in powers of the momentum transfer $t = (p'-p)^2$,

$$F(t) = 1 + \frac{1}{6} < r^2 > t + O(t^2) \tag{1.5}$$

where $< r^2 >$ is the mean square charge radius.

These examples illustrate the fact that, in the presence of a mass gap, the low energy structure of the theory can be analyzed in terms of an ordinary Taylor series expansion. The expansion parameters can be identified with the ratio of the momenta to the mass gap and the physical significance of the coefficients occurring in the expansion is well known.

The properties of the partition function at low temperatures are determined by the lightest excitations of the system. If the lightest particle carries mass M, the pressure, e.g., is proportional to $\exp(-M/T)$ and therefore becomes exponentially small if the temperature falls below M. Likewise, if the system is enclosed in a box of size L, the finite size effects generated by the walls of the box are of order $\exp(-ML)$ and are invisibly small if the box is large compared to the Compton wavelength $1/M$.

In the absence of a gap - i.e. if the theory contains massless particles - the low energy structure is quite different. The exchange of massless particles then generates poles at $p^2 = 0$ and cuts starting there, such that the Taylor series expansion in powers of the momenta fails. A well-known example is Coulomb scattering, which to lowest order in the electromagnetic interaction is described by the exchange of a photon. The corresponding scattering amplitude is proportional to e^2/t where $t = (p' - p)^2$ is the momentum transfer between the two charged particles. Clearly, the scattering is not dominated by the S-wave here, even if the momenta are very small. In the presence of massless particles, the thermal properties at low T are also quite different; the pressure then only tends to zero with a power of the temperature. In the case of the finite size effects generated by the walls of a box, the qualitative difference between theories which contain a mass gap and theories which don't is even more striking: for massless theories, some of the finite size effects even persist if the size of the box is sent to infinity.

The absence of a gap is the exception rather than the rule. For an analysis of the low energy structure of theories without mass gap, it is essential to understand why these theories contain massless particles in the first place. Three different mechanisms which can insure the absence of a gap are well-known:

(i) Local gauge invariance

$$A_\mu \rightarrow A_\mu + D_\mu \alpha$$

can protect spin 1 particles from picking up mass. The mass of the photon vanishes for this reason.

(ii) Global chiral symmetry

$$\psi \rightarrow e^{i\alpha\gamma_5}\psi$$

can protect fermions from becoming massive (possibly, neutrinos are massless for this reason).

(iii) Spontaneously broken symmetries give rise to massless particles - Goldstone bosons (phonons, spin waves, pions,...).

In these lectures, I focus on (iii), i.e. discuss the low energy structure of theories which contain Goldstone bosons.

2. Goldstone Bosons

I first give a rudimentary version of the Goldstone theorem [1]. Suppose the Hamiltonian of the model is invariant under some Lie group G and denote the generators of this group by Q_i, such that

$$[Q_i, H] = 0. \tag{2.1}$$

The symmetry is called spontaneously broken if the ground state of the theory is not invariant under G. Suppose, therefore, that, for some of the generators

$$Q_i \mid 0 > \neq 0. \tag{2.2}$$

This immediately implies that the vacuum is not the only state of zero energy: since H commutes with Q_i, the vector $Q_i \mid 0 >$ describes a state with the same energy as the vacuum. The subset formed by those generators which do leave the ground state invariant is a subalgebra: if Q_i and Q_k annihilate the vacuum, then this is also true of the commutator $[Q_i, Q_k]$. These operators therefore generate a subgroup $H \subset G$. Spontaneous symmetry breakdown thus involves two groups - the symmetry group G of the Hamiltonian and the symmetry group H of the vacuum. Denote the number of parameters required to label the elements of G by n_G such that there are n_G generators and suppose that $n_H < n_G$ is the number of parameters occurring in H. The $n_G - n_H$ generators which belong to the quotient G/H of the two groups do not annihilate the ground state. The corresponding vectors $Q_i \mid 0 >$ are linearly independent, because, otherwise, a suitable linear combination of these generators would leave the vacuum invariant and hence belong to H. Accordingly, spontaneous breakdown of the group G to the subgroup H requires the occurrence of $n_G - n_H$ independent states of zero energy - the spectrum of the theory contains $n_G - n_H$ different flavours of Goldstone bosons.

This "proof" of the Goldstone theorem requires mathematical massage, because the states $Q_i \mid 0 >$ occurring in the above argument have infinite norm (the vacuum is the only normalizable state invariant under translations). You can find a proper discussion of the theorem in the Coleman lectures on "Secret Symmetry" [2], where you can also see what happens if G is not an ordinary symmetry of the Hamiltonian, but a local gauge group. I do not discuss these phenomena here, but consider the spontaneous breakdown of ordinary symmetries where the Goldstone theorem applies and where the Goldstone bosons do occur as asymptotic states. For the following, it is essential that the spectrum does not contain other massless asymptotic states. In the case of QCD, it is crucial that the gauge group (colour) commutes with the symmetry group

G (flavour) and that confinement of colour prevents gluons and quarks from showing up as massless physical particles.

3. Linear σ-Model

As an illustration, consider the linear σ-model which involves n scalar fields $\phi = (\phi^0, \ldots, \phi^{n-1})$, coupled through a $\lambda\phi^4$ interaction,

$$\mathcal{L} = \frac{1}{2}\partial_\mu \phi \partial^\mu \phi - \frac{\lambda}{4}(\phi^2 - v^2)^2 \tag{3.1}$$

where v is a constant of dimension $(\text{mass})^{\frac{1}{2}(d-2)}$. Since this Lagrangian is invariant under rotations of the vector ϕ, the symmetry group of the model is $G = O(n)$. With $n = 4$, the model describes the Higgs sector of the Standard Model.

In four dimensions, the field theory specified in eq. (3.1) presumably describes a free field, the renormalized coupling constant being equal to zero. This indicates that the Higgs model only represents an effective theory, valid at low energies and that the basic degrees of freedom associated with mass generation yet need to be discovered. In our context, the problems afflicting $\lambda\phi^4$ in $d = 4$ are however not of central interest. Although the model only makes sense in the presence of an ultraviolet cutoff, the corresponding violation of locality does not manifest itself at large distances. The low energy expansion only concerns momenta which are small compared to the mass m_H of the Higgs particle and is not sensitive to the structure of the theory at distances which are small compared to $1/m_H$. Incidentally, the general low energy analysis also applies in three dimensions, where the Lagrangian (3.1) does specify a nontrivial local field theory.

The ground state of the model is characterized by a nonzero expectation value of the field. With a suitable choice of the field basis, $< 0 \mid \phi \mid 0 >$ points along the first axis such that the perturbative expansion starts with

$$< 0 \mid \phi^0 \mid 0 >= v\{1 + O(\lambda)\}$$

$$< 0 \mid \phi^i \mid 0 >= 0 i = 1, \ldots, n-1. \tag{3.2}$$

Visibly, the ground state is not symmetric under G, but breaks the symmetry down to the little group of the vector $< 0 \mid \phi \mid 0 >$, i.e. to $H = O(n-1)$. The number of parameters needed to label the elements of $O(n)$ is $\frac{1}{2}n(n-1)$. The spectrum of the theory must therefore contain $n_G - n_H = n - 1$ Goldstone bosons.

In the present case, the occurrence of massless modes can directly be confirmed within perturbation theory. Setting $\phi =< 0 \mid \phi \mid 0 > + \varphi$ and expanding the Lagrangian in powers of φ to order φ^2, one indeed finds that only one of the n fields, viz. φ^0, receives a mass term, $M^2 = 2\lambda v^2$, while the remaining $n-1$ fields are massless.

4. Quantum Chromodynamics

As a second example, consider QCD

$$\mathcal{L} = -\frac{1}{4g^2} G^a_{\mu\nu} G^{\mu\nu a} + \bar{q} i \gamma^\mu D_\mu q - \bar{q} \mathcal{M} q \tag{4.1}$$

where $\mathcal{M}$ is the quark mass matrix,

$$\mathcal{M} = \begin{pmatrix} m_u & & & \\ & m_d & & \\ & & m_s & \\ & & & \ddots \end{pmatrix} \tag{4.2}$$

Since the quark masses are different from one another, the symmetry group of this theory is $G = U(1) \times U(1) \times \ldots$. There is one conserved charge for every one of the quark flavours; the corresponding conserved currents are $\bar{u}\gamma^\mu u$, $\bar{d}\gamma^\mu d, \ldots$

If the quark masses are set equal to zero, the symmetry group of the theory becomes larger. The Lagrangian does then not contain terms which connect the left- and right-handed components of the quark fields and remains invariant under a set of chiral transformations of the type (ii) which are characteristic for massless fermions (see section 1). For N massless quark flavours, the Lagrangian is invariant under independent rotations of the left- and right-handed quark fields,

$$q_L \to V_L q_L \qquad q_R \to V_R q_R \tag{4.3}$$

where q_L, e.g., stands for $\frac{1}{2}(1-\gamma_5)q$ and V_L, V_R are unitary $N \times N$-matrices. One of the $2N^2$ currents whose charges generate these transformations is however anomalous: despite the symmetry of the Lagrangian, the Noether current $\bar{q}\gamma_\mu\gamma_5 q$ fails to be conserved, the divergence being proportional to the winding number density of the gluon field. The actual symmetry group of massless QCD consists of those pairs of elements $V_L, V_R \in U(N)$ which obey the constraint $\det(V_L V_R^{-1}) = 1$, i.e.

$$G = SU(N)_L \times SU(N)_R \times U(1)_{L+R}. \tag{4.4}$$

One generally assumes that (if N is not too large) this symmetry is spontaneously broken, the vacuum being invariant only under the subgroup generated by the vector currents $\bar{q}\gamma_\mu \frac{1}{2}\lambda_i q$,

$$H = SU(N)_{R+L} \times U(1)_{R+L}. \tag{4.5}$$

Accordingly, the spectrum of massless QCD must contain $N^2 - 1$ Goldstone bosons where N is the number of quark flavours. Their quantum numbers can be read off from those of the states $Q_i^A \mid 0>$ which result if the axial charge operators

$$Q_i^A = \int d^3x\, A_i^0(x); \qquad A_i^\mu = \bar{q}\gamma^\mu\gamma_5 \frac{1}{2}\lambda_i q \tag{4.6}$$

are applied to the vacuum: $J^P = 0^-$.

The symmetry group in eq. (4.4) pertains to the theoretical limit where the quark masses are set equal to zero. In reality, the symmetry is broken by the quark mass term occurring in the Lagrangian. Indeed, the observed spectrum of particles does not contain massless hadrons. Remarkably, however, the eight lightest hadrons ($\pi^\pm, \pi^0, K^\pm, K^0, \bar{K}^0, \eta$) are pseudoscalars and three of them are particularly light. This pattern is to be expected if

(i) m_u, m_d, m_s happen to be small, such that the Lagrangian is approximately invariant under $SU(3)_L \times SU(3)_R$. The spontaneous breakdown of this approximate symmetry to $SU(3)_{L+R}$ generates $3^2 - 1 = 8$ approximately massless particles.

(ii) m_u, m_d are small compared to m_s, such that the group $SU(2)_L \times SU(2)_R$ is an almost exact symmetry of the Lagrangian. This is why $2^2 - 1 = 3$ of the pseudoscalar mesons are particularly light.

We will discuss the manner in which the masses of the Goldstone bosons are affected by the symmetry breaking in detail below and show that the observed masses of the eight mesons listed above require $m_u : m_d : m_s \simeq 0.55 : 1 : 20$.

The masses of the other quarks occurring in the Standard Model, c, b, t, are not small. The lightest pseudoscalar bound state with the quantum numbers of $\bar{d}c$, e.g. sits at $M_{D^+} \simeq 1.87$ GeV. Even though there is a theoretical limit ($m_d = m_c = 0$) where this state is massless and plays the role of a Goldstone boson, this limit is not very useful, because the mass term $m_c\bar{c}c$ cannot be treated as a small perturbation. In the context of QCD, applications of chiral perturbation theory either concern the group $SU(2)_L \times SU(2)_R$ broken spontaneously to isospin symmetry or the group $SU(3)_L \times SU(3)_R$, where the symmetry of the vacuum is the eightfold way. In either case, the Lagrangian can be decomposed into two pieces $\mathcal{L}_0 + \mathcal{L}_1$ where $\mathcal{L}_0$ is the symmetric part while $\mathcal{L}_1$ contains the symmetry breaking terms. If $\mathcal{L}_1$ is identified with $-(m_u\bar{u}u + m_d\bar{d}d)$, then $\mathcal{L}_0$ is symmetric under $SU(2)_L \times SU(2)_R$ and the perturbative expansion in powers of $\mathcal{L}_1$ amounts to a power series in m_u and m_d. If the mass term of the strange quarks is included in $\mathcal{L}_1$, the symmetry group of $\mathcal{L}_0$ is increased and now contains the factor $SU(3)_L \times SU(3)_R$. The price to pay is that the expansion in powers of $\mathcal{L}_1$ now gives rise to powers of m_s - the chiral perturbation series converges less rapidly.

5. Effective Field Theory

In the context of the strong interactions, the main consequences of the hidden approximate symmetry were derived in the sixties from a direct analysis of the Ward identities, using current algebra and pion pole dominance. In the meantime, an alternative method has been developed which deals with the problem in a more systematic manner and which is considerably more efficient. I now turn to a discussion of this method [3] [4] [5].

For definiteness, consider QCD with two massless flavours, where three Goldstone bosons occur. In the Green functions of the theory, massless one-particle states manifest themselves as poles at $p^2 = 0$. The two-point function of the axial current, e.g., contains a pole term due to the exchange of a pion between the two currents. The residue of the pole is given by the square of the matrix element

$$< 0 \mid A_i^\mu \mid \pi^k(p) >= ip^\mu \delta_{ik} F. \tag{5.1}$$

The pole occurring in the two-point function entails corresponding singularities in other Green functions. The Ward identities obeyed by the three-point function with two axial and one vector current, e.g., require that this quantity contains a simultaneous pole in the two axial legs. Picturing the pole factors as pions which propagate in space, the leading low energy contributions to the two- and three-point functions are shown in Figs. 1a, b.

The Ward identities for the four-point function of the axial current relate this object to the three-point function just mentioned. They admit a solution only if the four-point function contains a simultaneous pole in all four legs as indicated in Fig. 1c. When analyzing these amplitudes in the framework of current algebra, one assumes that, at leading order in the low energy expansion,

(i) the only singularities which occur are poles originating in the progagation of pions and

(ii) the residues can be expanded in powers of the momenta.

Truncating the expansion at the first nontrivial order, the residues reduce to polynomials in the momenta. The coefficients occurring in these polynomials can then be worked out by explicitly solving the Ward identities.

The effective Lagrangian technique is based on the following heuristic considerations. The graphs shown in Fig. 1 may be viewed as tree graphs of a field theory which involves pion fields as basic variables. Vertices which exclusively emit pion lines - like the central vertex in Fig. 1c - represent interaction terms of the corresponding Lagrangian. If these vertices involve polynomials of the momenta, the corresponding interaction term contains derivatives of the pion field. The pion propagator originates in a kinetic term of the form $\frac{1}{2}\partial_\mu \pi \partial^\mu \pi$. The full Lagrangian is obtained by adding all interaction terms, involving arbitrarily many pion fields.

The vertex shown in Fig. 1a links the pion field to the axial current. In the language of the pion field theory, the occurrence of this vertex implies that the axial current contains a term linear in the pion field,

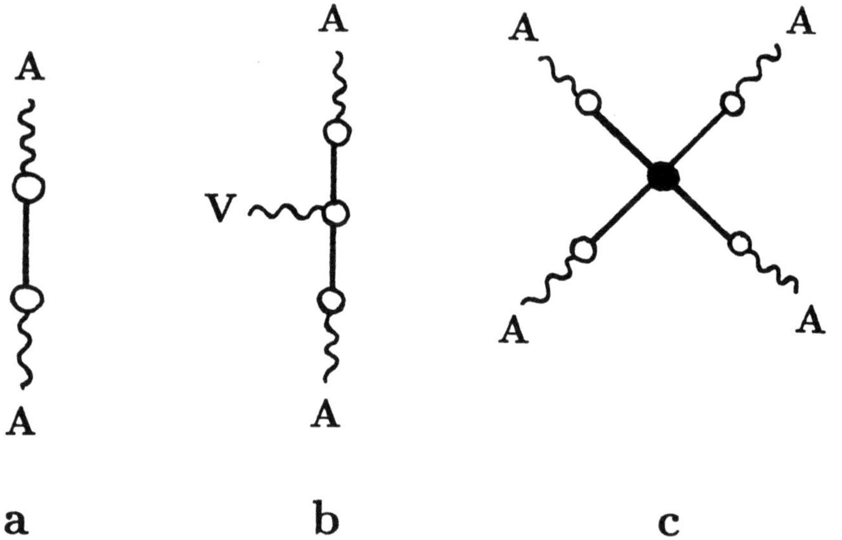

Fig. 1. Pion pole term occurring in the 2-, 3- and 4-point functions of the vector and axial currents

$$A_\mu^i = -F\partial_\mu \pi^i + \dots \tag{5.2}$$

while a term of the form $\partial_\mu \pi\pi\pi$, e.g., corresponds to a vertex where the axial current emits three pion lines. The Ward identities require the vector and axial currents to be conserved. According to the Noether theorem, the pion field theory does give rise to conserved vector and axial currents if the corresponding action is invariant under $G = SU(2) \times SU(2)$. This indicates that the solutions of the Ward identities which obey the conditions (i) and (ii) are in one-to-one correspondence with the tree graphs of a pion field theory for which

(a) the Lagrangian admits an expansion in powers of the derivatives of the field,

(b) the pion field transforms according to a nontrivial representation of the group G,

(c) the action is invariant under these transformations.

The above arguments are plausibility considerations. To my knowledge, a rigorous proof of the claim that the requirements (i) and (ii) imply the existence of an effective field theory with the properties (a)-(c) is still lacking. If you are looking for a decent problem in mathematical physics: this is one. Note also that (i) and (ii) are assumptions, within both the current algebra approach and the effective Lagrangian technique. When you are through with a proof of the above proposition, could you please also demonstrate that the Goldstone bosons necessarily dominate the low energy behaviour of the Green functions in this specific sense? In the following, I assume that they do, take it for granted that the low energy structure can be worked out by means of an effective Lagrangian and proceed by analyzing its specific form. As a first step in this direction, we need to look more closely at the transformation properties of the pion field.

6. Transformation Properties of the Goldstone Bosons

The pion field transforms according to a representation of the group G, i.e. every $g \in G$ induces a mapping of the form $\boldsymbol{\pi}' = \boldsymbol{f}(g, \boldsymbol{\pi})$. For this mapping to be a representation, the function $\boldsymbol{f}$ must obey the composition law

$$\boldsymbol{f}(g_1, \boldsymbol{f}(g_2, \boldsymbol{\pi})) = \boldsymbol{f}(g_1 g_2, \boldsymbol{\pi}). \tag{6.1}$$

Remarkably, this property determines the function $\boldsymbol{f}(g, \boldsymbol{\pi})$ essentially uniquely [6]. To verify this claim, we first consider the image of the origin, $\boldsymbol{f}(g, 0)$. The composition law shows that the set of elements h which map the origin onto itself forms a subgroup $H \subset G$. Moreover, $\boldsymbol{f}(gh, 0)$ coincides with $\boldsymbol{f}(g, 0)$ for any $g \in G$, $h \in H$. Hence the function $\boldsymbol{f}(g, 0)$ lives on the space G/H which is obtained from G by identifying elements g, g' which only differ by right multiplication with a member of H, $g' = gh$. The function $\boldsymbol{f}(g, 0)$ thus maps the elements of G/H into the space of pion field variables. The mapping is invertible, because $\boldsymbol{f}(g_1, 0) = \boldsymbol{f}(g_2, 0)$ implies $g_1^{-1} g_2 \in H$. One can therefore identify the values of the pion field with the elements of G/H: the Goldstone bosons live in this space. Next, choose a representative element n in each one of the equivalence classes $\{gh, h \in H\}$, such that every group element can be decomposed as $g = nh$. The composition law (6.1) then shows that the image n' of the element n under the action of $g \in G$ is obtained by decomposing the product gn into $n'h$ – the standard action of G on the space G/H. This shows that the geometry fully fixes the transformation law of the pion field, except for the freedom in the choice of coordinates on the manifold G/H.

In the case of $G = SU(2) \times SU(2)$ and $H = SU(2)$, the quotient G/H is the group SU(2). The pion field may therefore be represented as a 2×2 matrix field $U(x) \in SU(2)$. Since one needs three coordinates to parametrize the elements of SU(2), the matrix field $U(x)$ is equivalent to a set of 3 ordinary fields $\pi^1(x), \pi^2(x), \pi^3(x)$. One may use, e.g., canonical coordinates,

$$U(x) = \exp i\alpha\pi(x); \quad \pi(x) = \sum_{i=1}^{3} \pi^i(x)\tau_i \tag{6.2}$$

where τ_1, τ_2, τ_3 are the Pauli matrices. The constant α carries dimension $[\text{mass}]^{-1}$; we will fix it later on. I recommend it as an exercise to verify that, in the present case, the action of G on the pion field is given by

$$U'(x) = V_R U(x) V_L^+. \tag{6.3}$$

The matrix $U(x)$ thus transforms linearly. Note, however, that the corresponding transformation law for the pion field $\pi(x)$ is nonlinear. As indicated by the above general discussion, the occurrence of nonlinear realizations of the symmetry group is a characteristic feature of the effective Lagrangian technique.

7. Form of the Effective Lagrangian

I now turn to the requirements (a) and (c) of section 5. Lorentz invariance implies that the leading terms in the expansion of the effective Lagrangian in powers of derivatives are of the form

$$\mathcal{L}_{eff} = f_1(U) + f_2(U) \times U + f_3(U) \times \partial_\mu U \times \partial^\mu U + O(p^4)$$

where the symbol $O(p^4)$ indicates that the remainder contains four or more derivatives of the pion field. The crosses refer to the fact that the coefficients $f_2(U)$ and $f_3(U)$ carry indices which are contracted against those of the matrices U and $\partial_\mu U$. The first term does not contain derivatives. The corresponding action is invariant under $U \to V_R U V_L^+$ if and only if $f_1(U)$ is independent of U. Hence the first term is an irrelevant cosmological constant and can be dropped. Integrating by parts, the second term can be transformed into the third one, so that we can drop f_2, too. Without loss of generality we can then write the Lagrangian in the form $\tilde{f}_3(U) \times \Delta_\mu \times \Delta^\mu$ where Δ_μ stands for $-iU^{-1}\partial_\mu U$. The advantage of this manipulation is that Δ_μ is invariant under $U \to V_R U$, such that only $\tilde{f}_3(U)$ is affected by this operation. The requirement that the action must remain invariant therefore implies that $\tilde{f}_3(U)$ is independent of U. Finally, under the transformation $U \to U V_L^+$, the traceless quantity Δ_μ transforms according to the representation $D^{(1)}$. Since the product $D^{(1)} \times D^{(1)}$ contains the identity only once, there is a single invariant of order p^2,

$$\mathcal{L}_{eff} = g tr \Delta_\mu \Delta^\mu = g\, tr\, (\partial_\mu U \partial^\mu U^+). \tag{7.1}$$

This shows that the leading term in the derivative expansion of the effective Lagrangian contains only one free coupling constant, g.

Expanding the matrix $U(x) = \exp i\alpha\pi(x)$ in powers of the pion field, we obtain

$$\mathcal{L}_{eff} = 2g\alpha^2 \partial_\mu \pi \partial^\mu \pi + \frac{g\alpha^4}{12}\, tr\, ([\partial_\mu \pi, \pi][\partial^\mu \pi, \pi]) + \dots$$

where interactions involving six or more pion fields are omitted. In order for the first term to agree with the standard normalization of the kinetic energy, we fix the parameter α introduced in eq. (6.2) in terms of the coupling constant g

by setting $\alpha^2 = 1/4g$. The Noether currents associated with the SU(2)×SU(2) symmetry of the Lagrangian (7.1) are

$$V_i^\mu = ig\, tr\,(\tau_i[\partial^\mu U, U^+])$$
$$A_i^\mu = ig\, tr\,(\tau_i\{\partial^\mu U, U^+\}). \tag{7.2}$$

Comparing the expression for the axial current with eq. (5.1) or (5.2), we see that the coupling constant g is related to the pion decay constant F by $g = F^2/4$. At leading order in the derivative expansion, the effective Lagrangian therefore only involves the pion decay constant,

$$\mathcal{L}_{eff} = \frac{F^2}{4} tr\,(\partial_\mu U \partial^\mu U^+)$$
$$U(x) = \exp\left\{\frac{i}{F}\sum_{k=1}^{3}\pi^k(x)\tau_k\right\}. \tag{7.3}$$

The field theory characterized by this Lagrangian is referred to as the nonlinear σ-model.

It is well-known that for $d > 2$ this model is not renormalizable - taken by itself, it is not a decent theory. Are we running after a cloud of dust here? Actually, in the above analysis only the tree graphs of the effective Lagrangian played a role. Renormalizability is not an issue which concerns the tree graphs. We will have occasion to discuss the significance of loop graphs later on, when we consider the low energy expansion beyond leading order. Clearly, the effective Lagrangian must then also be worked out beyond the leading term in the derivative expansion. In the framework of the effective Lagrangian, the nonlinear σ-model only represents one building block of the construction - it does not occur by itself. As we will see, the effective theory as a whole is a perfectly renormalizable scheme.

8. Universality

In the preceding section, we arrived at the conclusion that the leading term in the derivative expansion of the effective Lagrangian associated with the low energy behaviour of QCD is an explicitly known expression which only involves the pion decay constant. This result is remarkable in more than one respect. First, it rigorously establishes that all of the Ward identities for all of the vector and axial current Green functions admit a solution which obeys the two requirements (i) and (ii) formulated in section 5. The solution is given by the tree graphs of the nonlinear σ-model. If the plausibility arguments underlying the effective Lagrangian technique are taken for granted, the uniqueness of the effective Lagrangian also implies that, at leading order in the low energy expansion, the solution of the Ward identities is unique and only contains the pion decay constant. Since the analysis applies to any theory for which $SU(2) \times SU(2)$ is spontaneously broken to SU(2), the specific properties of the underlying theory are irrelevant - the low energy structure is universal.

The extension to three massless flavours is straightforward. In this case, the Goldstone bosons live in $G/H = SU(3)$. Accordingly, there are eight pion fields which may be identified with the canonical coordinates on SU(3),

$$U(x) = \exp \frac{i}{F} \sum_{k=1}^{8} \pi^k(x) \lambda_k \tag{8.1}$$

where $\lambda_1, \ldots, \lambda_8$ are the Gell-Mann matrices. There is again only one invariant at order p^2, the one given in eq. (7.3).

In the case of $G = O(n)$, $H = O(n-1)$, the quotient G/H is the $(n-1)$-dimensional sphere. Accordingly, the pion field is a unit vector $U(x)$ with n components. The derivative expansion of the effective Lagrangian again starts with a term of order p^2 and involves a single coupling constant,

$$\mathcal{L}_{eff} = \frac{1}{2} F^2 \partial_\mu U \partial^\mu U. \tag{8.2}$$

In particular, if n is equal to four, the pion field lives in the three-dimensional sphere. This manifold can be mapped one-to-one onto the group SU(2), setting

$$U = U^0 + i \sum_{k=1}^{3} U^k \tau_k. \tag{8.3}$$

One readily checks that the Lagrangians (7.3) and (8.2) are identical. This is not a great surprise, because the groups $G = O(4)$ and $H = O(3)$ occurring in the spontaneous breakdown of the Higgs model are locally isomorphic to $SU(2) \times SU(2)$ and to SU(2), respectively. The equivalence of the two effective theories implies that, at low energies, the Green functions of the Higgs model and of QCD with two massless flavours are the same, except for the magnitude of the pion decay constant. For QCD, $F \simeq 93$ MeV, while in the case of the Higgs model, $F \simeq 245$ GeV.

9. Geometry

Quite generally, the leading term in the low energy expansion of the effective Lagrangian is determined by the geometry of the manifold G/H [7[[5]. Denote the coordinates on this manifold by $\omega^1, \ldots, \omega^r$ such that the Goldstone bosons are described by the fields $\omega^1(x), \ldots, \omega^r(x)$. Under the action of G, the coordinates transform in a nonlinear manner. There always exists a metric $g_{ab}(\omega)$ on G/H such that the line element

$$ds^2 = \sum_{a,b} g_{ab}(\omega) d\omega^a d\omega^b \tag{9.1}$$

is invariant under the action of G. In the examples considered above, this requirement fixes the metric uniquely up to a normalization constant. This is not always the case, however. In general, there are several independent quadratic forms on G/H which are invariant under G and the general G-invariant metric is given by a linear combination of these. Denote the general metric by $g_{ab}(\omega)$. The leading term in the derivative expansion of the effective Lagrangian is given by

$$\mathcal{L}_{eff} = \sum_{a,b} g_{ab}(\omega) \partial_\mu \omega^a \partial^\mu \omega^b. \tag{9.2}$$

The coefficients of the linear combination referred to above are the coupling constants of this Lagrangian.

An example where the effective Lagrangian involves more than one coupling constant at order p^2 is the following. Take $G = SU(N)$ and identify H with the trivial subgroup $H = \{e\}$. The quotient G/H then coincides with G and the action of the symmetry group on the pion field $U(x) \in SU(N)$ is of the form $U(x) \to VU(x)$. It is easy to see that in this case, there are several independent G-invariant quadratic forms. Consider two neighbouring elements U, $U + dU$ of G/H. The product $U^{-1}dU$ is invariant under G. Hence the linear forms Ω^a which occur if this product is expanded in the Gell-Mann basis

$$U^{-1}dU = i \sum_{a=1}^{N^2-1} \Omega^a \lambda_a. \tag{9.3}$$

are invariant under G. They can be used to construct the general G-invariant quadratic form as

$$ds^2 = \sum_{a,b} c_{ab} \Omega^a \Omega^b \tag{9.4}$$

where c_{ab} are arbitrary constants. In this case, the leading term in the derivative expansion of the effective Lagrangian thus involves $\frac{1}{2}N^2(N^2-1)$ coupling constants. Physically, this embarras de richesse reflects the fact that if the vacuum breaks the symmetry totally, then there is no symmetry left over which would require the pion matrix elements of the currents associated with G to be the same. Accordingly, the pole terms in the two-point functions of these currents need not have the same residue - several low energy constants are needed to specify the leading terms in the low energy expansion of the two-point functions. The Ward identities do not shed any light on the values of these constants - they merely require that Green functions involving more than two currents contain singularities which match those of the two-point functions.

10. Symmetry Breaking

In the preceding sections, we have dropped the masses of the u- and d-quarks. In their presence, the Lagrangian of the theory is not invariant under $SU(2) \times SU(2)$, because the mass term

$$\mathcal{L} = \mathcal{L}_0 - \bar{q}\mathcal{M}q \tag{10.1}$$

connects the right- and left-handed components of the quark fields,

$$\bar{q}\mathcal{M}q = \bar{q}_R \mathcal{M} q_L + h.c.. \tag{10.2}$$

Note that the mass terms associated with s, c, b, t are included in $\mathcal{L}_0$; the quark field q only contains u and d and $\mathcal{M}$ is the matrix

$$\mathcal{M} = \begin{pmatrix} m_u & 0 \\ 0 & m_d \end{pmatrix}. \tag{10.3}$$

It is instructive to compare the QCD Lagrangian with the Hamiltonian of a Heisenberg ferromagnet,

$$H = H_0 - \sum_a \mu \boldsymbol{s}_a \cdot \boldsymbol{H}. \tag{10.4}$$

Here, $\boldsymbol{s}_a$ is the spin associated with lattice site a, μ is the magnetic moment and $\boldsymbol{H}$ is an external magnetic field. The term H_0 is invariant under simultaneous $O(3)$-rotations of all spin variables, while the term which involves the external magnetic field breaks this symmetry. Clearly, the quark masses (m_u, m_d) play a role analogous to the external magnetic field and the quark condensate $< 0 \mid \bar{u}u \mid 0 >, < 0 \mid \bar{d}d \mid 0 >$ is analogous to the magnetization. In particular, spontaneous magnetization at zero external field corresponds to a nonzero value of the quark condensate in the chiral limit $m_u, m_d \to 0$.

In the case of the magnet, the symmetry breaking term transforms according to the spin 1 representation of $O(3)$. The decomposition of the quark mass term given in eq. (10.2) shows that this term transforms according to the representation $D^{\frac{1}{2},\frac{1}{2}}$ of $SU(2)_L \times SU(2)_R$. Equivalently, we may say that the QCD Lagrangian is invariant under the transformation (4.3) of the quark fields, provided the mass matrix is transformed accordingly,

$$\mathcal{M} \to V_R \mathcal{M} V_L^+. \tag{10.5}$$

The occurrence of a mass term of course modifies the form of the effective Lagrangian,

$$\mathcal{L}_{eff} = \mathcal{L}_{eff}(U, \partial U, \partial^2 U, \ldots, \mathcal{M}) \tag{10.6}$$

which now remains invariant under the transformation $U(x) \to V_R U(x) V_L^+$ of the pion field only if one simultaneously also transforms the quark mass matrix in the same manner. The modification of the Lagrangian generated by the quark masses can be analyzed by expanding in powers of $\mathcal{M}$. The first term in this expansion is the effective Lagrangian of the massless theory which we have considered in the preceding sections. The term linear in $\mathcal{M}$ is of the form

$$\mathcal{L}_{sb} = f(U, \partial U, \ldots) \times \mathcal{M}. \tag{10.6}$$

Next, we observe that derivatives of the pion field are suppressed by powers of the momenta. At leading order in an expansion in both, powers of $\mathcal{M}$ and powers of derivatives, the symmetry breaking term in the effective Lagrangian reduces to an expression of the form $f(U) \times \mathcal{M}$. Moreover, this expression must be invariant under simultaneous chiral transformations of the matrices U and $\mathcal{M}$. There are only two independent invariants: $tr(\mathcal{M}U^+)$ and its complex conjugate. Hence the leading symmetry breaking contribution is of the form

$$\mathcal{L}_{sb} = \frac{F^2}{2}\{B\, tr(\mathcal{M}U^+) + B^*\, tr(\mathcal{M}^+U)\} \tag{10.7}$$

where I have extracted a factor F^2 for later convenience. The symmetry breaking involves a new low energy constant, B, which need not be real. Since only the product $B\mathcal{M}$ matters, the phase of B occurs together with the phase of the quark mass matrix and is related to the possible occurrence of a parity violating term of the form $\Theta\, G_{\mu\nu}\tilde{G}^{\mu\nu}$ in the Lagrangian of QCD. The fact that the neutron dipole moment is very small implies that the strong interactions conserve parity to a very high degree of occuracy. Let us therefore require that the effective Lagrangian is parity invariant. Using the standard basis where the quark mass matrix is diagonal and real, this requirement implies that B is real (the parity operation sends π into $-\pi$ and hence interchanges U with U^+),

$$\mathcal{L}_{sb} = \frac{F^2 B}{2} tr \mathcal{M}(U + U^+). \tag{10.8}$$

Since U is an element of SU(2), the sum $U + U^+$ is proportional to the unit matrix. Accordingly, the leading contribution to the symmetry breaking part of the effective Lagrangian only involves the sum $m_u + m_d$ of the two quark masses and therefore conserves isospin - the breaking of isospin symmetry generated by the mass difference $m_u - m_d$ only shows up if the low energy expansion is carried beyond leading order.

Expanding $U = \exp(i\boldsymbol{\pi} \cdot \boldsymbol{\tau}/F)$ in powers of the pion field $\boldsymbol{\pi}$, the Lagrangian (10.8) gives rise to the following contributions:

$$\mathcal{L}_{sb} = (m_u + m_d)B\left\{F^2 - \frac{1}{2}\boldsymbol{\pi}^2 + \frac{\boldsymbol{\pi}^4}{24F^2} + \ldots\right\}. \tag{10.9}$$

Up to a sign, the first term represents the vacuum energy generated by the symmetry breaking. The second is quadratic in the pion field and therefore amounts to a pion mass term. The remaining contributions show that the symmetry breaking necessarily also modifies the interaction among the Goldstone bosons.

The derivative of the QCD Hamiltonian with respect to m_u is the operator $\bar{u}u$. The corresponding derivative of the vacuum energy therefore represents the vacuum expectation value of $\bar{u}u$. Evaluating this derivative with the first term in eq. (10.9), we obtain

$$< 0 \mid \bar{u}u \mid 0 >=< 0 \mid \bar{d}d \mid 0 >= -F^2 B\{1 + O(\mathcal{M})\}. \tag{10.10}$$

The low energy constant B is therefore related to the value of the quark condensate. In analyzing the form of the effective Lagrangian, we have retained only terms linear in the quark masses. The curly bracket in eq. (10.10) indicates that, in this relation, the higher terms generate contributions of order $\mathcal{M}$.

According to eq. (10.9), the pion mass is given by

$$M_\pi^2 = (m_u + m_d)B\{1 + O(\mathcal{M})\}. \tag{10.11}$$

If the quark masses are set equal to zero, the pion mass vanishes, as it should - $SU(2) \times SU(2)$ is then an exact symmetry and the Goldstone bosons are strictly massless. As long as the symmetry breaking is small, the Goldstone bosons only pick up a small mass which is proportional to the quare root of the symmetry breaking parameter $m_u + m_d$. In accord with the remarks made above, isospin breaking does not manifest itself at this order of the expansion - the masses of π^+, π^0 and π^- are the same.

Eliminating the low energy constant B, the relations (10.10) and (10.11) lead to the well-known result of Gell-Mann, Oakes and Renner [8],

$$F^2 M_\pi^2 = -(m_u + m_d) < 0 \mid \bar{u}u \mid 0 > +O(\mathcal{M}^2). \tag{10.12}$$

The extension to N quark flavours is straightforward - the above analysis goes through without any essential modifications and leads to an effective Lagrangian of the same form,

$$\mathcal{L}_{eff} = \frac{F^2}{4} tr\,\{\partial_\mu U \partial^\mu U^+ + 2B\mathcal{M}(U + U^+)\}. \tag{10.13}$$

The field $U(x)$ now is an element of SU(N) and describes $N^2 - 1$ Goldstone bosons; $\mathcal{M}$ is the diagonal matrix formed with the N quark masses $m_u, m_d, m_s, \ldots$.

I recommend it as an exercise to work out the kinetic part of the Lagrangian (10.13) for the case of three flavours and to show that the meson masses are given by

$$\begin{aligned} M_{\pi^+}^2 &= (m_u + m_d)B + O(\mathcal{M}^2) + O(e^2) \\ M_{K^+}^2 &= (m_u + m_s)B + O(\mathcal{M}^2) + O(e^2) \\ M_{K^0}^2 &= (m_d + m_s)B + O(\mathcal{M}^2) + O(e^2) \end{aligned} \tag{10.14}$$

The states π^0 and η mix with an angle of order $(m_u - m_d)/(m_s - \hat{m}) \ll 1$, such that the two levels repel. Show that this generates a mass difference between π^0 and π^+, given by

$$M_{\pi^0}^2 \simeq M_{\pi^+}^2 - \frac{1}{4}\left(\frac{m_u - m_d}{m_s - \hat{m}}\right)^2 (M_K^2 - M_\pi^2)$$

$$\hat{m} \equiv \frac{1}{2}(m_u + m_d). \tag{10.15}$$

Numerically, this effect is tiny - the observed mass difference mainly originates in the electromagnetic self-energy of the charged pion. Show that the squares of the masses obey the Gell-Mann-Okubo-formula.

The mass formulae (10.14) indicate that the Goldstone boson masses break SU(3) symmetry - the absence of isospin breaking at leading order observed in the case of $SU(2) \times SU(2)$ does not repeat itself here. In fact, none of the other multiplets shows SU(3) breaking effects comparable to those seen in the masses of the pseudoscalar octet. The reason is now evident: M_K^2 is 13 times larger than M_π^2, because m_s happens to be about 13 times larger than $m_u + m_d$ - the quark masses violently break SU(3) symmetry. Unlike the mass of say the nucleon which approaches a nonzero limit if the quark masses are turned off, the masses of the Goldstone bosons are exclusively due to symmetry breaking and are therefore just as asymmetric as the quark masses themselves.

Plenty of further exercises are available to familiarize yourself with the effective Lagrangian technique. You might, e.g., calculate the elastic $\pi\pi$ scattering amplitude by evaluating the terms of order π^4 in the effective $SU(2) \times SU(2)$ Lagrangian and derive the low energy theorems of Weinberg [7], e.g.

$$T_{\pi^+\pi^0 \to \pi^+\pi^0} = \frac{t - M_\pi^2}{F^2} + O(p^4, p^2\mathcal{M}, \mathcal{M}^2) \tag{10.16}$$

where $t = (p'_+ - p_+)^2$ is the invariant momentum transfer. Another instructive example is the decay $\eta \to 3\pi$. Again, it suffices to work out the terms of order π^4, except that in this case one needs to use the Lagrangian for three flavours where the field $U(x)$ includes the η. Also, one must be careful to use a field basis where the kinetic part of the Lagrangian is diagonal (in a basis where terms proportional to $\pi^0\eta$ occur, the η field does not exclusively create η-particles). Show that, to first order in $m_u - m_d$, the transition amplitude is given by

$$T_{\eta \to \pi^+\pi^-\pi^0} = \frac{\sqrt{3}}{4}\frac{m_u - m_d}{m_s - \hat{m}}\frac{s - \frac{4}{3}M_\pi^2}{F^2} + \ldots \tag{10.17}$$

11. Higher Order Terms

The expression (10.13) for the effective Lagrangian is the leading term in a double series: powers of derivatives and powers of the quark mass matrix. When working out pion matrix elements, one is considering momenta which obey

$p^2 = (m_u + m_d)B$. It is therefore convenient to count the quark masses like two powers of momentum and to order the double series accordingly. Both of the terms occurring in eq. (10.13) are then quantities of order p^2. Denoting this contribution by $\mathcal{L}^{(2)}$, the full effective Lagrangian is a series of the form

$$\mathcal{L}_{eff} = \mathcal{L}^{(2)} + \mathcal{L}^{(4)} + \mathcal{L}^{(6)} + \dots \tag{11.1}$$

The term $\mathcal{L}^{(4)}$ contains contributions with four derivatives of the pion field, terms with two derivatives and one power of $\mathcal{M}$ and terms with two powers of $\mathcal{M}$. Note that, in four dimensions, Lorentz invariance implies that the number of derivatives is even, because the tensors $g_{\mu\nu}$ and $\epsilon_{\mu\nu\rho\sigma}$ available to form invariants are of even rank. At order p^4, the following invariants occur [8]:

$$\begin{aligned}
P_0 &= < \partial_\mu U \partial_\nu U^+ \partial^\mu U \partial^\nu U^+ > \\
P_1 &= < \partial_\mu U \partial^\mu U^+ >^2 \\
P_2 &= < \partial_\mu U \partial_\nu U^+ >< \partial^\mu U \partial^\nu U^+ > \\
P_3 &= < \partial_\mu U \partial^\mu U^+ \partial_\nu U \partial^\nu U^+ > \\
P_4 &= < \partial_\mu U \partial^\mu U^+ >< \chi U^+ + U \chi^+ > \\
P_5 &= < \partial_\mu U \partial^\mu U^+ (\chi U^+ + U \chi^+) > \\
P_6 &= < \chi U^+ + U \chi^+ >^2 \\
P_7 &= < \chi U^+ - U \chi^+ >^2 \\
P_8 &= < \chi U^+ \chi U^+ + U \chi^+ U \chi^+ >
\end{aligned} \tag{11.2}$$

where $< A >$ stands for the trace of the matrix A and χ is proportional to the mass matrix:

$$\chi = 2B\mathcal{M}. \tag{11.3}$$

If there are less than four flavours, then not all of these invariants are independent. For $N = 3$, P_0 can be dropped because it can be represented as a linear combination of P_1, P_2 and P_3; for $N = 2$, only P_1, P_2, P_4, P_6 and P_7 are independent. The list (11.2) is however not complete. There are three additional categories of terms:

(i) total derivatives, i.e. contributions of the form $\partial_\mu \omega^\mu$,
(ii) expressions which contain $\Box U$,
(iii) terms involving the ϵ-tensor,

$$\mathcal{L}_{WZ} = \epsilon^{\mu\nu\rho\sigma} f(U) \times \partial_\mu U \times \partial_\nu U \times \partial_\rho U \times \partial_\sigma U. \tag{11.4}$$

The first category is irrelevant, because total derivatives do not affect the action. Terms involving the second derivative $\Box U$ differ from $P_0, \dots, P_8$ only by a multiple of the classical equation of motion associated with $\mathcal{L}^{(2)}$,

$$\partial_\mu \{\partial^\mu U U^+\} = \frac{1}{2}(\chi U^+ - U \chi^+) - \frac{1}{2N} < \chi U^+ - U \chi^+ > . \tag{11.5}$$

These terms can be removed from the effective Lagrangian with a suitable change of the field variables. Consider, e.g., the transformation

$$U' = U\{1 + \alpha(U^+ \Box U - \Box U^+ U) + O(p^4)\}, \tag{11.6}$$

If α is real, the map preserves unitarity and also keeps the determinant equal to one. Inserted in the leading term $\mathcal{L}^{(2)}$, this change of variables generates a contribution of the form $< \Box U^+ \Box U >$, proportional to the arbitrary constant α. With a suitable choice of α, we can therefore bring the effective Lagrangian to a form where such a contribution does not occur. Here it is essential that, in the framework of the effective Lagrangian, the pion field plays the role of a mere auxiliary variable devoid of physical significance - in QCD, a pion field does not occur. In fact, the matrix elements of the operator $\boldsymbol{\pi}(x)$ are ambiguous. In particular, they depend on the conventions chosen when introducing coordinates on SU(N). The change of variables occurring in eq. (11.6) amounts to a coordinate transformation which not only involves the field itself, but also the derivatives thereof,

$$\boldsymbol{\pi}' = \boldsymbol{f}(\boldsymbol{\pi}, \partial\boldsymbol{\pi}, \partial^2\boldsymbol{\pi}, \ldots, \mathcal{M}). \tag{11.7}$$

Since the particular change of variables specified in eq. (11.6) respects the transformation properties of the matrix U under $\mathrm{SU}(N)_L \times \mathrm{SU}(N)_R$, there is no reason to prefer the field $\boldsymbol{\pi}(x)$ to $\boldsymbol{\pi}(x)'$. I recommend it as an exercise in the technology of effective Lagrangians to verify that all of the terms of category (ii) can be eliminated with a change of variables of this type.

In contrast to the first two categories, (iii) is physically interesting. As pointed out by Wess and Zumino [9], the occurrence, of anomalies in the Ward identities for the currents of $\mathrm{SU}(N)_L \times \mathrm{SU}(N)_R$ implies that the effective Lagrangian contains specific vertices of order p^4 which involve the ϵ-tensor (note that I am not referring to the anomalous divergence of the flavour singlet axial current here, but to the anomalies contained in the three- and four-point functions of the non-singlet-currents, e.g., in $< TA^i_\lambda V^k_\mu V^l_\nu >$.) In fact, the function $f(U)$ occurring in eq. (11.4) turns out to be fully determined by the anomalies, which in turn are governed by the short distance singularities of the underlying theory. For QCD, $f(U)$ is proportional to the number of colours while, if the underlying theory only involves scalar fields, then $f(U)$ vanishes. The geometric significance of the Wess-Zumino term was uncovered by Witten [10] who pointed out that it can be represented in compact form if the four-dimensional base manifold is treated as boundary of a five-dimensional domain. I do not dwell on the structure of this term further as this would necessitate a discussion of the anomalies mentioned above. Suffice it to say that the effective Lagrangian machine can be equipped with the appropriate wheels (external vector and axial vector fields) to provide a systematic low energy expansion also for Green functions involving currents and that this extension leads to a rather elegant and powerful vehicle. Quite apart from the Wess-Zumino term which belongs to the standard equipment of this vehicle, it also features a number of physically interesting extras. If this advertisement prompts you to try it out, you can find it in the garage [8]. No strings attached.

At order p^4, the general effective Lagrangian is a linear combination of the terms listed in eq. (11.2), supplemented by the Wess-Zumino term,

$$\mathcal{L}^{(4)} = \sum_{i=0}^{8} L_i P_i + \mathcal{L}_{WZ}. \tag{11.8}$$

While the leading term $\mathcal{L}^{(2)}$ only contains two low energy constants (F and B), this expression involves nine such constants: $L_0, \ldots, L_8$ (as mentioned above, L_0 can be dropped if $N = 3$). The virtue of using the quantity $\chi = 2B\mathcal{M}$ instead of the mass matrix itself is that all of these constants are dimensionless.

If the derivative expansion is carried further, an ever growing number of low energy constants occurs - the effective theory contains infinitely many coupling constants. In this respect, the situation is the same as for theories containing a mass gap, where the low energy expansion is given by the Taylor series and also involves an infinite sequence of low energy parameters (scattering length, effective range, ...). As mentioned above, the effective Lagrangian can be extended in such a way that it also yields the low energy expansion of the Green functions. The price to pay is that the effective Lagrangian then contains even more terms. At first nonleading order, altogether thirteen constants are needed to specify the scattering amplitude and the Green functions of the vector, axial, scalar and pseudoscalar currents if $N \geq 4$ (twelve for $N = 3$ and ten for $N = 2$).

The underlying theory contains much less parameters. In the case of QCD, e.g., F and B are multiples of Λ_{QCD} which in principle, are calculable and the coupling constants L_i also represent calculable numbers. Indeed, lattice calculations do provide rough determinations of F and B which will become more accurate in the future. At the present time, the main source of information about the coupling constants is low energy phenomenology which allows one to pin down most of the coupling constants occurring in $\mathcal{L}^{(4)}$ to within rather narrow limits [8]. Quite independently, a rough a priori estimate of these constants can be obtained on the basis of the following observation. The effective Lagrangian explicitly describes the poles and cuts generated by the Goldstone bosons, but replaces the low energy singularities due to the exchange of other particles (such as ρ-mesons) by the corresponding Taylor series. In fact, in the case of SU(2)×SU(2), the size of all of the coupling constants occurring in $\mathcal{L}^{(4)}$ can be understood in a semi-quantitative manner if one assumes that they are dominated by the contribution from ρ-exchange [8]. More recently [11], it was shown that the values of the coupling constants occurring in the effective Lagrangian for $N = 3$ can be predicted by evaluating the pole graphs due to the exchange of the lightest massive mesons (octets with $J^P = 1^-, 1^+, 0^+$ and singlets with $J^P = 0^-, 0^+$). The result agrees remarkably well with the outcome of the phenomenological analysis. I will illustrate the method with a specific example below.

12. Unitarity, Loops, Renormalizability

In the preceding sections we were dealing only with tree graphs, i.e., with classical field theory. It is not legitimate, however, to simply disregard graphs which contain loops, because the tree graph contributions to the scattering amplitudes do not satisfy unitarity. As discussed by Lehmann at the Schladming Winter School in 1973 [12], unitarity requires that the low energy expansion of the $\pi\pi$ scattering amplitude contains specific contributions of order p^4 which are not polynomials in the invariants s, t but exhibit logarithmic branch points. In the language of the Feynman diagrams associated with the Lagrangian $\mathcal{L}^{(2)}$, these contributions are represented by the one-loop graph shown in Fig. 2a and its crossed variants. At one-loop order, the tadpole graphs of Fig. 2b also need to be taken into account, which renormalize the pion mass and the pion decay constant.

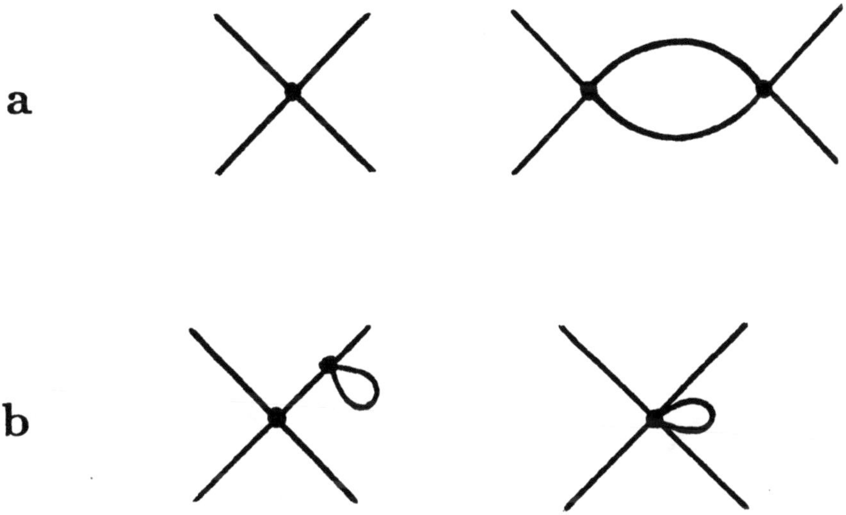

Fig. 2. Chiral perturbation theory graphs contributing to the low energy expansion of the $\pi\pi$ scattering amplitude to first nonleading order. The full circles represent vertices generated by the leading term in the derivative expansion of the effective Lagrangian, $\mathcal{L}^{(2)}$

Graphs involving two or more loops of course also occur - only the sum of all Feynman diagrams of a local field theory leads to a unitary S-matrix. Chiral perturbation theory is based on the fact that not all of these graphs are of the same order of magnitude. While the tree graph contribution to the $\pi\pi$

scattering amplitude is of order p^2 [see eq. (10.16)], the one-loop graphs are of order p^4. More generally, graphs containing a different number of loops occur at different orders of the low energy expansion: in d dimensions, graphs with ℓ loops are suppressed compared to the tree graphs by the power $[p^{d-2}]^\ell$. The rule is readily checked for individual graphs such as those shown in Figs. 2a or 2b. The loop integrals are homogeneous functions of the external momenta and of the pion mass occurring in the propagators. The degree of homogeneity is determined by the dimension of the integral which in turn is fixed by the overall power of the pion decay constant arising from the various vertices. A more thorough discussion of the issue can be found in ref.3.

As discussed above, the Lagrangian $\mathcal{L}^{(2)}$ is not the full story - graphs involving vertices of $\mathcal{L}^{(4)}, \mathcal{L}^{(6)}, \ldots$ also need to be taken into account. In the case of the $\pi\pi$ scattering amplitude, graphs containing ℓ loops are of order $p^{2+\ell(d-2)}$ only if they exclusively involve vertices of $\mathcal{L}^{(2)}$. Graphs containing one vertex of $\mathcal{L}^{(4)}$ ($\mathcal{L}^{(6)}$) are smaller by one (two) powers of p^2. Hence, to evaluate the scattering amplitude in four dimensions to order p^4, we need to work out the tree and one-loop graphs of $\mathcal{L}^{(2)}$ shown in Fig. 2 and add the tree graphs of Fig. 3 which involve one vertex from $\mathcal{L}^{(4)}$. Higher orders in the derivative expansion of the effective Lagrangian and two-loop graphs only start contributing at order p^6.

Fig. 3. Contributions to the $\pi\pi$ scattering amplitude generated by the first nonleading term in the derivative expansion of the effective Lagrangian: the full squares represent vertices of $\mathcal{L}^{(4)}$

Note that the graphs can be ordered by counting powers of the momentum only if $d > 2$. In two dimensions, the constant F is dimensionless and the degree of homogeneity is therefore independent of the number of loops. In $d = 2$, the Lagrangian $\mathcal{L}^{(2)}$ taken by itself specifies a decent, renormalizable theory, which moreover is asymptotically free and thus shares the qualitative properties of four-dimensional nonabelian gauge theories. In particular, the low energy structure of the theory cannot be analyzed perturbatively. [Incidentally, supplementing $\mathcal{L}^{(2)}$ by the Wess-Zumino term, one arrives at a two-dimensional field theory with very peculiar properties: the Wess-Zumino-Novikov-Witten model. In this model, the coupling constant F can be tuned in such a fashion that the β-function vanishes - the theory becomes conformally invariant.]

In $d = 4$, the Lagrangian $\mathcal{L}^{(2)}$ by itself is meaningless, but taken together with the infinite string of higher order terms $\mathcal{L}^{(4)}, \mathcal{L}^{(6)}, \ldots$ it does specify a renormalizable framework. It is convenient to regularize the loop integrals by means of dimensional regularization, because this method preserves the symmetries of the Lagrangian. The poles occurring at $d = 4$ then only require counter terms which are Lorentz invariant and symmetric under $SU(N) \times SU(N)$. By construction, the full effective Lagrangian contains all terms permitted by this symmetry. The divergences can therefore be absorbed in a renormalization of the coupling constants occurring in the Lagrangian. In particular, the divergences contained in the one-loop graphs are absorbed in a renormalization of the coupling constants $L_0, \ldots, L_8$ occurring in $\mathcal{L}^{(4)}$. Dimensional regularization also takes care of a technical complication connected with the fact that the effective Lagrangian contains derivative couplings. This property implies that the measure occurring in the functional integral does not coincide with the standard translation invariant measure on the space of the pion fields. In general, the measure generates additional contributions involving power divergences, such as $\delta(0) \sim \Lambda^4$. In dimensional regularization, however, power divergences do not occur (in particular, $\delta(0)$ vanishes) and the complications associated with the measure can simply be ignored [for a more detailed discussion and references to the literature, see, e.g. [8]].

13. Mass of the Goldstone Bosons to One Loop

As an illustration of the machinery, let us work out the pion mass to second order in the expansion in powers of the quark masses. For simplicity, I take the masses of the N quark flavours to be the same, $m_u = m_d = \ldots = m$, such that the spectrum of the theory consists of degenerate multiplets of SU(N). In particular, the $N^2 - 1$ Goldstone bosons then obtain the same mass M. The leading term in the expansion of M^2 in powers of m is given in eq. (10.11). Denote this term by M_1^2,

$$M_1^2 \equiv 2mB. \tag{13.1}$$

At order m^2, we need to evaluate the tree graphs of $\mathcal{L}^{(4)}$ and add the one-loop contributions generated by $\mathcal{L}^{(2)}$. The tree graph contributions to the mass are determined by the kinetic part of the Lagrangian, i.e. by the piece which is quadratic in the pion field $\pi(x)$. The invariants P_0, P_1, P_2, P_3 only involve vertices with four or more pion fields and do therefore not affect the mass, but the term P_4 does contain a quadratic piece,

$$P_4 = \frac{4N}{F^2} M_1^2 \partial_\mu \pi \partial^\mu \pi + O(\pi^4), \tag{13.2}$$

and P_5 is of a similar structure. In the case of P_6, the expansion in powers of π yields

$$P_6 = M_1^4 \left\{ 4N^2 - \frac{8N}{F^2} \pi^2 + O(\pi^4) \right\}. \tag{13.3}$$

The term P_7 only contains vertices with six or more pion fields and P_8 is similar to P_6. Collecting the various contributions, the kinetic part of the effective Lagrangian becomes

$$\mathcal{L}^{(2)} + \mathcal{L}^{(4)} = \frac{1}{2} \partial_\mu \pi \partial^\mu \pi \left\{ 1 + \frac{8M_1^2}{F^2} (NL_4 + L_5) \right\}$$
$$-\frac{1}{2} M_1^2 \pi^2 \left\{ 1 + \frac{16M_1^2}{F^2} (NL_6 + L_8) \right\}. \tag{13.4}$$

Graphically, this result corresponds to the tree graphs shown in Figs. 4a and 4b.

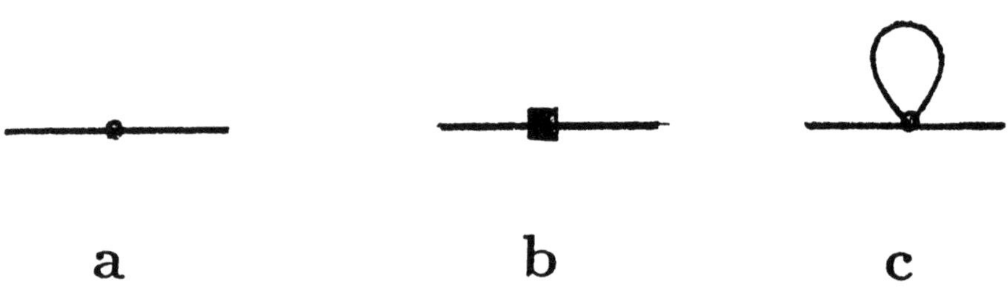

Fig. 4. Meson self energy graphs. At leading order in the low energy expansion only a tree graph generated by $\mathcal{L}^{(2)}$ contributes (a). At first nonleading order, a tree graph from $\mathcal{L}^{(4)}$ occurs (b) as well as a one-loop graph from $\mathcal{L}^{(2)}$ (c)

The one-loop graph of Fig. 4c stems from the vertex $\sim tr([\partial_\mu \pi, \pi][\partial^\mu \pi, \pi])$ contained in $\mathcal{L}^{(2)}$, two of the pion fields being contracted with

$$< 0 \mid T\pi^a(x)\pi^b(y) \mid 0 >= \frac{1}{i} \delta^{ab} \Delta(x-y)$$
$$\Delta(z) = \frac{1}{(2\pi)^d} \int d^d p \frac{e^{-ipz}}{M_1^2 - p^2 - i\epsilon}. \tag{13.5}$$

Since the arguments of the pion fields coincide, the contractions are proportional to the propagator or its derivatives at the origin. The term $\partial_\mu \Delta(z)$ vanishes at $z = 0$, as it is antisymmetric under $z \to -z$. The second derivative can be evaluated with the differential equation

$$\Box\,\Delta(z) + M_1^2\Delta(z) = \delta(z), \tag{13.6}$$

using the fact that, in dimensional regularization, $\delta(0)$ vanishes. The sum over the flavour of the meson circling around the loop leads to

$$\sum_a \lambda^a\lambda^a = \frac{2(N^2-1)}{N};\sum_a \lambda^a\lambda^b\lambda^a = -\frac{2}{N}\lambda^b. \tag{13.7}$$

The result for the mass of the Goldstone bosons then takes the form

$$M^2 = M_1^2\left\{1 - \frac{8M_1^2}{F^2}(NL_4 + L_5 - 2NL_6 - 2L_8) - \frac{i}{NF^2}\Delta(0)\right\}. \tag{13.8}$$

Performing a Wick rotation, the propagator at the origin can be represented as

$$\begin{aligned} -i\Delta(0) &= \frac{1}{(2\pi)^d}\int\frac{d^dk}{M_1^2+k^2} = \frac{1}{(2\pi)^d}\int_0^\infty d\lambda\int d^dk e^{-\lambda(M_1^2+k^2)} \\ &= \frac{1}{(4\pi)^{d/2}}\Gamma(1-\frac{d}{2})M_1^{d-2}. \end{aligned} \tag{13.9}$$

The Γ-function contains a pole at $d=4$, $\Gamma(1-d/2) = 2/(d-4)+\ldots$, generating a divergence in $\Delta(0)$ which is proportional to M_1^2. Eq. (13.8) shows that this divergence can indeed be absorbed with a suitable renormalization of the coupling constants $L_4,\ldots,L_8$, as claimed above. The singularity can be extracted by rewriting $\Delta(0)$ as

$$-i\Delta(0) = M_1^2\left\{c + \bar{c}(M_1)\right\} \tag{13.10}$$

where the divergent part c

$$c \equiv \frac{\Gamma(1-d/2)}{(4\pi)^{d/2}}\mu^{d-4} \tag{13.11}$$

is independent of M_1, but contains an arbitrary renormalization scale μ. The remainder

$$\bar{c}(M_1) = \frac{\Gamma(1-d/2)}{(4\pi)^{d/2}}(M_1^{d-4} - \mu^{d-4}) \tag{13.12}$$

approaches a finite limit as as $d\to 4$, given by

$$\bar{c}(M_1) = \frac{1}{8\pi^2}\ln\frac{M_1}{\mu}. \tag{13.3}$$

The Goldstone boson mass then becomes

$$M^2 = M_1^2\left\{1 - \frac{8M_1^2}{F^2}(NL_4^r + L_5^r - 2NL_6^r - 2L_8^r) + \frac{M_1^2}{8\pi^2NF^2}\ln\frac{M_1}{\mu}\right\} + O(m^3) \tag{13.14}$$

where $L_i^r = L_i^r(\mu)$ is the renormalized coupling constant at running scale μ (note that the curly bracket is independent of this scale). The formula shows that the expansion in powers of the quark mass is not an ordinary Taylor series, but contains a nonanalytic piece $\sim m^2 \log m$, referred to as a "chiral logarithm". The occurrence of such contributions is characteristic of chiral perturbation theory.

14. Masses of the Light Quarks

One of the many applications of chiral perturbation theory concerns the relative magnitude of m_u, m_d and m_s. In the following, I briefly discuss this example which adequately illustrates both the strengths and the weaknesses of the method.

Let me start with the estimate of the light quark masses we obtained in 1975 [13]:

$$m_u \simeq 4\,MeV,\ m_d \simeq 6\,MeV,\ m_s \simeq 135\,MeV. \tag{14.1}$$

This pattern is remarkable in several respects. First of all, in accordance with the qualitative discussion of section 4, m_s is large compared to m_u, m_d. The quark mass term $m_u\bar{u}u + m_d\bar{d}d + m_s\bar{s}s$ is therefore dominated by the strange quark. This implies that the chiral $SU(3)_L \times SU(3)_R$ symmetry which results if all three quark masses are set equal to zero is on the same footing as the eightfold way which results if m_u, m_d and m_s are set equal. Since the eightfold way is known to represent a decent approximate symmetry of the strong interaction, the same should also be true of $SU(3)_L \times SU(3)_R$. Furthermore, the mass term of the strange quark does not break chiral $SU(2)_L \times SU(2)_R$. This group should therefore represent an almost exact symmetry of the strong interaction. In fact, the estimate (14.1) indicates that the difference $m_u - m_d$ is by no means small compared to m_u, m_d. The breaking of $SU(2)_L \times SU(2)_R$ therefore ought to be comparable to isospin breaking which is known to be very small. This conclusion leads to a puzzle: the pion mass is an $SU(2)_L \times SU(2)_R$ symmetry breaking effect. Why do the pions not show comparable breaking of isospin symmetry? We have already solved this puzzle, in section 10. As noted there, the leading term in the derivative expansion of the effective Lagrangian associated with chiral $SU(2)_L \times SU(2)_R$ is isospin symmetric - despite the fact that the underlying theory is not. In the effective Lagrangian for *three* quark flavours, the mass difference $m_u - m_d$ does show up at leading order, generating a splitting between K^+ and K^0 [compare eq. (10.14)]. The relative magnitude of this isospin breaking effect is however also small, because it is of order $(m_u - m_d)/m_s$ and m_s happens to be large compared to $m_u - m_d$. In the case of the nucleon, isospin breaking is even less significant: the mass difference $M_p - M_n$ is of the same order of magnitude as $m_u - m_d$, but the scale to compare this with is the nucleon mass. To summarize: the magnitude of the isospin breaking effects is small compared to $(m_u - m_d)/(m_u + m_d)$, because the relevant scale is not given by $m_u + m_d$. In the pseudoscalar octet, the scale is set by m_s while for the remaining states, the magnitude of isospin breaking effects is determined by the ratio $(m_u - m_d)/\Lambda$ where Λ is a typical hadronic scale such as the mass of the nucleon.

Since 1975, the literature offers a considerable number of papers dealing with the pattern of the light quark masses, both concerning the ratios $m_u : m_d : m_s$ and the absolute magnitude, say of m_s. Chiral perturbation theory does not shed much light on the value of m_s, because the quark mass matrix only enters the effective Lagrangian through the product $B\mathcal{M}$ and direct phenomenological information on the low energy constant B is not available. Chiral symmetry merely relates the magnitude of the quark masses to the size of the quark condensate [see eq. (10.12)] - some of the methods proposed to estimate the light quark masses are based on this relation. I do not review the various estimates here, because chiral perturbation theory does not play an important role in this context. It does allow one, however, to determine the ratios $m_u : m_d : m_s$ to within rather narrow limits and I now turn to this topic.

The main source of information about the light quark masses is the mass spectrum of the pseudoscalar octet. Using the lowest order mass formulae (10.14) and ignoring the electromagnetic interaction, the experimental values of M_{π^+}, M_{K^+} and M_{K^0} imply $m_u : m_d : m_s = 0.66 : 1 : 20$. There are two categories of corrections to this result: electromagnetic contributions and higher order terms in the quark mass expansion.

A. Corrections of Order e^2

Dashen's theorem [14] states that, in the limit $m_u = m_d = m_s = 0$, the electromagnetic self-energies of the pseudoscalar mesons are proportional to the square of their charge Q_P,

$$M_P^2 = (M_P^2)_{QCD} + e^2 Q_P^2 C + O(e^2\mathcal{M}). \qquad (14.2)$$

The constant C can be expressed as an integral over the difference between the vector and axial vector spectral functions [15] which can be evaluated on the basis of τ-decay data [16]. Alternatively, C can be determined from the mass difference between π^+ and π^0. As mentioned above, in the absence of the electromagnetic interaction, this difference is tiny, of order $(m_u - m_d)^2$. Hence the observed mass difference must almost entirely be due to the electromagnetic self-energy, i.e. $e^2 C$ must approximately be equal to $M_{\pi^+}^2 - M_{\pi^0}^2$. The result of the spectral function analysis is in good agreement with this conclusion. The relation (14.2) therefore implies

$$M_{K^0}^2 - M_{K^+}^2 = (M_{K^0}^2 - M_{K^+}^2)_{QCD} - (M_{\pi^+}^2 - M_{\pi^0}^2) + O(e^2\mathcal{M}). \qquad (14.3)$$

Numerically, this amounts to

$$M_{K^0} - M_{K^+} = 5.3\,MeV - 1.3\,MeV$$

where the first term is the contribution generated by $m_u - m_d$, while the second term represents the electromagnetic self-energy.

Correcting the masses M_{π^+} and M_{K^+} for electromagnetic contributions in this manner and neglecting the terms of order $\mathcal{M}^2$, the mass formulae (10.14) imply $m_u : m_d : m_s = 0.55 : 1 : 20.1$, the ratios advocated by Weinberg [18].

B. Corrections of Order $\mathcal{M}^2$

We have evaluated the corrections of order $\mathcal{M}^2$ to the pseudoscalar masses in the preceding section, however only for the case of N quark flavours of equal mass. Since we are now considering $m_u \neq m_d \neq m_s$, the various mesons propagating in the loops carry different masses and the vertices generated by the symmetry breaking terms of $\mathcal{L}^{(2)}$ and $\mathcal{L}^{(4)}$ also distinguish between m_u, m_d and m_s. Otherwise, the calculation is the same. There are five different masses in the octet: $\pi^0, \pi^+, K^0, K^+, \eta$. Since we are using the observed difference between π^0 and π^+ to correct for the electromagnetic contributions, only four of these masses provide information about the mass spectrum in QCD. It is convenient to analyze this information in terms of the following three dimensionless ratios:

$$\begin{aligned} \Delta_{GMO} &\equiv (4M_K^2 - 3M_\eta^2 - M_\pi^2)/(M_\eta^2 - M_\pi^2) \\ Q_1 &\equiv M_K^2/M_\pi^2 \\ Q_2 &\equiv (M_{K^0}^2 - M_{K^+}^2 + M_{\pi^+}^2 - M_{\pi^0}^2)/(M_K^2 - M_\pi^2). \end{aligned} \tag{14.4}$$

At leading order in the expansion in powers of $\mathcal{M}$, the squares of the meson masses obey the Gell-Mann-Okubo formula. Hence Δ_{GMO} is a quantity of order $m_s - \hat{m}$. The one-loop calculation leads to the following explicit expression [8]

$$\Delta_{GMO} = \frac{8}{F^2}(M_K^2 - M_\pi^2)\{L_5^r - 12L_7^r - 6L_8^r\} + \chi\log \tag{14.5}$$

where the term "$\chi\log$" stands for chiral logarithms (compare eq. (13.14)). In the chiral expansion of Q_1 and Q_2, the leading term is different from zero - it is given by a ratio of quark masses. The contributions generated by the tree graphs of $\mathcal{L}^{(4)}$ and by the one-loop graphs of $\mathcal{L}^{(2)}$ amount to a correction of order $m_s - \hat{m}$:

$$\begin{aligned} Q_1 &= \frac{m_s + \hat{m}}{m_u + m_d}\{1 + \Delta_M + O(\mathcal{M}^2)\} \\ Q_2 &= \frac{m_d - m_u}{m_s - \hat{m}}\{1 + \Delta_M + O(\mathcal{M}^2)\}. \end{aligned} \tag{14.6}$$

Remarkably, the correction is the same in the two cases [8]:

$$\Delta_M = \frac{8}{F^2}(M_K^2 - M_\pi^2)(2L_8^r - L_5^r) + \chi\log. \tag{14.7}$$

The ratio $Q^2 \equiv Q_1/Q_2$ therefore agrees with the corresponding ratio of quark masses, up to corrections of second order

$$Q^2 = \frac{m_s^2 - \hat{m}^2}{m_d^2 - m_u^2}\{1 + O(\mathcal{M}^2)\}\,. \tag{14.8}$$

Neglecting the terms of order $\mathcal{M}^2$, this implies that the quark mass ratios are constrained to an ellipse

$$\left(\frac{m_u}{m_d}\right)^2 + \frac{1}{Q^2}\left(\frac{m_s}{m_d}\right)^2 = 1 \tag{14.9}$$

whose semi-axes are given by 1 and Q, respectively ($\hat{m}^2$ is negligibly small compared to m_s^2 - I have dropped this term). The observed values of the meson masses give $Q \simeq 24$. The corresponding ellipse is shown as a dashed line in Fig. 5, while the full lines represent an estimate of the uncertainties due to the higher order contributions occurring in eqs. (14.3) and (14.8) (for details, see ref.20).

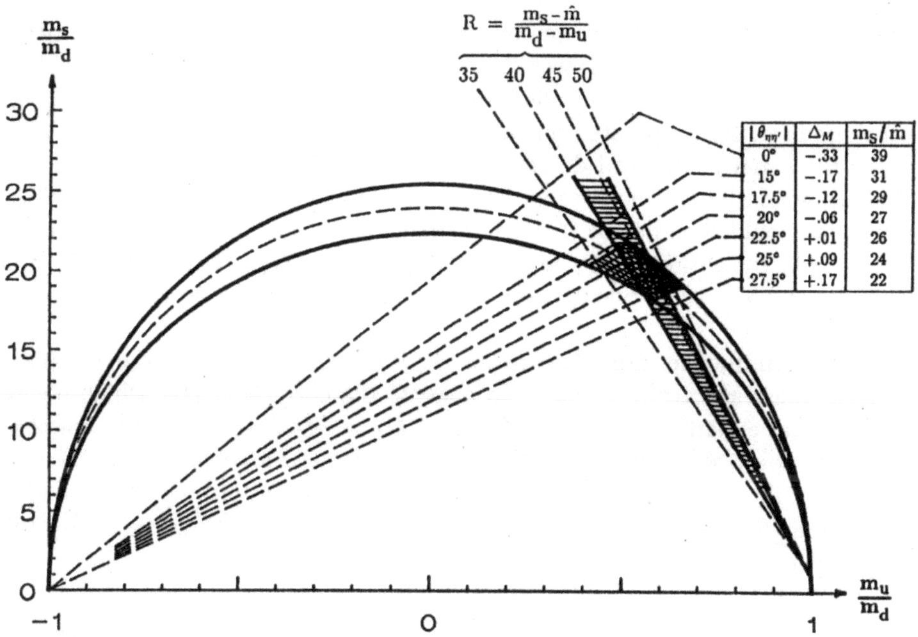

Fig. 5. The elliptic band indicates the range of quark mass ratios permitted by the low energy theorem (14.8). The cross-hatched area is the intersection of this band with the sector allowed by the phenomenology of $\eta\eta'$ mixing. The shaded wedge shows the constraint imposed on the ratio R by the mass splittings in the baryon octet, according to ref.21

Note that the result (14.9) does not contain any of the coupling constants occurring in the effective Lagrangian. This relation neatly illustrates the fact that, although the chiral Lagrangian contains a considerable number of parameters which chiral symmetry leaves unspecified, chiral perturbation theory nevertheless leads to parameter free predictions.

The elliptic constraint (14.9) does not suffice to determine the two individual mass ratios $m_u : m_d$ and $m_s : m_d$. In particular, as emphasized by Kaplan and Manohar [19], this constraint does not exclude the possibility that m_u vanishes (the quark mass matrix could then be brought to real, diagonal form by means of a chiral $\mathrm{SU}(N_f)_L \times \mathrm{SU}(N_f)_R$ rotation and QCD would therefore necessarily be invariant under P, C and T; if the determinant of the quark mass matrix is different from zero, this property of the strong interactions can be accommodated, but is not explained). The recent literature contains several papers [20] which deal with this possibility, pointing out that the mass ratios advocated by Weinberg [18] are based on lowest order chiral perturbation theory and are therefore subject to corrections of order $m_s - \hat{m}$ which could be substantial (the fact that the ratio $F_K/F_\pi = 1.22$ differs from one also originates in a correction of this type).

In our ancient treatise on quark masses [21], we arrived at a rather accurate value for the ratio $R = (m_s - \hat{m})/(m_d - m_u)$ which, together with the relation (14.9) pins the quark mass ratios down to within small uncertainties. This value was obtained from a variety of sources (mass splittings in the baryon octet, $K^0 - K^+$, $\rho\omega$-mixing) which were shown to lead to consistent results provided the terms of order $\mathcal{M}^{3/2}$ in the chiral expansion of the baryon masses are accounted for. As we pointed out at that time, our analysis excludes the value $m_u = 0$ by many standard deviations. Conversely, if m_u were to vanish, that analysis would be entirely wrong, because the higher order terms in the chiral expansion could then not be treated as small corrections. [At leading order and for $m_u = 0$, the ratios Q_1, Q_2 are related by $1/Q_2 = Q_1 - 1$. Numerically, this amounts to 43 = 12! Evidently, the chiral expansion would be in bad shape if m_u were zero.]

To determine the individual ratios $m_u : m_d$ and $m_s : m_d$ from the pseudoscalar mass spectrum, we need to analyze the correction Δ_M which occurs in the chiral expansion (14.6) of the ratio $Q_1 = M_K^2/M_\pi^2$. Here, the coupling constants don't do us the favour to drop out. We need an estimate of the magnitude of $2L_8 - L_5$. The value of L_5 can be determined on phenomenological grounds, exploiting the fact that the ratio F_K/F_π is also determined by this coupling constant,

$$\frac{F_K}{F_\pi} = 1 + 4\frac{(M_K^2 - M_\pi^2)}{F^2} L_5^r + \chi \log \tag{14.10}$$

and using the experimental value $F_K/F_\pi = 1.22$. In the case of L_8, however, phenomenological information is not available because of a hidden symmetry first pointed out in ref.19.

15. The Hidden Symmetry of Kaplan and Manohar

The symmetry originates in the fact that the effective Lagrangian exclusively incorporates chiral invariance. In particular, the only property of the quark mass matrix which the effective Lagrangian was told about is that it transforms according to $\mathcal{M} \to V_R \mathcal{M} V_L^+$. As pointed out by Kaplan and Manohar [19], the matrix

$$\mathcal{M}' = \alpha_1 \mathcal{M} + \alpha_2 (\mathcal{M}^+)^{-1} \det \mathcal{M} \tag{15.1}$$

transforms in the same manner, α_1 and α_2 being arbitrary constants. For a real diagonal mass matrix, this amounts to

$$m'_u = \alpha_1 m_u + \alpha_2 m_d m_s (cycl.\ u \to d \to s \to u). \tag{15.2}$$

Symmetry alone does therefore not distinguish $\mathcal{M}'$ from $\mathcal{M}$. If $\mathcal{L}(U, \partial U, \ldots, \mathcal{M})$ is an effective Lagrangian consistent with chiral symmetry, so is $\mathcal{L}(U, \partial U, \ldots, \mathcal{M}')$. Since only the product $B\mathcal{M}$ enters the Lagrangian, α_1 merely changes the value of the constant B. The term proportional to α_2 is a correction of order $\mathcal{M}^2$ which, upon insertion in $\mathcal{L}^{(2)}$ generates a contribution to $\mathcal{L}^{(4)}$. The contribution can again be removed by changing some of the coupling constants:

$$\begin{aligned} B' = B/\alpha_1 \ ; \quad L'_6 = L_6 - \alpha \\ L'_7 = L_7 - \alpha \ ; \quad L'_8 = L_8 + 2\alpha \end{aligned} \tag{15.3}$$

where $\alpha = \alpha_2 F^2 / 32 \alpha_1 B$. The effective Lagrangian is therefore invariant under a simultaneous change of the quark mass matrix and of the coupling constants. Accordingly, the meson masses and scattering amplitudes which one calculates with the effective Lagrangian are also invariant under this operation. The quantity Δ_{GMO}, e.g., only involves the combination $2L_7 + L_8$ which is indeed invariant (compare eq. (14.5)). I recommend to verify that the chiral representations for Q_1 and Q_2 and the elliptic constraint (14.8) also pass the test (up to terms of order $(m_u - m_d)^2$ which were neglected in the derivation of these formulae). The chiral representation for the Green functions of the vector and axial currents are also invariant [17]. Since there is experimental information only about masses, scattering amplitudes and matrix elements of the electromagentic or weak currents and since the chiral representation for these does not distinguish $\mathcal{M}, B, L_i$ form $\mathcal{M}', B', L'_i$, phenomenology does not allow us to determine the magnitude of the constants B, L_6, L_7, L_8.

I emphasize that we are not dealing with a hidden symmetry of QCD here - this theory is not invariant under the change (15.2) of the quark masses. The quark masses can be determined, e.g., by measuring the current correlation functions very accurately at short distances. Also, when simulating QCD by means of Monte Carlo calculations on a lattice, a symmetry of this sort does not occur - the mass spectra associated with the quark mass matrices $\mathcal{M}$ and $\mathcal{M}'$ are different. The symmetry arises because we are not making use of the explicit form of the QCD Lagrangian, but are only exploiting its symmetry properties under chiral rotations. Attempts at elevating the above symmetry to a basic property of QCD are futile - the symmetry pertains to the method by means of which we are analyzing the theory, not to QCD itself. We do however face the problem that phenomenological information about the constant L_8 which occurs in Δ_M is not available and we therefore have to resort to theoretical arguments to estimate its magnitude.

16. Singularities Generated by Low Lying States

Consider the nonet of pseudoscalar currents

$$P^A = \bar{q}\frac{\lambda^A}{2}i\gamma_5 q; \quad A = 0, \ldots, 8 \tag{16.1}$$

with $\lambda^0 = \sqrt{2/3}$. In the chiral limit, the corresponding two-point functions are described by two independent spectral functions $\overset{8}{\rho}_{PP}(s)$ and $\overset{1}{\rho}_{PP}(s)$:

$$i\int dx e^{iqx} < 0 \mid TP^i(x)P^k(0) \mid 0 >= \delta^{ik}\int_0^\infty \frac{ds}{s-q^2-i\epsilon}\,\overset{8}{\rho}_{PP}(s)$$

$$i\int dx e^{iqx} < 0 \mid TP^0(x)P^0(0) \mid 0 >= \int_0^\infty \frac{ds}{s-q^2-i\epsilon}\,\overset{1}{\rho}_{PP}(s) \tag{16.2}$$

The scalar and pseudoscalar currents represent the response of the QCD Lagrangian to an infinitesimal deformation of the quark mass matrix. The above two-point-functions can therefore be obtained by treating the quark mass matrix as an external field and taking the second derivative of the vacuum-to-vacuum amplitude with respect to this field. The same procedure can also be applied at the level of the effective theory, calculating the vacuum-to-vacuum amplitude with the effective Lagrangian specified in the preceding sections. Note however that chiral symmetry permits a term proportional to $\mathrm{tr}(\mathcal{M}\mathcal{M}^+)$. We did not include this term in eq. (11.8), because it does not affect thescattering amplitude or the meson masses, being independent of the pion field. In the present context, this term cannot be dropped; in the chiral representation for the above two-point-functions, it shows up as a contact contribution independent of q^2. It does drop out, however, in the difference between the singlet and the octet:

$$\frac{1}{B^2}\int_0^\infty \frac{ds}{s-q^2-i\epsilon}\left\{\overset{1}{\rho}_{PP}(s) - \overset{8}{\rho}_{PP}(s)\right\} = \frac{F^2}{q^2} - 48L_7 + O(q^2). \tag{16.3}$$

In the chiral limit, this representation is exact. The leading term of order $1/q^2$ originates in the one-pion intermediate state occurring in the two-point function of the octet while the contribution of order one is determined by the coupling constant L_7. The spectral functions do not receive contributions from $\pi\pi$ intermediate states, on account of parity. The continuum starts with three pions and only manifests itself at the two-loop level (i.e. through terms of order q^2 in eq. (16.3)).

The formula (16.3) unambiguously specifies the physical significance of the coupling constant L_7. The hidden symmetry of the effective Lagrangian discussed in the preceding section (under which $L_7 \rightarrow L_7 - \alpha$) does not apply to the chiral representation of the Green functions involving scalar or pseudoscalar currents.

The operator expansion of the product $P^i(x)P^k(0)$ starts with the unit operator. Up to logarithmic factors which reflect the anomalous dimension of P^i, the coefficient of the unit operator is proportional to x^{-6}. Since P^0 and P^i belong to the same chiral multiplet, the coefficient of the unit operator in the expansion of $P^0(x)P^0(0)$ is the same. The short distance singularity associated with the unit operator therefore drops out in the difference we are considering here. Chirality conservation also implies that the coefficient of the operator $\bar{q}q$ is proportional to the quark mass and hence vanishes in the chiral limit. In the case of the operator $G^a_{\mu\nu}G^{\mu\nu a}$, the singularity again drops out because this operator is a chiral singlet. This shows that the difference of the two correlation functions is less singular than x^{-2}. The spectral functions therefore obey the sum rule

$$\int_0^\infty ds \left\{\rho^1_{PP}(s) - \rho^8_{PP}(s)\right\} = 0. \tag{16.4}$$

The two relations (16.3) and (16.4) can be compared with the Weinberg sum rules obeyed by the spectral functions associated with the vector and axial currents [22]. In the chiral limit, these relations read

$$\int_0^\infty ds \left\{\rho^8_{VV}(s) - \rho^8_{AA}(s)\right\} = F^2$$
$$\int_0^\infty dss \left\{\rho^8_{VV}(s) - \rho^8_{AA}(s)\right\} = 0. \tag{16.5}$$

As it is well known, the contributions to the spectral functions generated by the lowest intermediate states (π, A_1, ρ) nearly saturate the Weinberg sum rules. In the case of the pseudoscalar spectral functions, the lowest states are π and η'. Assuming that the corresponding contributions saturate the integrals occurring in eqs. (16.3) and (16.4), we obtain [8] [7]

$$< 0 \mid P^i \mid \pi^k > \simeq \delta^{ik} < 0 \mid P^0 \mid \eta' >$$
$$L_7 \simeq -\frac{F^2}{48M^2_{\eta'}}. \tag{16.6}$$

In the large N_c limit, both of these relations are exact. The η' then also plays the role of a Goldstone boson which can be included in the effective Lagrangian.

A detailed discussion of the role of the η' in the context of chiral perturbation theory is given in ref.8.

The estimate (16.6) expresses the coupling constant L_7 in terms of the mass of the lowest lying bound state with the appropriate quantum numbers. As mentioned in section 11, analogous estimates can be given for all of the coupling constants occurring in the effective Lagrangian [11]. Both the signs and the magnitudes of the coupling constants can thus be understood theoretically, on the basis of the observed spectrum of low lying bound states: the coupling constants are not free parameters but are determined by the low energy singularities which remain, once the Goldstone boson poles and cuts are removed.

As an illustration, consider the coupling constant L_5. The corresponding estimate reads [11]

$$L_5 \simeq F^2 7over4M_S^2 \tag{16.7}$$

where M_S is the mass of the lowest lying scalar multiplet. Ignoring the chiral logarithm associated with the $\pi\pi$ continuum underneath the resonance, the formula (14.10) thus predicts

$$F_K 7over F_\pi \simeq 1 + \frac{M_K^2 - M_\pi^2}{M_S^2}.$$

This explains why SU(3) breaking effects only occur at the level of 20 or 30 %: the magnitude of these effects is determined by the ratio $(M_K^2 - M_\pi^2)/M_S^2 \simeq 0.25$.

Qualitatively, the chiral expansion in powers of m_u, m_d and m_s is a meaningful notion, because the corresponding eight Goldstone bosons are indeed the eight lightest hadrons. In chiral perturbation theory, this observation acquires quantitative meaning: the corrections of order $\mathcal{M}$ are given by M_P^2/M_S^2 or $M_P^2/M_{\eta'}^2$ where M_P stands for the mass of one of the Goldstone bosons while M_S or $M_{\eta'}$ are the masses of the lightest scalar or pseudoscalar non-Goldstone states. The two ratios are relatively small, since the masses $M_S \simeq M_{\eta'} \simeq 1$ GeV exceed the Goldstone boson masses by about a factor of two (for a more detailed discussion, see ref.17).

Let us now return to L_7. There is an independent phenomenological check on the estimate given in eq. (16.6). As pointed out in ref.8, the coupling constant L_7 also determines the $\eta\eta'$ mixing angle:

$$\sin^2 \Theta_{\eta\eta'} = -\frac{24(M_\eta^2 - M_\pi^2)^2}{F^2(M_{\eta'}^2 - M_\eta^2)} L_7. \tag{16.8}$$

The phenomenology of the decays $\eta \to \gamma\gamma$ and $\eta' \to \gamma\gamma$ indicates that the mixing angle is in the range $20^0 < \Theta_{\eta\eta'} < 25^0$ while the estimate (16.6) corresponds to $\Theta_{\eta\eta'} \simeq 16^0$.

Once L_7 is pinned down, it is a straightforward matter to determine the value of the correction Δ_M which occurs in the mass formulae for the pseudoscalar mesons. Using the experimental values for Δ_{GMO} and for F_K/F_π, one arrives at the numbers shown in the upper right of Fig. 5. Clearly, the phenomenological information about the mixing angle indicates that Δ_M is small, consistent with zero. The corresponding quark mass ratios are obtained by intersecting the straight lines which represent a given value of Δ_M with the elliptic band discussed before. The figure also shows that the independent information about the ratio $R = (m_s - \hat{m})/(m_d - m_u)$ extracted from the baryon masses in ref.21 is in perfect agreement with these results.

17. Final Remarks

There are plenty of further applications of chiral perturbation theory. Although I did not make an attempt at a comprehensive survey in these lectures, I did discuss a few additional examples. In particular, I described the present status of our knowledge about the σ-term in πN scattering, a recurrent theme in the literature on chiral perturbation theory from the very beginning. This material is described in detail in ref.23. I also discussed the extension of chiral perturbation theory needed to analyze thermal expectation values. In particular, I reviewed some work concerning the melting of the quark condensate with rising temperature [24] and the kinetic properties of the hadronic phase [25]. Finally, some predictions of chiral perturbation theory concerning finite size effects were briefly discussed. The beautiful data on the distribution of the magnetization obtained from Monte Carlo simulations of the Heisenberg model in $d = 3$ and of the Higgs model in $d = 4$ [26] are in perfect agreement with the calculated shape of the constraint effective potential [27]. When the lattice simulations of QCD will reach realistic pion masses at sufficiently small lattice spacings, chiral perturbation theory should also prove to be useful there. The finite size effects in several correlation functions have been calculated explicitly [28] in view of these applications.

I did not discuss K-decays which are currently under active investigation within chiral perturbation theory [29]. The planned Dafne accelerator at Frascati will allow several of these processes to be investigated experimentally - the comparison with the theoretical analysis should lead to a considerable improvement in our understanding of the low energy structure of QCD. Chiral perturbation theory can also be used to analyze the low energy structure of the Higgs sector and of its analogue in extended versions of the Standard Model - this topic is discussed in detail in the lectures of Bagger [30].

In conclusion, let me underline the limitations of the method:

(i) The chiral expansion is useful only at low energies, small quark masses, low temperatures and large volumes

$$p, m_{quark}, T, \frac{1}{L} \ll \quad \text{scaleoftheory}.$$

(ii) The effective Lagrangian involves many coupling constants. Although theoretical a priori estimates can be given, an accurate determination of these constants requires phenomenological input.

Despite these limitations, chiral perturbation theory is an unambiguous framework which leads to results of physical interest.

References

1 J. Goldstone: Nuovo Cimento **19**, 154 1961

2 S. Coleman: Erice Lectures 1973, in *Laws of hadronic matter*, Academic Press London and New York (1975); reprinted in S. Coleman, *Aspects of symmetry*. Cambridge Univ. Press (1985)

3 S. Weinberg: Physica **A96**, 327 (1979)

4 H. Georgi: Weak interactions and modern particle theory (Benjamin/Cummings, Menlo Park, 1984)

5 M. Bando, T. Kugo and K. Yamawaki: Phys. Rep.**164**, 217 (1988)

6 S. Coleman, J. Wess and B. Zumino: Phys. Rev. **177**, 2239 (1969)

7 G. Boulware, L.S. Brown: Ann. Phys. (N.Y.) **138**, 392 (1982)

8 M. Gell-Mann, R.J. Oakes and B. Renner: Phys. Rev. **175**, 2195 (1968).

9 S. Weinberg: Phys. Rev. Lett. **17**, 616 (1966)

10 J. Gasser and H. Leutwyler: Ann. Phys. (N.Y.) **158**, 142 (1984). Nucl. Phys. **B250**, 465 (1985)

11 J. Wess and B. Zumino: Phys. Lett. **B37**, 95 (1971)

12 E. Witten: Nucl. Phys. B223, **422**, 433 (1983)

13 G. Ecker, J. Gasser, A. Pich and E. de Rafael: Nucl. Phys. **B321**, 311 (1989)

14 H. Lehmann: Acta Physica Austriaca Supp. **11**, 139 (1973)

15 J. Gasser and H. Leutwyler: Nucl. Phys. **B94**, 269 (1975)

16 R. Dashen: Phys. Rev. **183**, 1245 (1969)

17 T. Das, G. Guralnik, U. Mathur, F. Low and J. Young: Phys. Rev. Lett. **18**, 759 (1967)

18 R. Peccei and J. Sola: Nucl. Phys. **B281**, 1 (1987)

19 S. Weinberg: in A Festschrift for I.I.Rabi, ed. L. Motz: (New York Acad Sci., (1977) p.185

20 H. Leutwyler: Nucl. Phys. **B337**, 108 (1990)

21 D.B. Kaplan and A.V. Manohar: Phys. Rev. Lett. **56**, 2004 (1986)

22 K. Choi, C.W. Kim and W.K. Sze: Phys. Rev. Lett. **61**, 794 (1988); K. Choi and C.W. Kim: Phys. Rev. **D40**, 890 (1989); J.M. Cline: Phys. Rev. Lett. **63**, 1338 (1969); J.F. Donoghue: Ann. Rev. Nucl. Part. Sci. **39**, 1 (1989)

23 J. Gasser and H. Leutwyler: Phys. Rep. **87**, 77 (1982)

24 S. Weinberg: Phys. Rev. Lett. **18**, 188, 507 (1967)

25 J. Gasser, H. Leutwyler and M. Sainio: Phys. Lett. **B253**, 252, 260 (1991)

26 J. Gasser and H. Leutwyler: Phys. Lett. **B184**, 83 (1987) P. Gerber and H. Leutwyler: Nucl. Phys. **B321**, 387 (1989); For a review of chiral perturbation theory at finite temperatures and finite volumes, see H. Leutwyler: Nucl. Phys. B (Proc. Suppl.) **4**, 248 (1988)

27 J. Goity and H. Leutwyler: Phys. Lett. **228**, 517 (1989); H. Leutwyler and A. Smilga: Nucl. Phys. **B342**, 302 (1990); A.Schenk: Absorption and dispersion of pions at finite temperature. Nucl. Phys. in print; H. Bebie, P. Gerber, J. Goity and H. Leutwyler: The role of the entropy in an expanding hadronic gas, preprint University of Bern BUTP-91/14

28 I. Dimitrovic, J. Nager, K. Jansen and T. Neuhaus: Shape of the constraint effective potential: a Monte Carlo study, preprint BUTP-91/22; M. Göckeler, K. Jansen and T. Neuhaus: Constraint effective potential and the σ-mass in the $O(4)\phi^4$ theory, preprint HLRZ Zürich **228**, (1991)

29 M. Göckeler and H. Leutwyler: Nucl. Phys. **B350**, 228 (1991)

30 J. Gasser and H. Leutwyler: Phys. Lett. **B188**, 477 (1987); Nucl. Phys. **B397**, 763 (1988); P. Hasenfratz and H. Leutwyler: Nucl. Phys. **B343**, 241 (1990); F. Hansen: Nucl. Phys. **B345**, 685 (1990); F. Hansen and H. Leutwyler: Nucl. Phys. **B350**, 201 (1991)

31 J.F. Donoghue and B.R. Holstein: Phys. Rev. **D40**, 3700 (1989); B.R. Holstein: Phys. Rev. **D41**, 2829 (1990); J. Bijnens: Nucl. Phys. **B337**, 635 (1990); C. Riggenbach et al.: Phys. Rev. **D43**, 127 (1991). For a survey of rare kaon decays, see Pich A. Nucl. Phys. B (Proc. Suppl.) **16**, 378 (1990); G. Ecker talks given at topical meeting on CP violation, Calcutta (1990) and at XXIV Int. symp. on the theory of elementary particles, Gosen, Germany (1990), preprints University of Vienna UWTH 1990-6 and 1990-56; R. Battiston, D. Cocolicchio, G.L. Fogli and N. Paver: in Proposal for a ϕ-factory, preprint Frascati L.N.F.-90/031 (R) and CERN 5664/90

32 J. Bagger: Lectures given at the Theoretical Acavanced Study Institute in Elementary Particle Physics, Boulder (1991)

The Spins Inside the Proton and Chiral Symmetry Breaking

H. Fritzsch

Max-Planck-Institut für Physik und Astrophysik
Werner Heisenberg Institut für Physik
Föhringer Ring 6, München, Germany

1. Introduction

Using electron, muon and neutrino beams, the experimentalists have been able to investigate the internal structure of the nucleons. At high energies the nucleons look like composite systems consisting of the three valence quarks, of gluons and of the sea of $\bar{q}q$-pairs. The valence quarks determine the quantum numbers of the nucleons like the baryon number and the electric charge which are additive in the corresponding quantum numbers of the valence quarks. Of course, the $\bar{q}q$-pairs and the gluons do not contribute to these quantum numbers.

The question arises whether also the spin quantum number of the nucleon can be regarded as additive in the spins of the quarks. A priori, there is no reason why the $\bar{q}q$-pairs and the gluons should not contribute to the spin. However many properties of the baryons, e.g. their magnetic moments, can be described fairly well in a naive constituent model of static quarks, bound in a spin-independent potential. The underlying SU(6) symmetry implies that the spin of a baryon is indeed given by the spins of the three constituent quarks. In such a model there is no room for the $(\bar{q}q)$-pairs or the gluons to contribute to the spin.

Recently the results of the experiments in deep inelastic scattering of polarized leptons off polarized nucleons have led to speculations about the internal spin structure of the nucleon. In a naive interpretation of the experimental results one is led to the conclusion that the portion of the nucleon spin carried by the quarks is much smaller than expected — the data do not even exclude the possibility that this portion is zero [1].

Before we turn to a discussion of the phenomenon, we discuss the main physical results. We define the distribution functions of the quarks of flavor g and helicity $+1/2(-1/2)$ by $q_+(q_-)$. The lowest moment of the structure function g_1, measured in the deep inelastic scattering of polarized leptons off hadronic targets, is given by the moments of these quark densities Δq:

$$\int_0^1 g_1 dx = \frac{1}{2}\left(\frac{4}{9}\Delta u + \frac{1}{9}\Delta d + \frac{1}{9}\Delta s\right),$$
$$\left(\Delta u = \int_0^1 dx\,(u_+ + \bar{u}_+ - u_- - \bar{u}_-)\,etc.\right) \tag{1}$$

The spin density moments Δq are determined by the nucleon matrix elements of the associated axial vector currents (s_μ: spin vector):

$$\Delta q \cdot s_\mu = < p, s \mid \bar{q}\gamma_\mu\gamma_5 q | p, s > \tag{2}$$

The experimental data give [1]:

$$\int_0^1 g_1 dx = 0.114 \pm 0.012 \pm 0.026 \tag{3}$$

The Bjorken sum rule which follows from the algebra of currents in QCD, relates the difference of the u/d - moments to the axial vector coupling constant measured in β-decay:

$$\Delta u - \Delta d = g_A \tag{4}$$

(we neglect radiative corrections of the order of α_s/π).

Using SU_3 one finds:

$$g_A = F + D, \Delta u + \Delta d - 2\Delta s = 3F - D \tag{5}$$

Here F and D are defined by the axial vector matrix elements of the members of the baryon octet. An analysis of the hyperon decays gives:[2]

$$\begin{aligned} F &= 0.47 \pm 0.04 \quad D = 0.81 \pm 0.03 \\ \Delta u &= 0.74 \pm 0.10 \quad \Delta d = -0.54 \pm 0.10 \\ \Delta s &= -0.20 \pm 0.11 \\ \Delta\Sigma &= \Delta u + \Delta d + \Delta s = 0.01 \pm 0.29 \end{aligned} \tag{6}$$

An essential feature of the data is that the "sea" of the $\bar{s}s$-pairs in the nucleon appears to be highly polarized; it contributes significantly to the axial singlet charge. This implies that the "Zweig" rule does not seem to work for the matrix elements of the axialvector currents.

The quantity $\Delta\Sigma$ is the nucleon matrix element of the axial baryonic current and can be defined as the axial baryon charge of the nucleon:

$$\Delta\Sigma \cdot s_\mu = < p \mid \bar{u}\gamma_\mu\gamma_5 u + \bar{d}\gamma_\mu\gamma_5 d + \bar{s}\gamma_\mu\gamma_5 s \mid p > = < p \mid j_\mu^{05} \mid p > \tag{7}$$

In a naive wave function picture of the nucleon the axial baryon number corresponds to the portion of the nucleon spin carried by the quarks. Independent of a specific wave function model we can <u>define</u> $\Delta\Sigma$ as the relative amount of the nucleon spin carried by the intrinsic spins of the quarks. In the sim plest SU_6 type model of the nucleon this quantity is one. In reality it may depart significantly from one, due to the contributions of orbital momenta and of the $\bar{q}q$-pairs or the gluons to the nucleon spin. Nevertheless it is surprising to observe that $\Delta\Sigma$ seems to be small compared to one. However we emphasize that the experiments give solely an information about the axial baryonic charge of the nucleon, and not about the spin. Only in a nonrelativistic $SU(6)$ type model, in which the quarks move in an s-wave, the axial baryonic charge and the spin of the nucleon, multiplied by two, are both equal to one. There is no reason why $\Delta\Sigma$ could not be much less than one, or even zero, if we doubt the validity of the "naive" $SU(6)$ model.

Of course, a possible vanishing of the axial singlet nucleon charge must be discussed in view of the fact that the octet axial charges are, of course, different from zero. Nevertheless they depart substantially from the values one obtains in a nonrelativistic $SU(6)$ approach, which, for example, predicts $g_A/g_V = 5/3$, while in reality one has $g_A/g_V \approx 1,27$.

Furthermore the octet charges obey the Goldberger-Treiman relations, which relate the mass of the nucleon and the axial charges to the coupling and decay constants of the pseudoscalar mesons. The latter act is massless Nambu-Goldstone bosons in the chiral limit of QCD [3]. This suggests that also the value of the singlet axial charge is not unrelated to the chiral symmetry of QCD and its dynamical breaking. For this reason it is useful to examine the nucleon matrix element of the axial singlet current in this respect. First we consider it in the chiral limit of $SU(3)_L \times SU(3)_R$, in which $m_u = m_d = m_s = 0$. In this limit the octet of axial vector currents is conserved, while the singlet current is not conserved due to the gluonic anomaly:

$$\begin{aligned} \partial^\mu j_\mu^{i5} &= 0 \quad (i = 1, 2 \cdots 8) \\ \partial^\mu j_\mu^{i5} &= 3 \cdot \frac{\alpha_s}{2\pi} \cdot \text{ tr } G\tilde{G} = a \end{aligned} \tag{8}$$

It is known that this limit, in which the masses of the three light quark flavors are neglected, is not far away from the real world of hadrons. In the limit there exist eight massless pseudoscalar mesons, serving as the Goldstone bosons. However the ninth pseudoscalar, the η' – meson, remains massive and has a mass not far from its physical mass, i.e. about 900 MeV. The axialvector charges of the baryons are related to the coupling constants of the pseudoscalar mesons with the baryons by the Goldberger — Treiman relations, e.g. those for the pions (f_π : pion decay constant, M: nucleon mass):

$$2Mg_A = 2f_\pi g_{\pi NN} \tag{9}$$

We remind the reader how these relations are obtained. The matrix element of the axial vector current in the octet channel can be described by two form factors:

$$\bar{u}(p)[G_1(q^2)\gamma_\mu\gamma_5 - G_2(q^2)q_\mu\gamma_5]u(p), \quad q = p - p' \tag{10}$$

The induced pseudoscalar formfactor G_2 acquires a pole at $q^2 = 0$, since the pion mass vanishes in the chiral limit:

$$G_2(q^2) = \frac{2f_\pi g_{\pi NN}}{q^2} \tag{11}$$

Due to the conservation of the current one finds $2M \cdot G_1(0) = 2Mg_A = 2f_\pi \cdot g_{\pi NN}$. We stress that this relation follows as the result of an interplay between the axial vector form factors G_1 and G_2. It is the latter, which contains the pion pole. But the conservation of the current leads to the constraint about G_1, i.e. to a condition about the axial charge — the Goldberger — Treiman relation. In other words: the chiral symmetry allows us to convert a statement about the divergence of the axial vector current into a statement about the matrix element of the current. Due to the pole in G_2 one finds a nonzero matrix element, even though the current is conserved. In the absence of the pole the chiral symmetry would be trivially fulfilled — the nucleon mass would have to vanish.

Let us consider the nucleon matrix element of the axial baryonic current:

$$< p \mid j_\mu^{05} \mid p' >= \bar{u}(p)(G_1^0 \gamma_\mu \gamma_5 - G_2^0 q_\mu \gamma_5) u(p') \tag{12}$$

Here the induced pseudoscalar form factor does not have a Goldstone pole at $q^2 = 0$. Instead of the Goldberger Treiman relation one finds after taking the divergence and setting $q = 0$ [4,5]:

$$G_1^0(0) = \Delta\Sigma = A(0) \tag{13}$$

where A is the formfactor of the anomalous divergence:

$$< p \mid 3 \cdot \frac{\alpha_s}{2\pi} \text{ tr } G\tilde{G} \mid p >= 2MA(q^2)\bar{u}(p) i\gamma_5 u(p') \tag{14}$$

We conclude: The axial baryonic charge $\Delta\Sigma$ (the spin of the nucleon) is nothing but the nucleon matrix element of the anomalous divergence, i.e. a purely gluonic quantity. This quantity is essentially unknown. Eventually it may be calculated in the future within the lattice approach to QCD.

It is interesting to note that the fact that the singlet quantity $\Delta\Sigma$ is a gluonic quantity while the octet spin densities, e.g. $\Delta u + \Delta d - 2\Delta s$, are determined by the nucleon matrix elements of quark bilinears, indicates a substantial violation of the Zweig rule for the axial vector nonet. The latter would imply $\Delta s = 0$, and we would have $\Delta\Sigma = \Delta u + \Delta d + \Delta s = \Delta u + \Delta d - 2\Delta s$. Thus the matrix element of the anomalous divergence, a gluonic quantity, would have to be equal to the matrix element of the eighth component of the axial vector, a quark bilinear. Of course, there is no reason why this should be the case.

We conclude: The violation of the Zweig rule in the pseudoscalar channel, which is well-known and caused by the QCD anomaly, implies via the mechanism of spontaneous symmetry breaking another violation of this rule for the nucleon matrix elements of the axial vector current. The strength of this violation is given by the magnitude of the spin density moment Δs. Therefore it is not surprising that in particular this spin density moment appears to be large.

Apparently the violation of the Zweig rule is such that the axial singlet charge $\Delta\Sigma$ is rather small, perhaps even zero. Thus the constituent quark model needs a revision which must take into account this effect, being a consequence of the dynamics of chiral symmetry and its breaking. Below we shall discuss such a revision, which is able to combine both chiral dynamics and the naive constituent quark model [6,7,8].

First we consider a simplified case, namely the one of QCD with the two flavors u and d only. The strange quarks and the heavy flavors c, b and t are disregarded. Furthermore we assume $m_u = m_d = 0$, i.e. the chiral symmetry $SU(2)_L \times SU(2)_R$ is exactly fulfilled. The pions are massless. Due to the QCD anomaly the singlet pseudoscalar η (quark composition $(\bar{u}u + \bar{d}d)/2$) has a mass of the order of the nucleon mass M. The Goldberger - Treiman relation is exactly valid:

$$2M_n g_A = 2F_\pi g_{\pi NN} \tag{15}$$

In the SU(6) type constituent quark model the axialvektor coupling constant g_A is given by the nucleon expectation value of the quark spin operator $\frac{1}{2}\sigma_z$:

$$g_A =< \sigma_z(u) > - < \sigma_z(d) >= 5/3 \tag{16}$$

where one has:

$$1/2\sigma_z(u) = 2/3, \quad 1/2\sigma_z(d) = -1/6 \tag{17}$$

$1/2(\sigma_z(u) + \sigma_z(d)) = 1/2$ (= nucleon spin).

In reality g_A is not equal to 5/3, but about 1.27, i.e. the prediction of the "constituent model" is violated by about 24 %. This violation can be understood without giving up the simple ideas of the constituent quark model, as an effect due to orbital motions and relativistic effects. Thus in the isovector channel both the chiral dynamics and the constituent quark model do not in contradict, but rather supplement each other. This observation encourages us to consider the "constituent quarks" as separate entities. In a "Gedankenexperiment" we could consider a polarized "constituent quark" $Q(Q = U, D)$ and study its coupling constant. They would also obey a Goldberger - Treiman relation [9]:

$$2M_q \tilde{g}_A = 2F_\pi g_{\pi QQ} \tag{18}$$

(M_q: constituent quark mass, $\tilde{g}_A$ axialvector coupling constant of the constituent quark, $q_{\pi QQ}$: pion — quark coupling constant).

Suppose we consider the corresponding matrix elements of the vector and axialvector currents and relate them to the various moments of the quark density functions. One finds naively:

$$
\begin{aligned}
< U \mid \bar{u}\gamma_\mu u \mid U > &= p_\mu / M_U = \left[\frac{p_\mu}{M_U}\right] \int_0^1 (u_+ + u_- - \bar{u}_+ - \bar{u}_-) dx \\
< U \mid \bar{d}\gamma_\mu d \mid U > &= 0 \\
< U \mid \bar{u}\gamma_\mu\gamma_5 u \mid U > &= s_\mu \cdot \Delta u = s_\mu \cdot \int_0^1 (u_+ + \bar{u}_+ - u_- - \bar{u}_-) dx \\
&= s_\mu \cdot 1 \\
< U \mid \bar{d}\gamma_\mu\gamma_5 d \mid U > &= 0
\end{aligned}
\tag{19}
$$

(s_μ: spin vector, p_μ: four-momentum, the quark density functions refer to the U-quark and should carry an index u, which is not explicitly denoted here.) These relations reflect the expectation that in a constituent U-quark the quark density functions must be arranged such that the correct flavor structure is obtained and that its total spin is carried by the u-flavor. The d-flavor is not supposed to contribute to the spin.

We could go further and be more specific about the structure of the quark density functions. The success of the "Zweig" rule relies on the assumption that ($\bar{q}q$)-pairs contribute very little to the hadronic wave functions. Correspondingly we could consider a limit in which the ($\bar{q}q$)-pairs are neglected ("valence quark dominance"). In this limit we find for a U-quark:

$$
\begin{aligned}
\bar{u}_+ = \bar{u}_- = u_- = 0 \\
d_+ = d_- = \bar{d}_+ = 0
\end{aligned}
\tag{20}
$$

Only the density function u_+ is different from zero. This is easily understood if we consider the free quark model, in which the "constituent quarks" and the "current quarks" are identical and we have not only the relations (2), but in addition the function u_+ is known: $u_+ = \delta(x-1)$.

Thus the essential difference between a "constituent quark" inside a hadron and a free quark lies in the shape of the density function u_+. The confinement forces merely cause this function to depart from a δ-function and to spread out over the available x-range.

It turns out that the picture of a constituent quark described above is not consistent with the constraints given by the chiral symmetry. In the constituent model we have $\bar{d}_\pm = 0$. This implies for a U-quark that both the isoscalar and the isovector combinations of the spin density moments are equal to one: $\Delta u - \Delta d = \Delta u + \Delta d = 1$. The isovector part is determined by the pion pole. If the isosinglet η-meson would also be a Goldstone particle, the associated coupling constants would conspire such that the isovector and isoscalar spin density moments would be equal, and the results of the "naive" constituent model would be obtained. However due to the QCD anomaly the isosinglet spin density function does not receive a Goldstone pole contribution. Instead it is given by the constituent quark matrix element of the anomalous divergence:

$$
\Delta u + \Delta d = A
$$

$$< u \mid 2 \cdot \frac{\alpha_s}{2\pi} \text{ tr } G\tilde{G} \mid u > = 2M_u A\bar{u}i\gamma_5 u$$
$$\Delta u = (1 + A)/2 \quad \Delta d = (A - 1)/2 \tag{21}$$

There is no dynamical reason why A should be equal to one. If it were, the spin density moments would indeed reproduce the constituent quark model result. In particular Δd would vanish. This is not ruled out a priori, but if it were, it would be a miraculous coincidence.

For all other values of A the spin density moment Δd does not vanish. We conclude that for $A \neq 1$ the constituent U quark must contain $(\bar{q}q)$-pairs. Thus a violation of the "Zweig rule" is automatically implied. It is interesting that these pairs are generated by the same nonperturbative mechanism due to the gluon anomaly which causes the η-meson to acquire a mass and not to act as a Goldstone particle in the chiral limit of $SU(2)_L \times SU(2)_R$.

Intuitively one can understand the violation of the "Zweig rule" discussed above as follows. The chiral dynamics of a "constituent quark" would obey the "Zweig rule" if it were surrounded by a cloud of π and η Goldstone bosons. The Goldstone poles of the axialvector current matrix elements would imply, via the Goldberger — Treiman relations in the isovector and isoscalar channel, that the matrix elements obey the constraints given by the "Zweig rule" (in particular $\Delta d = 0$ for a u-quark etc.). However the QCD anomaly causes the η-pole at $q^2 = 0$ to disappear. As a result the "Goldstone cloud" of an U-quark consists only of π-mesons. Thus the dynamical structure of the constituent quark is drastically changed. In particular $(\bar{u}u)$ and $(\bar{d}d)$ pairs are generated.

We note that these pairs cannot simply be regarded as the pairs inside virtual π-mesons. Their presence is caused by the chiral dynamics, in particular by the Goldberger — Treiman relations for the axialvector matrix elements. Their appearance is a nonperturbative phenomenon just like the generation of the η-mass due to the gluonic anomaly.

In ref.10 the author has argued that the nucleon matrix element of the anomalous divergence should be very small. The argument is based on the observation that the "constituent quark model" requires the quarks in a nucleon to be in an s-wave. In particular the gluonic components of the wave function should also be dominantly in an s-wave. This implies that a pseudoscalar density like $\alpha_s \cdot G_{\mu\nu}\tilde{G}^{\mu\nu}$ is not expected to have a sizeable matrix element, in accordance with the experimental results.

It is easy to apply similar considerations to the constituent quarks. Also for them we should have $A = 0$, at least to a good approximation. Thus we find for a constituent U quark that the "Zweig rule" for the density moments is maximally violated:

$$\Delta u = 1/2, \quad \Delta d = -1/2. \tag{22}$$

We can go further and specify the various density moments. if the "Zweig rule" were valid (both π and η Goldstone modes present), we would have

$$\int u_+ dx = 1, \quad u_- = \bar{u}_+ = \bar{u}_- = 0 \quad d_+ = d_- = \bar{d}_+ = \bar{d}_- = 0 \tag{23}$$

Such a constraint which is not invariant under the renormalization group can only be imposed for a particular value of the energy scale μ, which is expected to be the characteristic hadronic energy scale. The removal of the η Goldstone pole causes a shift in the density moments, which we can parametrize by two functions h_+ and h_-:

$$\begin{aligned} u_+ &= u_+^v h_+, \quad \bar{u}_+ = d_+ = \bar{d}_+ = h_+, \\ u_- &= \bar{u}_- = d_- = \bar{d}_- = h_-. \end{aligned} \tag{24}$$

(u_+^v:intrinsic density function of U-quark in the absence of the anomaly, $\int u_+^v dx = 1$). We find:

$$\begin{aligned} \Delta\Sigma = \Delta u + \Delta d &= 1 + 4\int_0^1 (h_+ - h_-)dx = 0 \\ \Delta u - \Delta d &= 1 \end{aligned} \tag{25}$$

It follows:

$$\int_0^1 (h_+ - h_-)d_+ = -1/4. \tag{26}$$

We observe that $\Delta\Sigma$ vanishes because the constituent U quark contribution to $\Delta\Sigma$ is cancelled by the pairs. A cancellation is only possible, if the density function h_- is different from zero. On the other hand h_+ can be zero, in accordance with the sum rule (26). The simplest model obeying the constraints discussed above is one in which we have

$$\begin{aligned} &h_+ = 0, \qquad \int h_- dx = 1/4, \\ &u_+ = u_+^v, \qquad \int u_- dx = \int \bar{u}_- dx = \int d_- dx = \int \bar{d}_- dx = 1/4 \\ &d_+ = \bar{d}_+ = \bar{u}_+ = 0 \end{aligned} \tag{27}$$

Thus we obtain in the case $A = 0$ the following picture of a polarized constituent U quark in the $SU(2)_L \times SU(2)_R$ limit: The density function u_+, which describes the density of u-quarks polarized in the same direction as the U-quark, is unaffected by the QCD anomaly. The latter causes a large violation of the "Zweig rule"in the sense that $(\bar{q}q)$-pairs are generated. We shall refer to this "cloud" of $(\bar{q}q)$-pairs as the "anomaly-cloud". The density functions $\overset{(-)}{u}_-$ and $\overset{(-)}{d}_-$ are different from zero, i.e. the pairs are polarized oppositely to the original constituent quark. The sum of all (anti) quark spins is zero. Thus for $A = 0$ the quarks do not contribute to the spin of the constituent quark. The latter is provided by the orbital angular momentum of the pairs. This can be seen as follows. If we would turn off the QCD anomaly (e.g. formally by setting $n_c = \infty$), the "naive" picture should hold, i.e. the spin of the U-quark is carried continues to contribute its spin, but the $(\bar{q}q)$ pairs cancel the latter. Their total angular momentum must be zero. Otherwise the introduction of the anomaly would violate the conservation of angular momentum. Thus we have:

$$\begin{aligned} J_z(U) = +1/2 &= J_z(u_v) + J_z(cloud) + L_z(cloud) \\ &= +1/2 + (-1/2) + (+1/2) \end{aligned} \tag{28}$$

In the case $A \neq 0$ the cancellation between the spin of the valence quark and the spins of the "anomaly cloud" would not be complete, but the sum of the spins and of the orbited angular momenta of the pairs in the "anomaly cloud" would still be zero.

Thus far we have disregarded the polarization effects due to gluons. We find it unlikely, but not impossible that polarized gluons contribute also to the total angular momentum of the anomaly cloud, and the gluonic term would appear also in eq. (28).

Finally we consider the case of the three light flavors u, d, s. In the chiral limit of $SU(3)_L \times SU(3)_R$ we obtain for a constituent U quark in analogy to eq.(27):

$$\Delta U = 2/3 \quad \Delta d = -1/3 \quad \Delta s = -1/3. \tag{29}$$

In the symmetry limit the "anomaly cloud" is, of course, $SU(3)$ symmetric. In reality symmetry breaking will be present. The result will be that the effects of the $(\bar{s}s)$ pairs are somewhat reduced compared to those of the $(\bar{u}u)$ and $(\bar{d}d)$ pairs. For example, in a U-quark we expect: $\Delta d > \Delta s$. The actual spin density momenta of the U, D constituent quarks will lie between the extreme case of $SU(2) \times SU(2) (\Delta d = -1/2$ for a U-quark) and of $SU(3) \times SU(3) (\Delta d = -1/3$ for a U-quark). However we note that for $A = 0$ the limit of $SU(2)_L \times SU(2)_R$ is not ruled out experimentally. It would imply for a proton:

$$\Delta u + \Delta d = 0 \tag{30}$$

$$\Delta u - \Delta d = | g_A/g_V | = 1.26$$

$$\Delta u = 0.63 \quad \Delta d = -0.63$$

$$\int_0^1 g_1 dx = 1/2(4/9 \Delta u + 1/9 \Delta d) = 0.105,$$

while the experimental value for this integral is

$$\int_0^1 g_1 dx = 0.114 \pm 0.038 \tag{31}$$

In this limiting case the $SU(3)$ symmetry for the axialvector currents would be very badly broken. It is known that the $SU(3)$ breaking in the axialvector channel is sizeable, but certainly not as large as to allow the case $\Delta s = 0$. Details of the symmetry breaking cannot be presented here.

Recently it was argued that the anomaly could contribute to the axial singlet charge of gluons are highly polarized in a polarized nucleon. In this case their contribution to the singlet charge could be calculated perturbatively [11,12]. In our approach we see no reason for a large gluonic polarization. Thus the effect discussed in ref.11,12 would be negligible in comparison to the nonperturbative phenomenon discussed here.

The smallness of the axialsinglet charge, parametrized above by the parameter A, follows also within the Skyrme type model, discussed in ref.13. The connection of this model to the scheme discussed here remains unclear, although some common features exist. In our approach we would also expect that in the case of one flavor the spin of a constituent quark is cancelled partially or fully by the "anomaly cloud". Thus we see no qualitative difference between the cases of one or two (three, ...) flavors. On the other hand in the Skyrme model the case of one flavor is not defined.

The picture of "constituent quarks", carrying a polarized "anomaly cloud", described here, implies that many aspects of hadronic physics, especially those in which polarization and spin aspects are relevant, must be reconsidered. Among them are the magnetic moments of the baryons, the polarization phenomena of hyperons in hadronic processes and the spin asymmetries observed in strong interaction processes. Many further tests of the ideas presented here can be envisaged, once spin asymmetries can be measured in electroweak lepton-hadron reactions at high energies. The generation of a cloud of $(\bar{q}q)$-pairs by the QCD anomaly reminds us of the "Cooper pairs" in the BCS-theory of superconductivity. Indeed there are some analogies between superconductivity and hadronic physics in the chiral limit, e.g. the appearance of the mass gap, which in QCD is related to the anomaly as well as to the dynamical breaking of scale invariance and the chiral symmetry, and the presence of pairing forces, which in QCD are responsible for the removal of the Goldstone pole in the singlet axialvector channel.

In this lecture I have described why polarized constituent quarks should be surrounded by a cloud of polarized quark-antiquark pairs. Our reasoning was entirely based on phenomenological arguments. It would be interesting to see how these polarized pairs are generated dynamically, via those nonperturbative effects, due to instantons etc., which are responsible also for the QCD mass gap and the breaking of the chiral symmetry in the axial singlet channel. An explicit dynamical model along these lines has not yet evolved.

Acknowledgement: It is a pleasure to thank the organizers of the Schladming school, especially Prof. H. Mitter, for their hospitality in ths wonderful part of the Tauern region in the Alps.

References

1 G. Baum et al.: Phys. Rev. Lett. **51**, 1135 (1983) J. Ashman et al.: Phys. Lett. **B 206**, 364 (1988) U.W. Hughes et al.: Phys. Lett. **B 212**, 511 (1988)
2 M. Bourquin et al.: Zeitschrift für Physik **C21**, 27 (1983). R.L. Jaffe and A. Manohar: MIT-preprint (1989).
3 See e.g.: H. Pagels, Phys. Rep. **16**, 219 (1975)
4 H. Fritzsch: Phys. Lett. **229B**, 1605 (1989)
5 G. Veneziano: Mod. Phys. Lett. **A 17**, 1605 (1989)
6 H. Fritzsch: Phys. Lett. **256 B**, 75 (1991)
7 H. Fritzsch: Mod. Phys. Lett. **A5**, 625 (1990)
8 U. Ellwanger and B.U. Stech: of Heidelberg preprint 1990.
9 M. Goldberger and S.B: Treiman: Phys. Rev. **110**, 1178 (1958)
10 H. Fritzsch: Phys. Lett. **229B**, 122 (1989)
11 G. Altarelli and G. Roos: Phys. Lett. **B 212**, 391 (1988)
12 R. Carlitz, J. Collins and A. Mueller: Phys. Lett. **B 214**, 229 (1988)
13 S. Brodsky, J. Ellis and M. Karliner: Phys. Lett. **B 206**, 309 (1988)

Light-Cone Quantization of Quantum Chromodynamics

St.J. Brodsky[1], *H.-Ch. Pauli*[2]

[1] Stanford Linear Accelerator Center, Stanford University, Stanford, California, USA
[2] Max-Planck-Institut für Kernphysik, D-6900 Heidelberg 1, Germany

1. Abstract

We discuss the light-cone quantization of gauge theories from two perspectives: as a calculational tool for representing hadrons as QCD bound-states of relativistic quarks and gluons, and also as a novel method for simulating quantum field theory on a computer. The light-cone Fock state expansion of wavefunctions at fixed light cone time provides a precise definition of the parton model and a general calculus for hadronic matrix elements. We present several new applications of light-cone Fock methods, including calculations of exclusive weak decays of heavy hadrons, and intrinsic heavy-quark contributions to structure functions. A general non-perturbative method for numerically solving quantum field theories, "discretized light-cone quantization," is outlined and applied to several gauge theories, including QCD in one space and one time dimension, and quantum electrodynamics in physical space-time at large coupling strength. The DLCQ method is invariant under the large class of light-cone Lorentz transformations, and it can be formulated such that ultraviolet regularization is independent of the momentum space discretization. Both the bound-state spectrum and the corresponding relativistic light-cone wavefunctions can be obtained by matrix diagonalization and related techniques. We also discuss the construction of the light-cone Fock basis, the structure of the light-cone vacuum, and outline the renormalization techniques required for solving gauge theories within the light-cone Hamiltonian formalism.

2. Introduction

In quantum chromodynamics, hadrons are relativistic bound states of confined quark and gluon quanta. Although the momentum distributions of quarks in nucleons are well-determined experimentally from deep inelastic lepton scattering measurements, there has been relatively little progress in computing the basic wavefunctions of hadrons from first principles in QCD. The most interesting progress has come from lattice gauge theory [1,2] and QCD sum rule calculations, [3] both of which have given predictions for the lowest moments $\langle x_i^n \rangle$ of the proton's distribution amplitude, $\phi_p(x_i, Q)$. The distribution amplitude is the fundamental gauge invariant wavefunction which describes the fractional longitudinal momentum distributions of the valence quarks in a hadron integrated over transverse momentum up to the scale Q. [4] However, the results from the two analyses are in strong disagreement: The QCD sum rule analysis predicts a strongly asymmetric three-quark distribution (See Fig. 1), whereas the lattice results, [2] obtained in the quenched approximation, favor a symmetric distribution in the x_i. Models of the proton distribution amplitude based on a quark–di-quark structure suggest strong asymmetries and strong spin-correlations in the baryon wavefunctions.[5] Even less is known from first principles in non-perturbative QCD about the gluon and non-valence quark contributions to the proton wavefunction, although data from a number of experiments now suggest non-trivial spin correlations, a significant strangeness content, and a large x component to the charm quark distribution in the proton.[6]

There are many reasons why knowledge of hadron wavefunctions, particularly at the amplitude level, will be necessary for future progress in particle physics. For example, in electroweak theory, the central unknown required for reliable calculations of weak decay amplitudes are the hadronic matrix elements. The coefficient functions in the operator product expansion needed to compute many types of experimental quantities are essentially unknown and can only be estimated at this point. The calculation of form factors and exclusive scattering processes, in general, depend in detail on the basic amplitude structure of the scattering hadrons in a general Lorentz frame. Even the calculation of the magnetic moment of a proton requires wavefunctions in a boosted frame. We thus need a practical computational method for QCD which not only determines its spectrum, but also a method which can provide the non-perturbative hadronic matrix elements needed for general calculations in hadron physics.

It is clearly a formidable task to calculate the structure of hadrons in terms of their fundamental degrees of freedom in QCD. Even in the case of abelian quantum electrodynamics, very little is known about the nature of the bound state solutions in the large α, strong-coupling, domain. A calculation of bound state structure in QCD has to deal with many complicated aspects of the theory simultaneously: confinement, vacuum structure, spontaneous breaking of chiral symmetry (for massless quarks), while describing a relativistic many-body system which apparently has unbounded particle number.

The first step is to find a language in which one can represent the hadron in terms of relativistic confined quarks and gluons. The Bethe-Salpeter formalism has been the central method for analyzing hydrogenic atoms in QED, providing a completely covariant procedure for obtaining bound state solutions. However, calculations using this method are extremely complex and appear to be intractable much beyond the ladder approximation. It also appears im-

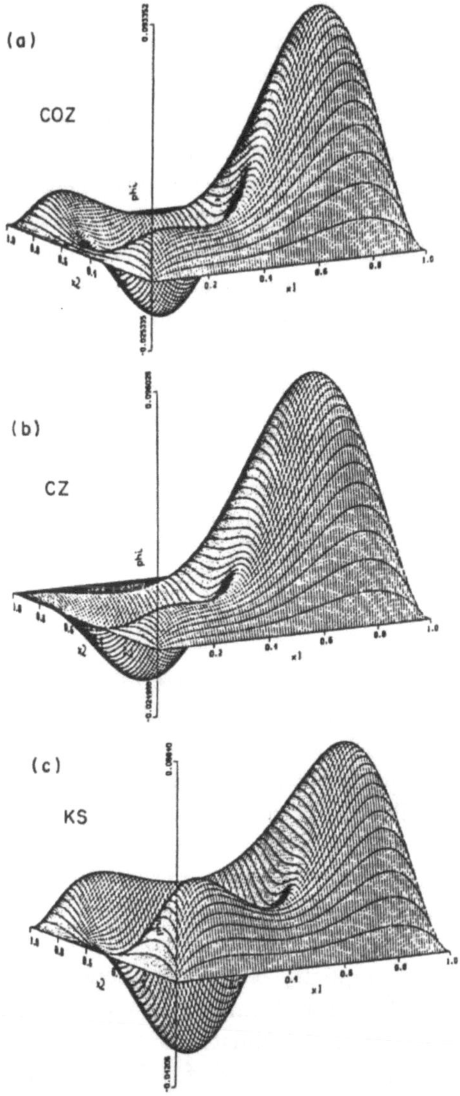

Fig. 1. The proton distribution amplitude $\phi_P(x_i, \mu)$ evaluated at the scale $\mu \sim 1\ GeV$ from QCD sum rules.[3] The enhancement at large x_1 correspond to a strong correlation between the a high momentum u quark with spin parallel to the proton spin.

practical to extend this method to systems with more than a few constituent particles.

An intuitive approach for solving relativistic bound-state problems would be to solve the Hamiltonian eigenvalue problem

$$H\,|\psi\rangle = \sqrt{\vec{P}^{\,2} + M^2}\,|\psi\rangle$$

for the particle's mass, M, and wavefunction, $|\psi\rangle$. Here, one imagines that $|\psi\rangle$ is an expansion in multi-particle occupation number Fock states, and that the operators H and $\vec{P}$ are second-quantized Heisenberg picture operators. Unfortunately, this method, as described by Tamm and Dancoff, [7] is severely complicated by its non-covariant structure and the necessity to first understand its complicated vacuum eigensolution over all space and time. The presence of the square root operator also presents severe mathematical difficulties. Even if these problems could be solved, the eigensolution is only determined in its rest system; determining the boosted wavefunction is as complicated as diagonalizing H itself.

Fortunately, "light-cone" quantization, the Lorentz-frame-independent method we shall emphasize in these lectures, offers an elegant avenue of escape.[8] The square root operator does not appear in light-cone formalism, and the vacuum structure is relatively simple; for example, there is no spontaneous creation of massive fermions in the light-cone quantized vacuum.

2.1 Quantization on the Light-Cone

There are, in fact, many reasons to quantize relativistic field theories at fixed light-cone time $\tau = t + z/c$. Dirac,[9] in 1949, showed that a maximum number of Poincare generators become independent of the dynamics in the "front form" formulation, including certain Lorentz boosts. In fact, unlike the traditional equal-time Hamiltonian formalism, quantization on the light-cone can be formulated without reference to the choice of a specific Lorentz frame; the eigensolutions of the light-cone Hamiltonian thus describe bound states of arbitrary four-momentum, allowing the computation of scattering amplitudes and other dynamical quantities. However, the most remarkable feature of this formalism is the apparent simplicity of the light-cone vacuum. In many theories the vacuum state of the free Hamiltonian is an eigenstate of the total light-cone Hamiltonian. The Fock expansion constructed on this vacuum state provides a complete relativistic many-particle basis for diagonalizing the full theory.

2.2 General Features of Light-Cone Quantization

In general, the Hamiltonian is the "time" evolution operator $H = i\frac{\partial}{\partial\tau}$ which propagates fields from one space-like surface to another. As emphasized by Dirac,[9] there are several choices for the evolution parameter τ. In the "Instant Form" $\tau = t$ is the ordinary Cartesian time. In the "Front Form," or light-cone quantization, one chooses $\tau = t + z/c$ as the light-cone coordinate with boundary conditions specified as a function of x, y, and $z^- = ct - z$. Another possible choice is the "point form," where $\tau = \sqrt{c^2t^2 - \vec{x}^2}$. Notice that all three forms become equivalent in the non-relativistic limit where, effectively, $c \to \infty$. A comparison of light-cone quantization with equal-time quantization is shown in Table 1.

	Instant Form	Front Form
Hamiltonian	$H = \sqrt{\overrightarrow{P}^2 + M^2} + V$	$P^- = \frac{P_\perp^2 + M^2}{P^+} + V$
Conserved quantities	$E, \ \overrightarrow{P}$	$P^-, \ P^+, \ \overrightarrow{P}_\perp$
Momenta	$P_z <> 0$	$P^+ > 0$
Bound state equation	$H\psi = E\psi$	$P^+P^-\psi = M^2\psi$
Vacuum	Complicated	Trivial

Table 1. A comparison of light-cone and equal-time quantization.

Although the instant form is the conventional choice for quantizing field theory, it has many practical disadvantages. For example, given the wavefunction of an n-electron atom, $\psi_n(\vec{x}_i, t = 0)$, at initial time $t = 0$, then, in principle, one can use the Hamiltonian H to evolve $\psi_n(\vec{x}_i, t)$ to later times t. However, an experiment which could specify the initial wavefunction would require the simultaneous measurement of the positions of all of the bound electrons, such as by the simultaneous Compton scattering of n independent laser beams on the atom. In contrast, determining the initial wavefunction at fixed light-cone time $\tau = 0$ only requires an experiment which scatters one plane-wave laser beam, since the signal reaching each of the n electrons is received along the light front at the same light-cone time $\tau = t_i + z_i/c$.

As we shall discuss in these lectures, light cone quantization allows a precise definition of the notion that a hadron consists of confined quarks and gluons. In light-cone quantization, a free particle is specified by its four momentum $k^\mu = (k^+, k^-, k_\perp)$ where $k^\pm = k^0 \pm k^3$. If the particle is on its mass shell and has positive energy, its light-cone energy is also positive: $k^- = (k_\perp^2 + m^2)/k^+ > 0$. In perturbation theory, transverse momentum $\sum k_\perp$ and the plus momentum $\sum k^+$ are conserved at each vertex. The light-cone bound-state wavefunction thus describes constituents which are on their mass shell, but off the light-cone energy shell: $P^- < \sum k^- i$.

As we shall show explicitly, one can construct a complete basis of free Fock states (eigenstates of the free light-cone Hamiltonian) $|n\rangle \langle n| = I$ in the usual way by applying products of free field creation operators to the vacuum state $|0\rangle$:

$$|0\rangle$$
$$|q\bar{q}:\underline{k}_i\lambda_i\rangle = b^\dagger(\underline{k}_1\lambda_1)\,d^\dagger(\underline{k}_2\lambda_2)\;|0\rangle$$

$$|q\bar{q}g:\underline{k}_i\lambda_i\rangle = b^\dagger(\underline{k}_1\lambda_1)\,d^\dagger(\underline{k}_2\lambda_2)\,a^\dagger(\underline{k}_3\lambda_3)\;|0\rangle$$

$$\vdots$$

where $b^\dagger$, $d^\dagger$ and $a^\dagger$ create bare quarks, antiquarks and gluons having three-momenta $\underline{k}_i$ and helicities λ_i.

Note, however, that in principle In the case of a theory such as QED, with massive fermions, all states containing particles have quanta with positive k^+, and the zero-particle state cannot mix with the other states in the basis. [10] The free vacuum in such theories is thus an exact eigenstate of H_{LC}. However, as we shall discuss in later sections, the vacuum in QCD is undoubtedly more complicated due to the possibility of color-singlet states with $P^+ = 0$ built on four or more zero-mode massless gluon quanta.

The restriction $k^+ > 0$ for massive quanta is a key difference between light-cone quantization and ordinary equal-time quantization. In equal-time quantization, the state of a parton is specified by its ordinary three-momentum $\vec{k}_\perp = (k^1, k^2, k^3)$. Since each component of $\vec{k}_\perp$ can be either positive or negative, there exist zero total momentum Fock states of arbitrary particle number, and these will mix with the zero-particle state to build up the ground state. However, in light-cone quantization each of the particles forming a zero-momentum state must have vanishingly small k^+. Such a configuration represents a point of measure zero in the phase space, and therefore such states can usually be neglected.

Actually some care must be taken here, since there are operators in the theory that are singular at $k^+ = 0$—*e.g.* the kinetic energy $(\vec{k}_\perp^{\,2} + M^2)/k^+$. In certain circumstances, states containing $k^+ \to 0$ quanta can significantly alter the ground state of the theory. One such circumstance is when there is spontaneous symmetry breaking. Another is the complication due to massless gluon quanta in a non-Abelian gauge theory. Nevertheless, the space of states that can play a role in the vacuum structure is much smaller for light-cone quantization than for equal-time quantization. This suggests that vacuum structure may be far simpler to analyze using the light-cone formulation.

Even in perturbation theory, light-cone quantization has overwhelming advantages over standard time-ordered perturbation theory. For example, in order to calculate a Feynman amplitude of order g^n in TOPTH one must suffer the calculation of the sum of n time-ordered graphs, each of which is a non-covariant function of energy denominators which, in turn, consist of sums of complicated square roots $p_i^0 = \sqrt{\vec{p}_i^{\,2} + m_i^2}$. On the other hand, in light-cone perturbation theory (LCPTH), only a few graphs give non-zero contributions, and those that are non-zero have light-cone energy denominators which are simple sums of rational forms $p^- = (\vec{p}_{\perp i}^{\,2} + m_i^2)/p_i^+$.

Probably the worst problem in TOPTH are the contributions from vacuum creation graphs, as illustrated for QED in Fig. 2(a). In TOPTH, all intermediate states contribute to the total amplitude as long as three-momentum is conserved; in this case $\vec{p_1} + \vec{p_2} + \vec{k} = \overrightarrow{0}$. The existence of vacuum creation and annihilation graphs implies that one cannot even compute any current matrix element without considering the effect of the currents arising from pair production from the vacuum. This is illustrated in Fig. 2(b). In contrast, in light-cone perturbation theory (LCPTH), an intermediate state contributes only if the total $\vec{p}_\perp$ and p^+ are conserved. In the case of vacuum creation graphs in QED, this implies $\vec{p}_{1\perp} + \vec{p}_{2\perp} + \vec{p}_{3\perp} = \overrightarrow{0}_\perp$ and $p_1^+ + p_2^+ + k_3^+ = 0$. However, the latter condition cannot be satisfied since each massive fermion has strictly positive $p_i^+ > 0$. Thus aside from theories which permit zero modes, there are no vacuum creation graphs in LCPTH.

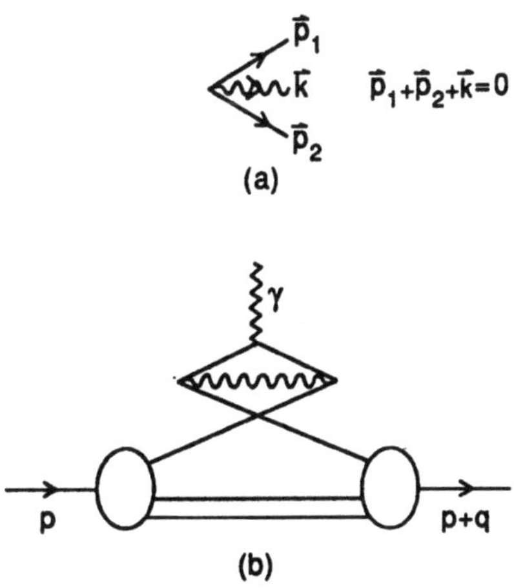

Fig. 2. (a) Illustration of a vacuum creation graph in time-ordered perturbation theory. A corresponding contribution to the form factor of a bound state is shown in figure (b).

In fact, light-cone perturbation theory is sufficiently simple that it provides in many cases a viable alternative to standard covariant (Feynman) perturbation theory. Each loop of a τ-ordered diagram requires a three-dimensional integration over the transverse momentum $d^2\vec{k}_{i\perp}$ and light-cone momentum fraction $x_i = k_i^+/p^+$ with $(0 < x_i < 1.)$ For example, the lowest order Schwinger contribution to the electron anomalous magnetic moment, $a = \frac{1}{2}(g-2) = \frac{\alpha}{2\pi}$, is easily computed from just one LCPTH diagram. (See Fig. 3). Calculations of the higher order terms in α require renormalization in the context of light-cone Hamiltonian field theory. As shown in Ref. 11 renormalization in LCPTH can be carried out in close correspondence to Lagrangian methods. In the case of QED one can use the Pauli-Villars method to regulate the ultra-violet divergences. Then for each τ-ordered diagram with divergent subgraphs, the required local counter-term can be computed using the method of "alternating denominators." [11] A simple example for one LCPTH graph for Compton scattering is

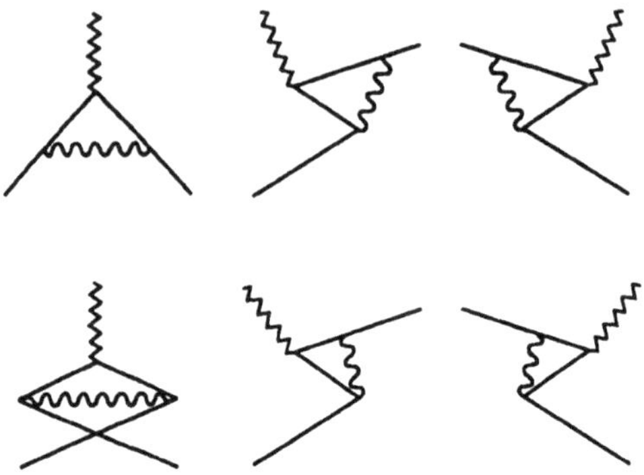

Fig. 3. Time-ordered contributions to the electron's anomalous magnetic moment. In light-cone quantization with $q^+ = 0$, only graph (a) needs to be computed to obtain the Schwinger result.

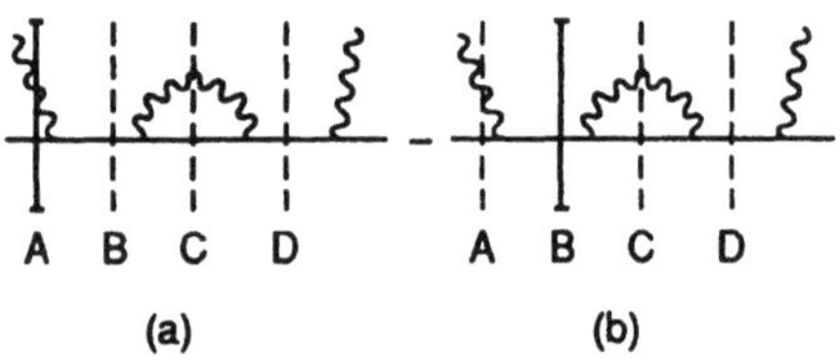

Fig. 4. Construction of a renormalized amplitude in LCPTH using the method of alternating denominators.[11] The mass renormalization counterterm is constructed locally in momentum space in graph (b) by substituting the light-cone energy difference $P_C^- = P_B^-$ rather than $P_C^- - P_A^-$.

shown in Fig. 4. Additional divergences which occur due to the γ^- couplings (in covariant gauges) can be eliminated by subtraction of the divergent amplitude subgraph at $p^+ = 0$.[12]

One of the most interesting applications of LCPTH would be the perturbative calculation of the annihilation cross section $R_{e^+e^-}$, since one would automatically calculate, to the same order in perturbation theory, the quark and gluon jet distributions appearing in the final state. It is advantageous to use the light-cone gauge $A^+ = 0$ since one wants to describe gluon distributions with physical polarization. The extra complications in the renormalization procedure induced by a non-covariant axial gauge have recently been discussed by Langnau and Burkardt. [12] A non-perturbative light-cone quantization calculation of $R_{e^+e^-}$ for QED in one space and one time has been given by Hiller.[13] We will return to these developments in later sections.

2.3 Representation of Hadrons on the Light-Cone Fock Basis

One of the most important advantages of light-cone quantization is that the light-cone Fock expansion can be used as the basis for representing the physical states of QCD. For example, a pion with momentum $\underline{P} = (P^+, \overrightarrow{P}_\perp)$ is described by the expansion,

$$|\pi : \underline{P}\rangle = \sum_{n,\lambda_i} \int \overline{\prod_i} \frac{dx_i d^2\vec{k}_{\perp i}}{\sqrt{x_i} 16\pi^3} \left| n : x_i P^+, x_i \overrightarrow{P}_\perp + \vec{k}_{\perp i}, \lambda_i \right\rangle \psi_{n/\pi}(x_i, \vec{k}_{\perp i}, \lambda_i)$$

where the sum is over all Fock states and helicities, and where

$$\overline{\prod_i} dx_i \equiv \prod_i dx_i \, \delta \left(1 - \sum_j x_j \right)$$

$$\overline{\prod_i} d^2\vec{k}_{\perp i} \equiv \prod_i d^2\vec{k}_{\perp i} \, 16\pi^3 \, \delta^2 \left(\sum_j \vec{k}_{\perp j} \right) .$$

The wavefunction $\psi_{n/\pi}(x_i, \vec{k}_{\perp i}, \lambda_i)$ is thus the amplitude for finding partons in a specific light-cone Fock state n with momenta $(x_i P^+, x_i \overrightarrow{P}_\perp + \vec{k}_{\perp i})$ in the pion. The Fock state is off the light-cone energy shell: $\sum k_i^- > P^-$. The light-cone momentum coordinates x_i, with $\sum_{i=1}^n x_i$ and $\vec{k}_{\perp i}$, with $\sum_{i=1}^n \vec{k}_{\perp i} = \overrightarrow{0}_\perp$, are actually relative coordinates; *i.e.* they are independent of the total momentum P^+ and $P_\perp$ of the bound state. The special feature that light-cone wavefunctions do not depend on the total momentum is not surprising, since x_i is the longitudinal momentum fraction carried by the ith-parton $(0 \le x_i \le 1)$, and $\vec{k}_{\perp i}$ is its momentum "transverse" to the direction of the meson. Both of these are frame independent quantities. The ability to specify wavefunctions simultaneously in any frame is a special feature of light-cone quantization.

In the light-cone Hamiltonian quantization of gauge theories, one chooses the light-cone gauge, $\eta \cdot A = A^+ = 0$, for the gluon field. The use of this gauge results in well-known simplifications in the perturbative analysis of light-cone dominated processes such as high-momentum hadronic form factors. It is indispensable if one desires a simple, intuitive Fock-state basis since there are neither negative-norm gauge boson states nor ghost states in $A^+ = 0$ gauge. Thus each term in the normalization condition

$$\sum_{n,\lambda_i} \int \overline{\prod_i} \frac{dx_i d^2\vec{k}_{\perp i}}{16\pi^3} \left|\psi_{n/\pi}(x_i, \vec{k}_{\perp i}, \lambda_i)\right|^2 = 1$$

is positive.

The coefficients in the light-cone Fock state expansion are the parton wavefunctions $\psi_{n/H}(x_i, \vec{k}_{\perp i}, \lambda_i)$ which describe the decomposition of each hadron in terms of its fundamental quark and gluon degrees of freedom. The light-cone variable $0 < x_i < 1$ is often identified with the constituent's longitudinal momentum fraction $x_i = k_i^z / P_z$, in a frame where the total momentum

$P^z \to \infty$. However, in light-cone Hamiltonian formulation of QCD, x_i is the boost-invariant light cone fraction,

$$x_i \equiv \frac{k_i^+}{P^+} = \frac{k_i^0 + k_i^z}{P^0 + P^z} ,$$

independent of the choice of Lorentz frame.

2.4 Calculation of Hadronic Processes from Light-Cone Wavefunctions

Given the light-cone wavefunctions, $\psi_{n/H}(x_i, \vec{k}_{\perp i}, \lambda_i)$, one can compute virtually any hadronic quantity by convolution with the appropriate quark and gluon matrix elements. For example, the leading-twist structure functions measured in deep inelastic lepton scattering are immediately related to the light-cone probability distributions:

$$2M\, F_1(x,Q) = \frac{F_2(x,Q)}{x} \approx \sum_a e_a^2\, G_{a/p}(x,Q)$$

where

$$G_{a/p}(x,Q) = \sum_{n,\lambda_i} \int \overline{\prod_i} \frac{dx_i d^2\vec{k}_{\perp i}}{16\pi^3} |\psi_n^{(Q)}(x_i, \vec{k}_{\perp i}, \lambda_i)|^2 \sum_{b=a} \delta(x_b - x)$$

is the number density of partons of type a with longitudinal momentum fraction x in the proton. This follows from the observation that deep inelastic lepton scattering in the Bjorken-scaling limit occurs if x_{bj} matches the light-cone fraction of the struck quark. (The $\sum_b$ is over all partons of type a in state n.) However, the light cone wavefunctions contain much more information for the final state of deep inelastic scattering, such as the multi-parton distributions, spin and flavor correlations, and the spectator jet composition.

As was first shown by Drell and Yan, [14] it is advantageous to choose a coordinate frame where $q^+ = 0$ to compute form factors $F_i(q^2)$, structure functions, and other current matrix elements at spacelike photon momentum. With such a choice the quark current cannot create pairs, and $\langle p'|j^+|p\rangle$ can be computed as a simple overlap of Fock space wavefunctions; all off-diagonal terms involving pair production or annihilation by the current or vacuum vanish. In the interaction picture one can equate the full Heisenberg current to the quark current described by the free Hamiltonian at $\tau = 0$. Accordingly, the form factor is easily expressed in terms of the pion's light cone wavefunctions by examining the $\mu = +$ component of this equation in a frame where the photon's momentum is transverse to the incident pion momentum, with $\vec{q}_\perp^{\,2} = Q^2 = -q^2$. The spacelike form factor is then just a sum of overlap integrals analogous to the corresponding nonrelativistic formula: [14] (See Fig. 5.)

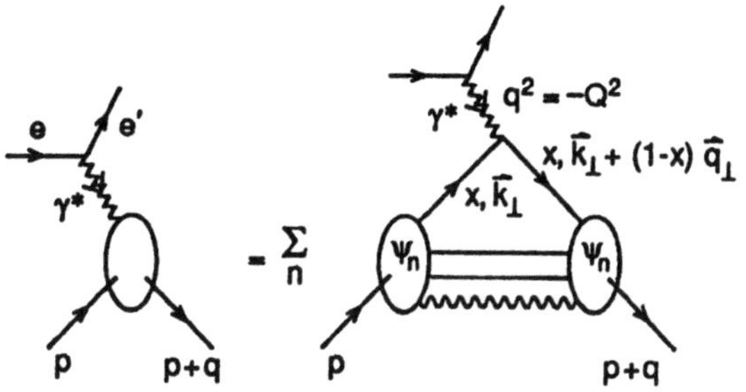

Fig. 5. Calculation of the form factor of a bound state from the convolution of light-cone Fock amplitudes. The result is exact if one sums over all ψ_n.

$$F(Q^2) = \sum_{n,\lambda_i} \sum_a e_a \int \prod_i \frac{dx_i\, d^2\vec{k}_{\perp i}}{16\pi^3}\, \psi_n^{(\Lambda)*}(x_i, \vec{\ell}_{\perp i}, \lambda_i)\, \psi_n^{(\Lambda)}(x_i, \vec{k}_{\perp i}, \lambda_i).$$

Here e_a is the charge of the struck quark, $\Lambda^2 \gg \vec{q}_\perp^{\,2}$, and

$$\vec{\ell}_{\perp i} \equiv \begin{cases} \vec{k}_{\perp i} - x_i \vec{q}_\perp + \vec{q}_\perp & \text{for the struck quark} \\ \vec{k}_{\perp i} - x_i \vec{q}_\perp & \text{for all other partons.} \end{cases}$$

Notice that the transverse momenta appearing as arguments of the first wavefunction correspond not to the actual momenta carried by the partons but to the actual momenta minus $x_i \vec{q}_\perp$, to account for the motion of the final hadron. Notice also that $\vec{\ell}_\perp$ and $\vec{k}_\perp$ become equal as $\vec{q}_\perp \to 0$, and that $F_\pi \to 1$ in this limit due to wavefunction normalization. All of the various form factors of hadrons with spin can be obtained by computing the matrix element of the plus current between states of different initial and final hadron helicities.[15]

As we have emphasized above, in principle, the light-cone wavefunctions determine all properties of a hadron. The general rule for calculating an amplitude involving wavefunction $\psi_n^{(\Lambda)}$, describing Fock state n in a hadron with $\underline{P} = (P^+, \vec{P}_\perp)$, has the form[4] (see Fig. 6):

$$\sum_{\lambda_i} \int \prod_i \frac{dx_i d^2\vec{k}_{\perp i}}{\sqrt{x_i} 16\pi^3}\, \psi_n^{(\Lambda)}(x_i, \vec{k}_{\perp i}, \lambda_i)\; T_n^{(\Lambda)}(x_i P^+, x_i \vec{P}_\perp + \vec{k}_{\perp i}, \lambda_i)$$

where $T_n^{(\Lambda)}$ is the irreducible scattering amplitude in LCPTh with the hadron replaced by Fock state n. If only the valence wavefunction is to be used, $T_n^{(\Lambda)}$ is irreducible with respect to the valence Fock state only; *e.g.* $T_n^{(\Lambda)}$ for a pion has no $q\bar{q}$ intermediate states. Otherwise contributions from all Fock states must be summed, and $T_n^{(\Lambda)}$ is completely irreducible.

The leptonic decay of the $\pi^\pm$ is one of the simplest processes to compute since it involves only the $q\bar{q}$ Fock state. The sole contribution to π^- decay is from

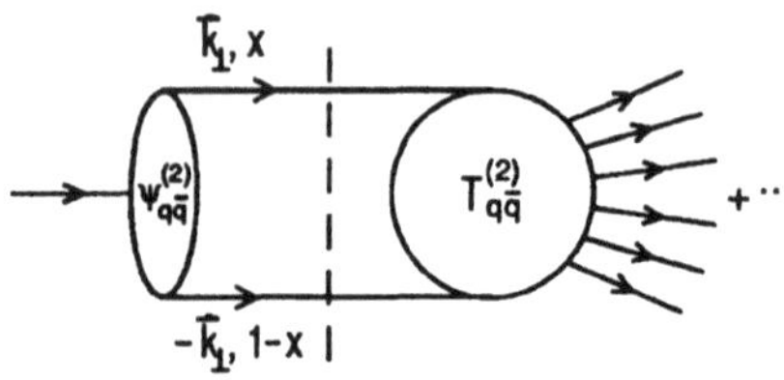

Fig. 6. Calculation of hadronic amplitudes in the light-cone Fock formalism.

$$\langle 0 \mid \bar{\psi}_u \gamma^+ (1-\gamma_5) \psi_d \mid \pi^- \rangle = -\sqrt{2} P^+ f_\pi$$

$$= \int \frac{dx\, d^2\vec{k}_\perp}{16\pi^3}\, \psi^{(\Lambda)}_{d\bar{u}}(x, \vec{k}_\perp) \frac{\sqrt{n_c}}{\sqrt{2}} \left\{ \frac{\bar{v}_\downarrow}{\sqrt{1-x}} \gamma^+ (1-\gamma_5) \frac{u_\uparrow}{\sqrt{x}} \;+\; (\uparrow \leftrightarrow \downarrow) \right\}$$

where $n_c = 3$ is the number of colors, $f_\pi \approx 93$ MeV, and where only the $L_z = S_z = 0$ component of the general $q\bar{q}$ wavefunction contributes. Thus we have

$$\int \frac{dx\, d^2\vec{k}_\perp}{16\pi^3}\, \psi^{(\Lambda)}_{d\bar{u}}(x, \vec{k}_\perp) = \frac{f_\pi}{2\sqrt{3}}.$$

This result must be independent of the ultraviolet cutoff Λ of the theory provided Λ is large compared with typical hadronic scales. This equation is an important constraint upon the normalization of the $d\bar{u}$ wavefunction. It also shows that there is a finite probability for finding a π^- in a pure $d\bar{u}$ Fock state.

The fact that a hadron can have a non-zero projection on a Fock state of fixed particle number seems to conflict with the notion that bound states in QCD have an infinitely recurring parton substructure, both from the infrared region (from soft gluons) and the ultraviolet regime (from QCD evolution to high momentum). In fact, there is no conflict. Because of coherent color-screening in the color-singlet hadrons, the infrared gluons with wavelength longer than the hadron size decouple from the hadron wavefunction.

The question of parton substructure is related to the resolution scale or ultraviolet cut-off of the theory. Any renormalizable theory must be defined by imposing an ultraviolet cutoff Λ on the momenta occurring in theory. The scale Λ is usually chosen to be much larger than the physical scales μ of interest; however it is usually more useful to choose a smaller value for Λ, but at the expense of introducing new higher-twist terms in an effective Lagrangian:[16]

$$\mathcal{L}^{(\Lambda)} = \mathcal{L}_0^{(\Lambda)}(\alpha_s(\Lambda), m(\Lambda)) + \sum_{n=1}^{N} \left(\frac{1}{\Lambda}\right)^n \delta\mathcal{L}_n^{(\Lambda)}(\alpha_s(\Lambda), m(\Lambda)) + \mathcal{O}\left(\frac{1}{\Lambda}\right)^{N+1}$$

where

$$\mathcal{L}_0^{(\Lambda)} = -\frac{1}{4} F^{(\Lambda)}_{a\mu\nu} F^{(\Lambda)a\mu\nu} + \bar{\psi}^{(\Lambda)} \left[i\gamma \cdot D^{(\Lambda)} - m(\Lambda) \right] \psi^{(\Lambda)} .$$

The neglected physics of parton momenta and substructure beyond the cutoff scale has the effect of renormalizing the values of the input coupling constant

$g(\Lambda^2)$ and the input mass parameter $m(\Lambda^2)$ of the quark partons in the Lagrangian.

One clearly should choose Λ large enough to avoid large contributions from the higher-twist terms in the effective Lagrangian, but small enough so that the Fock space domain is minimized. Thus if Λ is chosen of order 5 to 10 times the typical QCD momentum scale, then it is reasonable to hope that the mass, magnetic moment and other low momentum properties of the hadron could be well-described on a Fock basis of limited size. Furthermore, by iterating the equations of motion, one can construct a relativistic Schrödinger equation with an effective potential acting on the valence lowest-particle number state wavefunction.[4] Such a picture would explain the apparent success of constituent quark models for explaining the hadronic spectrum and low energy properties of hadron.

It should be emphasized that infinitely-growing parton content of hadrons due to the evolution of the deep inelastic structure functions at increasing momentum transfer, is associated with the renormalization group substructure of the quarks themselves, rather than the "intrinsic" structure of the bound state wavefunction.[17] The fact that the light-cone kinetic energy $\left\langle \frac{\vec{k}_\perp^2+m^2}{x} \right\rangle$ of the constituents in the bound state is bounded by Λ^2 excludes singular behavior of the Fock wavefunctions at $x \to 0$. There are several examples where the light-cone Fock structure of the bound state solutions is known. In the case of the super-renormalizable gauge theory, $QED(1+1)$, the probability of having non-valence states in the light-cone expansion of the lowest lying meson and baryon eigenstates to be less than 10^{-3}, even at very strong coupling.[18] In the case of QED(3+1), the lowest state of positronium can be well described on a light-cone basis with two to four particles, $|e^+e^-\rangle$, $|e^+e^-\gamma\rangle$, $|e^+e^-\gamma\gamma\rangle$, and $|e^+e^-e^+e^-\rangle$; in particular, the description of the Lamb-shift in positronium requires the coupling of the system to light-cone Fock states with two photons "in flight" in light-cone gauge. The ultraviolet cut-off scale Λ only needs to be taken large compared to the electron mass. On the other hand, a charged particle such as the electron does not have a finite Fock decomposition, unless one imposes an artificial infrared cut-off.

We thus expect that a limited light-cone Fock basis should be sufficient to represent bound color-singlet states of heavy quarks in QCD(3+1) because of the coherent color cancellations and the suppressed amplitude for transversely-polarized gluon emission by heavy quarks. However, the description of light hadrons is undoubtedly much more complex due to the likely influence of chiral symmetry breaking and zero-mode gluons in the light-cone vacuum. We return to this problem later.

Even without solving the QCD light-cone equations of motion, we can anticipate some general features of the behavior of the light-cone wavefunctions. Each Fock component describes a system of free particles with kinematic invariant mass squared:

$$\mathcal{M}^2 = \sum_i^n \frac{\vec{k}_{\perp i}^2 + m_i^2}{x_i},$$

On general dynamical grounds, we can expect that states with very high $\mathcal{M}^2$ are suppressed in physical hadrons, with the highest mass configurations computable from perturbative considerations. We also note that $\ell n\ x_i = \ell n\ \frac{(k^0+k^z)_i}{(P^0+P^z)} = y_i - y_P$ is the rapidity difference between the constituent with light-cone fraction x_i and the rapidity of the hadron itself. Since correlations between particles rarely extend over two units of rapidity in hadron physics, this argues that constituents which are correlated with the hadron's quantum numbers are primarily found with $x > 0.2$.

The limit $x \to 0$ is normally an ultraviolet limit in a light-cone wavefunction. Recall, that in any Lorentz frame, the light-cone fraction is $x = k^+/p^+ = (k^0+k^z)/(P^0+P^z)$. Thus in a frame where the bound state is moving infinitely fast in the positive z direction ("the infinite momentum frame"), the light-cone fraction becomes the momentum fraction $x \to k^z/p^z$. However, in the rest frame $\overrightarrow{P} = \overrightarrow{0}$, $x = (k^0 + k^z)/M$. Thus $x \to 0$ generally implies very large constituent momentum $k^z \to -k^0 \to -\infty$ in the rest frame; it is excluded by the ultraviolet regulation of the theory—unless the particle has strictly zero mass and transverse momentum.

If a particle has non-relativistic momentum in the bound state, then we can identify $k^z \sim xM - m$. This correspondence is useful when one matches physics at the relativistic/non-relativistic interface. In fact, any non-relativistic solution to the Schrödinger equation can be immediately written in light-cone form by identifying the two forms of coordinates. For example, the Schrödinger solution for particles bound in a harmonic oscillator potential can be taken as a model for the light-cone wavefunction for quarks in a confining linear potential: [19]

$$\psi(x_i, \vec{k}_{\perp i}) = A \exp(-b\mathcal{M}^2) = \exp - \left(b \sum_i^n \frac{k_{\perp i}^2 + m_i^2}{x_i} \right) .$$

This form exhibits the strong fall-off at large relative transverse momentum and at the $x \to 0$ and $x \to 1$ endpoints expected for soft non-perturbative solutions in QCD. The perturbative corrections due to hard gluon exchange give amplitudes suppressed only by power laws and thus will eventually dominate wavefunction behavior over the soft contributions in these regions. This *ansatz* is the central assumption required to derive dimensional counting perturbative QCD predictions for exclusive processes at large momentum transfer and the $x \to 1$ behavior of deep inelastic structure functions. A review is given in Ref. 20. A model for the polarized and unpolarized gluon distributions in the proton which takes into account both perturbative QCD constraints at large x and coherent cancellations at low x and small transverse momentum is given in Ref. 17.

2.5 The Light-Cone Hamiltonian Eigenvalue Problem

In principle, the problem of computing the spectrum in QCD and the corresponding light-cone wavefunctions for each hadron can be reduced to diagonalizing the QCD light cone Hamiltonian in Heisenberg quantum mechanics: Any hadron state must be an eigenstate of the light-cone Hamiltonian. For convenience we will work in the "standard" frame where $\underline{P}_\pi \equiv (P^+, P_\perp) = (1, 0_\perp)$ and $P_\pi^- = M_\pi^2$. Then the state $|\pi\rangle$ satisfies an equation

$$(M_\pi^2 - H_{LC})\,|\pi\rangle = 0.$$

Projecting this onto the various Fock states $\langle q\bar{q}|$, $\langle q\bar{q}g| \ldots$ results in an infinite number of coupled integral eigenvalue equations,[4]

$$\left(M_\pi^2 - \sum_i \frac{\vec{k}_{\perp i}^2 + m_i^2}{x_i}\right) \begin{bmatrix} \psi_{q\bar{q}/\pi} \\ \psi_{q\bar{q}g/\pi} \\ \vdots \end{bmatrix}$$

$$= \begin{bmatrix} \langle q\bar{q}|\, V \,|q\bar{q}\rangle & \langle q\bar{q}|\, V \,|q\bar{q}g\rangle & \cdots \\ \langle q\bar{q}g|\, V \,|q\bar{q}\rangle & \langle q\bar{q}g|\, V \,|q\bar{q}g\rangle & \cdots \\ \vdots & \vdots & \ddots \end{bmatrix} \begin{bmatrix} \psi_{q\bar{q}/\pi} \\ \psi_{q\bar{q}g/\pi} \\ \vdots \end{bmatrix}$$

where V is the interaction part of H_{LC}. Diagrammatically, V involves completely irreducible interactions—*i.e.* diagrams having no internal propagators—coupling Fock states. (See Fig. 7.) We will give the explicit forms of each matrix element of V in a later section. In principle, these equations determine the hadronic spectrum and wavefunctions. However, even though the QCD potential is essentially trivial on the light-cone momentum space basis, the many channels required to describe a hadronic state make these equations very difficult to solve. For example, Fock states with two or more gluons are required just to represent the effects of the running coupling constant of QCD.

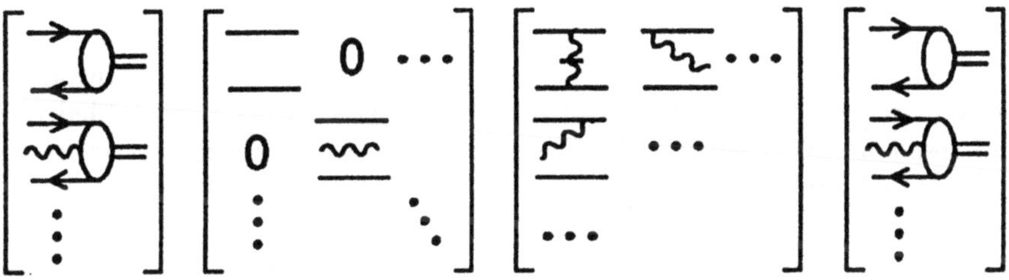

Fig. 7. Coupled eigenvalue equations for the light-cone wavefunctions of a pion.

In the case of gauge theories in one space and one time dimension, there are no physical gluon degrees of freedom in light-cone gauge. The computational problem is thus much more tractable, and it is possible to explicitly diagonalize the light-cone Hamiltonian and thus solve these theories numerically. In this method, "discretized light-cone quantization" (DLCQ) the light-cone Fock state basis is rendered discrete by imposing periodic (or anti-periodic) boundary conditions.[21]

A central emphasis of these lectures will be the use of DLCQ methods to solve non-perturbative problems in gauge theory. This method was first used to obtain the mass spectrum and wavefunctions of Yukawa theory, $\overline{\psi}\psi\phi$, in one space and one time dimensions.[21] This success led to further applications including QED(1+1) for general mass fermions and the massless Schwinger model by Eller *et al.*,[22] ϕ^4 theory in 1+1 dimensions by Harindranath and Vary,[23] and QCD(1+1) for N_C = 2,3,4 by Hornbostel *et al.*. [18] Complete numerical solutions have been obtained for the meson and baryon spectra as well as their respective light cone Fock state wavefunctions for general values of the coupling constant, quark masses, and color. Similar results for QCD(1+1) were also obtained by Burkardt [24] by solving the coupled light-cone integral equation in the low particle number sector. Burkardt was also able to study non-additive nuclear effects in the structure functions of nuclear states in QCD(1+1). In each of these applications, the mass spectrum and wavefunctions were successfully obtained, and all results agree with previous analytical and numerical work, where they were available. More recently, Hiller [13] has used DLCQ and the Lanczos algorithm for matrix diagonalization method to compute the annihilation cross section, structure functions and form factors in 1+1 theories. Although these are just toy models, they do exhibit confinement and are excellent tests of the light-cone Fock methods.

In addition to the above work on DLCQ, Wilson and his colleagues at Ohio State have developed a complimentary method, the Light-Front Tamm Damcoff approach.[25,26] which uses a fixed number Fock basis to truncate the theory. Wilson has also emphasized the potential advantages of using a Gaussian basis similar to that used in many-electron molecular systems, rather than the plane wave basis used in the DLCQ work.

The initial successes of DLCQ provide the hope that one can use this method for solving 3+1 theories. The application to higher dimensions is much more involved due to the expansion of the degrees of freedom and the need to introduce ultraviolet and infrared regulators and truncation procedures which minimize violations of gauge invariance and Lorentz invariance. This is in addition to the work involved implementing two extra dimensions with their added degrees of freedom. In these lectures, we will discuss some initial attempts to apply DLCQ to gauge theories in 3+1 dimensions.[27,28,29,30] We return to these applications in later sections.

The striking advantages of quantizing gauge theories on the light-cone have been realized by a number of authors, including Klauder, Leutwyler, and Streit,[31] Kogut and Soper,[32] Rohrlich,[33] Leutwyler,[34] Casher,[35] Chang, Root, and Yan,[36] Lepage and Brodsky,[4] Brodsky and Ji,[37] Lepage, Brodsky, Huang, and Mackenzie,[19] and McCartor.[38] Leutwyler recognized the utility of defining quark wavefunctions on the light-cone to give an unambiguous meaning to concepts used in the parton model. Casher gave the first construction of the light-cone Hamiltonian for non-Abelian gauge theory and gave an overview of important considerations in light-cone quantization. Chang, Root, and Yan demonstrated the equivalence of light-cone quantization with standard covariant Feynman analysis.

Franke, [39,40,41] Karmanov,[42,43] and Pervushin [44] have also done important work on light-cone quantization. The question of whether boundary conditions can be consistently set in light-cone quantization has been discussed by McCartor [45] and Lenz.[46] They have also shown that for massive theories that the energy and momentum derived using light-cone quantization are not only conserved, but also are equivalent to the energy and momentum one would normally write down in an equal-time theory.

The approach that we use in these lectures is closely related to the light-cone Fock methods used in Ref.4 in the analysis of exclusive processes in QCD. The renormalization of light-cone wavefunctions and the calculation of physical observables in the light-cone framework is also discussed in that paper. The analysis of light-cone perturbation theory rules for QED in light-cone gauge used here is similar to that given in Ref.19. A number of other applications of QCD in light-cone quantization are reviewed in Ref.20.

A mathematically similar but conceptually different approach to light-cone quantization is the "infinite momentum frame" formalism. This method involves observing the system in a frame moving past the laboratory close to the speed of light. The first developments were given by Weinberg.[47] Although light-cone quantization is similar to infinite momentum frame quantization, it differs since no reference frame is chosen for calculations, and it is thus manifestly Lorentz covariant. The only aspect that "moves at the speed of light" is the quantization surface. Other works in infinite momentum frame physics include Drell, Levy, and Yan, [48] Susskind and Frye, [49] Bjorken, Kogut, and Soper, [59] and Brodsky, Roskies, and Suaya. [11] This last reference presents the infinite momentum frame perturbation theory rules for QED in Feynman gauge, calculates one-loop radiative corrections, and demonstrates renormalizability.

3. Light-Cone Wavefunctions and High Momentum-Transfer Exclusive Processes

One of the major advantages of the light-cone formalism is that many properties of large momentum transfer exclusive reactions can be calculated without explicit knowledge of the form of the non-perturbative light-cone wavefunctions. The main ingredients of this analysis are asymptotic freedom, and the power-law scaling relations and quark helicity conservation rules of perturbative QCD. For example, consider the light-cone convolution formula given in the previous section for a meson form factor at high momentum transfer Q^2. If the internal momentum transfer is large then one can iterate the gluon-exchange term in the effective potential for the light-cone wavefunctions. The result is the hadron form factors can be written in a factorized form as a convolution of quark "distribution amplitudes" $\phi(x_i, Q)$, one for each hadron involved in the amplitude, with a hard-scattering amplitude T_H. [4,51] The pion's electromagnetic form factor, for example, can be written as [4,51,52]

$$F_\pi(Q^2) = \int_0^1 dx \int_0^1 dy\, \phi_\pi^*(y,Q)\, T_H(x,y,Q)\, \phi_\pi(x,Q) \left(1 + \mathcal{O}\left(\frac{1}{Q}\right)\right).$$

Here T_H is the scattering amplitude for the form factor but with the pions replaced by collinear $q\bar{q}$ pairs—*i.e.* the pions are replaced by their valence partons. We can also regard T_H as the free particle matrix element of the order $1/Q^2$ term in the effective Lagrangian for $\gamma^* q\bar{q} \to q\bar{q}$. [6]

The process-independent distribution amplitude [4] $\phi_\pi(x,Q)$ is the probability amplitude for finding the $q\bar{q}$ pair in the pion with $x_q = x$ and $x_{\bar{q}} = 1-x$. It is directly related to the light-cone valence wavefunction:

$$\begin{aligned}
\phi_\pi(x,Q) &= \int \frac{d^2\vec{k}_\perp}{16\pi^3}\, \psi^{(Q)}_{q\bar{q}/\pi}(x,\vec{k}_\perp) \\
&= P_\pi^+ \int \frac{dz^-}{4\pi}\, e^{ixP_\pi^+ z^-/2}\, \langle 0|\, \bar{\psi}(0)\, \frac{\gamma^+\gamma_5}{2\sqrt{2n_c}}\, \psi(z)\, |\pi\rangle^{(Q)} \Big|_{z^+ = \vec{z}_\perp = 0}
\end{aligned}$$

The $\vec{k}_\perp$ integration in the above equation is cut off by the ultraviolet cutoff $\Lambda = Q$ implicit in the wavefunction; thus only Fock states with invariant mass squared $\mathcal{M}^2 < Q^2$ contribute. We will return later to the discussion of ultraviolet regularization in the light-cone formalism.

It is important to note that the distribution amplitude is gauge invariant. In gauges other than light-cone gauge, a path-ordered "string operator" $P\exp(\int_0^1 ds\, ig\, A(sz)\cdot z)$ must be included between the $\bar{\psi}$ and ψ. The line integral vanishes in light-cone gauge because $A\cdot z = A^+ z^-/2 = 0$ and so the factor can be omitted in that gauge. This (non-perturbative) definition of ϕ uniquely fixes the definition of T_H which must itself then be gauge invariant.

The above result is in the form of a factorization theorem; all of the non-perturbative dynamics is factorized into the non-perturbative distribution amplitudes, which sums all internal momentum transfers up to the scale Q^2. On the other hand, all momentum transfers higher than Q^2 appear in T_H, which, because of asymptotic freedom, can be computed perturbatively in powers of the QCD running coupling constant $\alpha_s(Q^2)$.

Given the factorized structure, one can read off a number of general features of the PQCD predictions; *e.g.* the dimensional counting rules, hadron helicity conservation, color transparency, etc. [20] In addition, the scaling behavior of the exclusive amplitude is modified by the logarithmic dependence of the distribution amplitudes in $\ell n\, Q^2$ which is in turn determined by QCD evolution equations. [4]

An important application of the PQCD analysis is exclusive Compton scattering and the related cross process $\gamma\gamma \to \bar{p}p$. Each helicity amplitude for $\gamma p \to \gamma p$ can be computed at high momentum transfer from the convolution of the proton distribution amplitude with the $\mathcal{O}(\alpha_s^2)$ amplitudes for $qqq\gamma \to qqq\gamma$. The result is a cross section which scales as

$$\frac{d\sigma}{dt}(\gamma p \to \gamma p) = \frac{F(\theta_{CM}, \ell n\, s)}{s^6}$$

if the proton helicity is conserved. The helicity-flip amplitude and contributions involving more quarks or gluons in the proton wavefunction are power-law suppressed. The nominal s^{-6} fixed angle scaling follows from dimensional counting rules.[53] It is modified logarithmically due to the evolution of the proton distribution amplitude and the running of the QCD coupling constant.[4] The normalization, angular dependence, and phase structure are highly sensitive to the detailed shape of the non-perturbative form of $\phi_p(x_i, Q^2)$. Recently Kronfeld and Nizic [54] have calculated the leading Compton amplitudes using model forms for ϕ_p predicted in the QCD sum rule analyses;[3] the calculation is complicated by the presence of integrable poles in the hard-scattering subprocess T_H. The results for the unpolarized cross section are shown in Fig. 8.

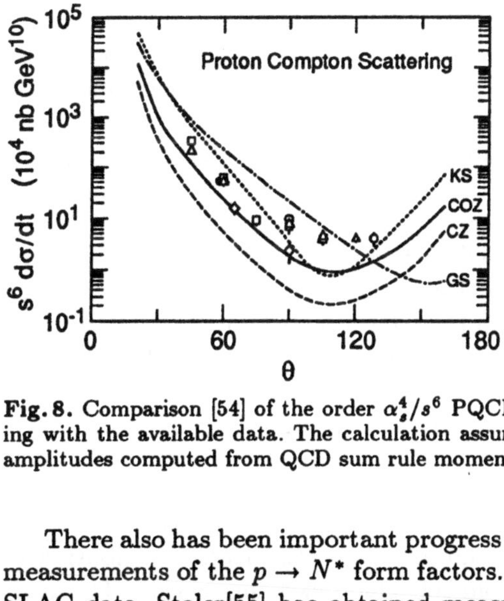

Fig. 8. Comparison [54] of the order α_s^4/s^6 PQCD prediction for proton Compton scattering with the available data. The calculation assumes PQCD factorization and distribution amplitudes computed from QCD sum rule moments.[3]

There also has been important progress testing PQCD experimentally using measurements of the $p \to N^*$ form factors. In a recent new analysis of existing SLAC data, Stoler[55] has obtained measurements of several transition form factors of the proton to resonances at $W = 1232, 1535$, and $1680\ MeV$. As is the case of the elastic proton form factor, the observed behavior of the transition form factors to the $N^*(1535)$ and $N^*(1680)$ are each consistent with the Q^{-4} fall-off and dipole scaling predicted by PQCD and hadron helicity conservation over the measured range $1 < Q^2 < 21\ GeV^2$. In contrast, the $p \to \Delta(1232)$ form factor decreases faster than $1/Q^4$ suggesting that non-leading processes are dominant in this case. Remarkably, this pattern of scaling behavior is what is expected from PQCD and the QCD sum rule analyses,[3] since, unlike the case of the proton and its other resonances, the distribution amplitude $\phi_{N^*}(x_1, x_2, x_3, Q)$ of the Δ resonance is predicted to be nearly symmetric in the x_i, and a symmetric distribution leads to a strong cancellation [56] of the leading helicity-conserving terms in the matrix elements of the hard scattering amplitude for $qqq \to \gamma^* qqq$.

These comparisons of the proton form factor and Compton scattering predictions with experiment are very encouraging, showing agreement in both the fixed-angle scaling behavior predicted by PQCD and the normalization predicted by QCD sum rule forms for the proton distribution amplitude. Assuming one can trust the validity of the leading order analysis, a systematic series of polarized target and beam Compton scattering measurements on proton and neutron targets and the corresponding two-photon reactions $\gamma\gamma \to p\bar{p}$ will strongly constrain a fundamental quantity in QCD, the nucleon distribution amplitude $\phi(x_i, Q^2)$. It is thus imperative for theorists to develop methods to calculate the shape and normalization of the non-perturbative distribution amplitudes from first principles in QCD.

3.1 Is PQCD Factorization Applicable to Exclusive Processes?

One of the concerns in the derivation of the PQCD results for exclusive amplitudes is whether the momentum transfer carried by the exchanged gluons in the hard scattering amplitude T_H is sufficiently large to allow a safe application of perturbation theory. [57] The problem appears to be especially serious if one assumes a form for the hadron distribution amplitudes $\phi_H(x_i, Q^2)$ which has strong support at the endpoints, as in the QCD sum rule model forms suggested by Chernyak and Zhitnitskii and others. [3]

This problem has now been clarified by two groups: Gari *et al.*. [58] in the case of baryon form factors, and Mankiewicz and Szczepaniak,[59] for the case of meson form factors. Each of these authors has pointed out that the assumed non-perturbative input for the distribution amplitudes must vanish strongly in the endpoint region; otherwise, there is a double-counting problem for momentum transfers occurring in the hard scattering amplitude and the distribution amplitudes. Once one enforces this constraint, (*e.g.* by using exponentially suppressed wavefunctions [19]) on the basis functions used to represent the QCD moments, or uses a sufficiently large number of polynomial basis functions, the resulting distribution amplitudes do not allow significant contribution to the high Q^2 form factors to come from soft gluon exchange region. The comparison of the PQCD predictions with experiment thus becomes phenomenologically and analytically consistent. An analysis of exclusive reactions on the effective Lagrangian method [60] is also consistent with this approach. In addition, as discussed by Botts,[61] potentially soft contributions to large angle hadron-hadron scattering reactions from Landshoff pinch contributions [62] are strongly suppressed by Sudakov form factor effects.

The empirical successes of the PQCD approach, together with the evidence for color transparency in quasi-elastic pp scattering [20] gives strong support for the validity of PQCD factorization for exclusive processes at moderate momentum transfer. It seems difficult to understand this pattern of form factor behavior if it is due to simple convolutions of soft wavefunctions. Thus it should be possible to use these processes to empirically constrain the form of the hadron distribution amplitudes, and thus confront non-perturbative QCD in detail.

3.2 Light-Cone Quantization and Heavy Particle Decays

One of the most interesting applications of the light-cone PQCD formalism is to large momentum transfer exclusive processes to heavy quark decays. For example, consider the decay $\eta_c \to \gamma\gamma$. If we can choose the Lagrangian cutoff $\Lambda^2 \sim m_c^2$, then to leading order in $1/m_c$, all of the bound state physics and virtual loop corrections are contained in the $c\bar{c}$ Fock wavefunction $\psi_{\eta_c}(x_i, k_{\perp i})$. The hard scattering matrix element of the effective Lagrangian coupling $c\bar{c} \to \gamma\gamma$ contains all of the higher corrections in $\alpha_s(\Lambda^2)$ from virtual momenta $|k^2| > \Lambda^2$. Thus

$$\mathcal{M}(\eta_c \to \gamma\gamma) = \int d^2k_\perp \int_0^1 dx\, \psi_{\eta_c}^{(\Lambda)}(x, k_\perp)\, T_H^{(\Lambda)}(c\bar{c} \to \gamma\gamma)$$

$$\Rightarrow \int_0^1 dx\, \phi(x, \Lambda)\, T_H^{(\Lambda)}(c\bar{c} \to \gamma\gamma)$$

where $\phi(x, \Lambda^2)$ is the η_c distribution amplitude. This factorization and separation of scales is shown in Fig. 9. Since the η_c is quite non-relativistic, its distribution amplitude is peaked at $x = 1/2$, and its integral over x is essentially equivalent to the wavefunction at the origin, $\psi(\vec{r} = \vec{0})$.

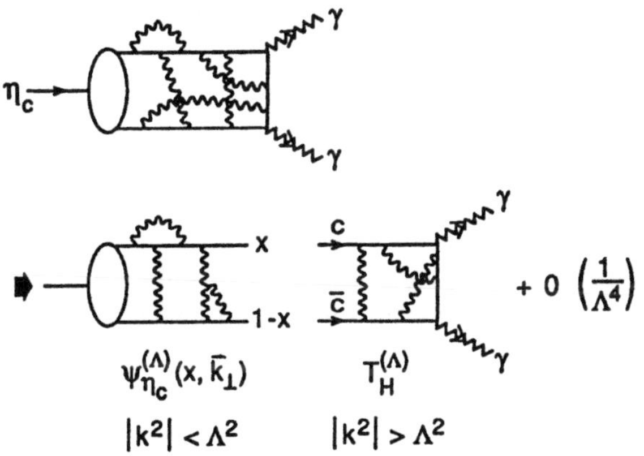

Fig. 9. Factorization of perturbative and non-perturbative contributions to the decay $\eta_c \to \gamma\gamma$.

Another interesting calculational example of quarkonium decay in PQCD is the annihilation of the J/ψ into baryon pairs. The calculation requires the convolution of the hard annihilation amplitude $T_H(c\bar{c} \to ggg \to uud\; uud)$ with the J/ψ, baryon, and anti-baryon distribution amplitudes.[4,3] (See Fig. 10.) The magnitude of the computed decay amplitude for $\psi \to \bar{p}p$ is consistent with experiment assuming the proton distribution amplitude computed from QCD sum rules.[3] The angular distribution of the proton in $e^+e^- \to J/\psi \to p\bar{p}$ is

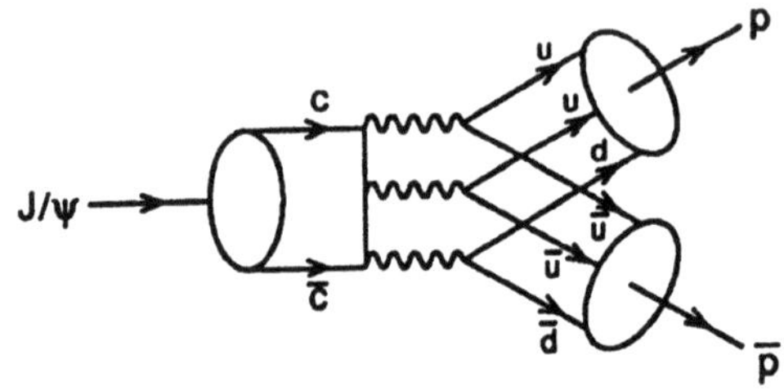

Fig. 10. Calculation of $J/\psi \to p\bar{p}$ in PQCD.

also consistent with the hadron helicity conservation rule predicted by PQCD; *i.e.* opposite proton and anti-proton helicity.

The effective Lagrangian method was used by Lepage, Caswell, and Thacker [16] to systematically compute the order $\alpha_s(\widehat{Q})$ corrections to the hadronic and photon decays of quarkonium. The scale $\widehat{Q}$ can then be set by incorporating vacuum polarization corrections into the running coupling constant. [63] A summary of the results can be found in Ref. 64.

3.3 Exclusive Weak Decays of Heavy Hadrons

An important application of the PQCD effective Lagrangian formalism is to the exclusive decays of heavy hadrons to light hadrons, such as $B^0 \to \pi^+\pi^-$, K^+, K^-. [65] To a good approximation, the decay amplitude $\mathcal{M} = \langle B|H_{Wk}|\pi^+\pi^-\rangle$ is caused by the transition $\bar{b} \to W^+\bar{u}$; thus $\mathcal{M} = f_\pi p_\pi^\mu \frac{G_F}{\sqrt{2}} \times \langle\pi^-|J_\mu|B^0\rangle$ where J_μ is the $\bar{b} \to \bar{u}$ weak current. The problem is then to recouple the spectator d quark and the other gluon and possible quark pairs in each B^0 Fock state to the corresponding Fock state of the final state π^-. (See Fig. 11.) The kinematic constraint that $(p_B - p_\pi)^2 = m_\pi^2$ then demands that at least one quark line is far off shell: $p_{\bar{u}}^2 = (yp_B - p_\pi)^2 \sim -\mu m_B \sim -1.5\ GeV^2$, where we have noted that the light quark takes only a fraction $(1-y) \sim \sqrt{(k_\perp^2 + m_d^2)}/m_B$ of the heavy meson's momentum since all of the valence quarks must have nearly equal velocity in a bound state. In view of the successful applications [55] of PQCD factorization to form factors at momentum transfers in the few GeV^2 range, it is reasonable to assume that $\langle|p_{\bar{u}}^2|\rangle$ is sufficiently large that we can begin to apply perturbative QCD methods.

The analysis of the exclusive weak decay amplitude can be carried out in parallel to the PQCD analysis of electroweak form factors [66] at large Q^2. The first step is to iterate the wavefunction equations of motion so that the large momentum transfer through the gluon exchange potential is exposed. The heavy quark decay amplitude can then be written as a convolution of the hard scattering amplitude for $Q\bar{q} \to W^+q\bar{q}$ convoluted with the B and π distribution amplitudes. The minimum number valence Fock state of each hadron gives the leading power law contribution. Equivalently, we can choose the ultraviolet cut-off scale in the Lagrangian at $(\Lambda^2 < \mu m_B)$ so that the hard scattering amplitude $T_H(Q\bar{q} \to W^+q\bar{q})$ must be computed from the matrix elements of

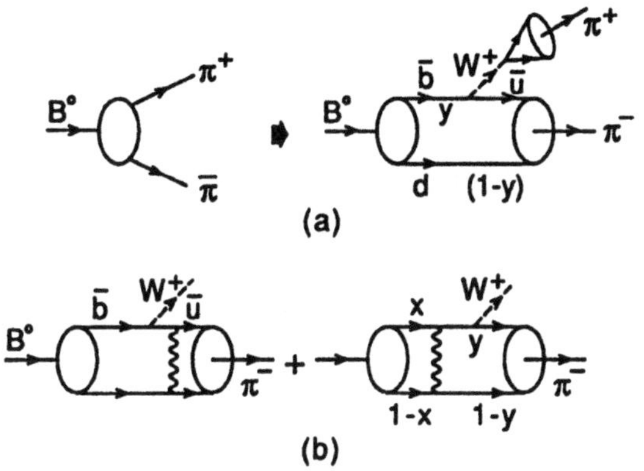

Fig. 11. Calculation of the weak decay $B \to \pi\pi$ in the PQCD formalism of Ref. 65. The gluon exchange kernel of the hadron wavefunction is exposed where hard momentum transfer is required.

the order $1/\Lambda^2$ terms in $\delta\mathcal{L}$. Thus T_H contains all perturbative virtual loop corrections of order $\alpha_s(\Lambda^2)$. The result is the factorized form:

$$\mathcal{M}(B \to \pi\pi) = \int_0^1 dx \int_0^1 dy \phi_B(y, \Lambda) T_H \phi_\pi(x, \Lambda)$$

which is expected to be correct up to terms of order $1/\Lambda^4$. All of the non-perturbative corrections with momenta $|k^2| < \Lambda^2$ are summed in the distribution amplitudes.[67]

In order to make an estimate of the size of the $B \to \pi\pi$ amplitude, in Ref. 65 we have taken the simplest possible forms for the required wavefunctions

$$\phi_\pi(y) \propto \gamma_5 \gamma \cdot p_\pi y(1-y)$$

for the pion and

$$\phi_B(x) \propto \frac{\gamma_5[\gamma \cdot p_B + m_B g(x)]}{\left[1 - \frac{1}{x} - \frac{\epsilon^2}{(1-x)}\right]^2}$$

for the B, each normalized to its meson decay constant. The above form for the heavy quark distribution amplitude is chosen so that the wavefunction peaks at equal velocity; this is consistent with the phenomenological forms used to describe heavy quark fragmentation into heavy hadrons. We estimate $\epsilon \sim 0.05$ to 0.10. The functional dependence of the mass term $g(x)$ is unknown; however, it should be reasonable to take $g(x) \sim 1$ which is correct in the weak binding approximation.

One now can compute the leading order PQCD decay amplitude

$$\mathcal{M}(B^0 \to \pi^- \pi^+) = \frac{G_F}{\sqrt{2}} V_{ud}^* V_{ub} P_{\pi^+}^\mu \langle \pi^- | V^\mu | B^0 \rangle$$

where

$$\langle \pi^- \,|\, V^\mu \,|\, B^0 \rangle = \frac{8\pi\alpha_s(Q^2)}{3} \int_0^1 dx \int_0^{1-\epsilon} dy\, \phi_B(x)\, \phi_\pi(y)$$

$$\times \frac{\mathrm{Tr}[\gamma \cdot P_{\pi^-} \gamma_5 \gamma^\nu \gamma \cdot k_1 \gamma^\mu (\gamma \cdot P_B + M_B g(x)) \gamma_5 \gamma_\nu]}{k_1^2 Q^2}$$

$$+ \frac{\mathrm{Tr}[\gamma \cdot P_{\pi^-} \gamma_5 \gamma^\nu (\gamma \cdot k_2 + M_B) \gamma^\nu (\gamma \cdot P_B + M_B g(x)) \gamma_5 \gamma_\nu]}{(k_2^2 - M_B^2) Q^2}$$

Numerically, this gives the branching ratio

$$BR(B^0 \to \pi^+\pi^-) \sim 10^{-8} \xi^2 N$$

where $\xi = 10|V_{ub}/V_{cb}|$ is probably less than unity, and N has strong dependence on the value of g: $N = 180$ for $g = 1$ and $N = 5.8$ for $g = 1/2$. The present experimental limit [68] is

$$BR(B^0 \to \pi^+\pi^-) < 3 \times 10^{-4}.$$

A similar PQCD analysis can be applied to other two-body decays of the B; the ratios of the widths will not be so sensitive to the form of the distribution amplitude, allowing tests of the flavor symmetries of the weak interaction.

4. Light-Cone Quantization of Gauge Theory

In this section we will outline the canonical quantization of QCD in $A^+ = 0$ gauge, following the discussion in Refs. 4 and 19. The quantization proceeds in several steps. First we identify the independent dynamical degrees of freedom in the Lagrangian. The theory is quantized by defining commutation relations for these dynamical fields at a given light-cone time $\tau = t + z$ (we choose $\tau = 0$). These commutation relations lead immediately to the definition of the Fock state basis. Expressing the dependent fields in terms of the independent fields, we then derive a light-cone Hamiltonian, which determines the evolution of the state space with changing τ. Finally we derive the rules for τ-ordered perturbation theory.

The purpose of this exercise is to illustrate the origins and nature of the Fock state expansion, and of light-cone perturbation theory in QCD. In this section we will ignore the subtleties to the zero-mode large scale structure of non-Abelian gauge fields. Although these have a profound effect on the structure of the vacuum, the theory can still be described with a Fock state basis and some sort of effective light-cone Hamiltonian. At the least, this procedure should be adequate to describe heavy quark systems. Furthermore, the short distance interactions of the theory are unaffected by this structure, according to the central ansatz of perturbative QCD.

The Lagrangian (density) for QCD can be written

$$\mathcal{L} = -\frac{1}{2}\,\mathrm{Tr}\,(F^{\mu\nu}\,F_{\mu\nu}) + \overline{\psi}\,(i\,\not{D} - m)\,\psi$$

where $F^{\mu\nu} = \partial^\mu A^\nu - \partial^\nu A^\mu + ig[A^\mu, A^\nu]$ and $iD^\mu = i\partial^\mu - gA^\mu$. Here the gauge field A^μ is a traceless 3×3 color matrix ($A^\mu \equiv \sum_a A^{a\mu}T^a$, $\mathrm{Tr}(T^aT^b) = 1/2\delta^{ab}$, $[T^a, T^b] = ic^{abc}T^c, \ldots$), and the quark field ψ is a color triplet spinor (for simplicity, we include only one flavor). In order to maintain charge conjugation symmetry in the construction of the Hamiltonian, it is understood that this expression is averaged with its Hermetian conjugate.

Given the Lagrangian density, one can calculate the energy momentum tensor and stress tensor in the usual way from the independent dynamical fields and their conjugate momenta. At a given light-cone time, say $\tau = 0$, the independent dynamical fields are $\psi_\pm \equiv \Lambda_\pm\psi$ and $A^i_\perp$ with conjugate fields $i\psi^\dagger_+$ and $\partial^+ A^i_\perp$, where $\Lambda_\pm = \gamma^o\gamma^\pm/2$ are projection operators ($\Lambda_+\Lambda_- = 0$, $\Lambda^2_\pm = \Lambda_\pm$, $\Lambda_+ + \Lambda_- = 1$) and $\partial^\pm = \partial^0 \pm \partial^3$. Using the equations of motion, the remaining fields in $\mathcal{L}$ can be expressed in terms of ψ_+, $A^i_\perp$:

$$\begin{aligned}
\psi_- \equiv \Lambda_-\psi &= \frac{1}{i\partial^+}\left[i\vec{D}_\perp\cdot\vec{\alpha}_\perp + \beta m\right]\psi_+ \\
&= \widetilde{\psi}_- - \frac{1}{i\partial^+}\, g\vec{A}_\perp\cdot\vec{\alpha}_\perp\psi_+ \ , \\
A^+ &= 0 \ , \\
A^- &= \frac{2}{i\partial^+}\, i\vec{\partial}_\perp\cdot\vec{A}_\perp + \frac{2g}{(i\partial^+)^2}\left\{\left[i\partial^+A^i_\perp, A^i_\perp\right] + 2\psi^\dagger_+\,T^a\,\psi_+\,T^a\right\} \\
&\equiv \widetilde{A}^- + \frac{2g}{(i\partial^+)^2}\left\{\left[i\partial^+A^i_\perp, A^i_\perp\right] + 2\psi^\dagger_+\,T^a\,\psi_+\,T^a\right\} \ ,
\end{aligned}$$

with $\beta = \gamma^o$ and $\vec{\alpha}_\perp = \gamma^o\vec{\gamma}$.

To quantize, we expand the fields at $\tau = 0$ in terms of creation and annihilation operators,

$$\begin{aligned}
\psi_+(x) = \int_{k^+>0} \frac{dk^+\,d^2k_\perp}{k^+\,16\pi^3} \sum_\lambda \Big\{ & b(\underline{k},\lambda)\; u_+(\underline{k},\lambda)\; e^{-ik\cdot x} \\
& + d^\dagger(\underline{k},\lambda)\; v_+(\underline{k},\lambda)\; e^{ik\cdot x}\Big\} \ , \quad \tau = x^+ = 0
\end{aligned}$$

$$A^i_\perp(x) = \int_{k^+>0} \frac{dk^+\,d^2k_\perp}{k^+\,16\pi^3} \sum_\lambda \left\{ a(\underline{k},\lambda)\; \epsilon^i_\perp(\lambda)\; e^{-ik\cdot x} + c{\cdot}c{\cdot}\right\} \ , \quad \tau = x^+ = 0 \ ,$$

with commutation relations ($\underline{k} = (k^+, \vec{k}_\perp)$):

$$\begin{aligned}
\left\{b(\underline{k},\lambda),\ b^\dagger(\underline{p},\lambda)\right\} &= \left\{d(\underline{k},\lambda),\ d^\dagger(\underline{p},\lambda')\right\} \\
&= \left[a(\underline{k},\lambda),\ a^\dagger(\underline{p},\lambda')\right] \\
&= 16\pi^3\, k^+\, \delta^3(\underline{k} - \underline{p})\, \delta_{\lambda\lambda'} \ , \\
\{b, b\} = \{d, d\} &= \ldots = 0 \ ,
\end{aligned}$$

where λ is the quark or gluon helicity. These definitions imply canonical commutation relations for the fields with their conjugates ($\tau = x^+ = y^+ = 0$, $\underline{x} = (x^-, x_\perp), \ldots$):

$$\left\{\psi_+(\underline{x}),\ \psi_+^\dagger(\underline{y})\right\} = \Lambda_+\, \delta^3(\underline{x}-\underline{y})\ ,$$
$$\left[A^i(\underline{x}),\ \partial^+ A_\perp^j(\underline{y})\right] = i\delta^{ij}\, \delta^3(\underline{x}-\underline{y})\ .$$

It should be emphasized that these commutation relations are not new; they are the usual commutation relation for free fields evaluated at fixed light-cone rather than ordinary time.

The creation and annihilation operators define the Fock state basis for the theory at $\tau = 0$, with a vacuum $|0\rangle$ defined such that $b\,|0\rangle = d\,|0\rangle = a\,|0\rangle = 0$. The evolution of these states with τ is governed by the light-cone Hamiltonian, $H_{LC} = P^-$, conjugate to τ. The Hamiltonian can be readily expressed in terms of ψ_+ and $A_\perp^i$:

$$H_{LC} = H_0 + V\ ,$$

where

$$\begin{aligned} H_0 &= \int d^3x \left\{ \mathrm{Tr}\left(\partial_\perp^i A_\perp^j \partial_\perp^i A_\perp^j\right) + \psi_+^\dagger\,(i\partial_\perp\cdot\alpha_\perp + \beta m)\,\frac{1}{i\partial^+}\,(i\partial_\perp\cdot\alpha_\perp + \beta m)\,\psi_+ \right\} \\ &= \sum_{\substack{\lambda \\ \text{colors}}} \int \frac{dk^+\, d^2k_\perp}{16\pi^3\, k^+} \left\{ a^\dagger(\underline{k},\lambda)\, a(\underline{k},\lambda) \frac{k_\perp^2}{k^+} + b^\dagger(\underline{k},\lambda)\, b(\underline{k},\lambda) \right. \\ &\quad \left. \times\, \frac{k_\perp^2+m^2}{k^+} + d^\dagger(\underline{k},\lambda)\, b(\underline{k},\lambda)\, \frac{k_\perp^2+m^2}{k^+} \right\} + \text{constant} \end{aligned}$$

is the free Hamiltonian and V the interaction:

$$\begin{aligned} V = \int d^3x\ &\left\{ 2g\ \mathrm{Tr}\left(i\partial^\mu \widetilde{A}^\nu \left[\widetilde{A}_\mu, \widetilde{A}_\nu\right]\right) - \frac{g^2}{2}\,\mathrm{Tr}\left(\left[\widetilde{A}^\mu, \widetilde{A}^\nu\right]\left[\widetilde{A}_\mu, \widetilde{A}_\nu\right]\right) \right. \\ &+ g\overline{\widetilde{\psi}}\ \widetilde{\not{A}}\,\widetilde{\psi} + g^2\,\mathrm{Tr}\left(\left[i\partial^+\widetilde{A}^\mu, \widetilde{A}_\mu\right]\frac{1}{(i\partial^+)^2}\left[i\partial^+\widetilde{A}^\nu, \widetilde{A}_\nu\right]\right) \\ &+ g^2\overline{\widetilde{\psi}}\ \widetilde{\not{A}}\,\frac{\gamma^+}{2i\partial^+}\,\widetilde{\not{A}}\,\widetilde{\psi} - g^2\overline{\widetilde{\psi}}\gamma^+\left(\frac{1}{(i\partial^+)^2}\left[i\partial^+\widetilde{A}^\nu, \widetilde{A}_\nu\right]\right)\widetilde{\psi} \\ &\left. + \frac{g^2}{2}\ \overline{\psi}\gamma^+\ T^a\psi\,\frac{1}{(i\partial^+)^2}\,\overline{\psi}\gamma^+\ T^a\psi \right\}\ , \end{aligned}$$

with $\widetilde{\psi} = \widetilde{\psi}_- + \psi_+$ ($\to \psi$ as $g \to 0$) and $\widetilde{A}^\mu = (0, \widetilde{A}^-, A_\perp^i)$ ($\to A^\mu$ as $g \to 0$). The Fock states are obviously eigenstates of H_0 with

$$H_0\left|n : k_i^+, k_{\perp i}\right\rangle = \sum_i \left(\frac{k_\perp^2+m^2}{k^+}\right)_i \left|n : k_i^+, k_{\perp i}\right\rangle\ .$$

It is equally obvious that they are not eigenstates of V, though any matrix element of V between Fock states is trivially evaluated.

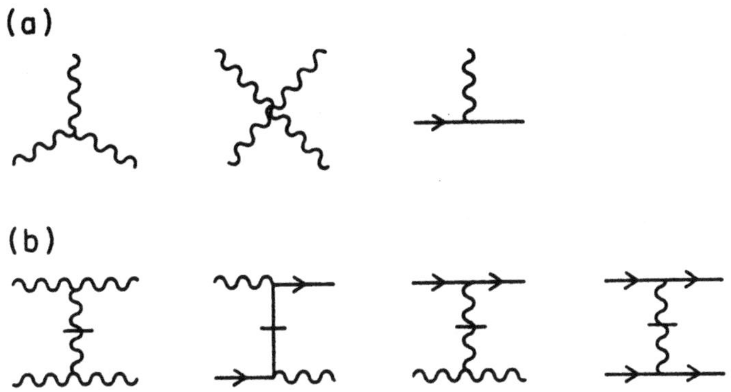

Fig. 12. Diagrams which appear in the interaction Hamiltonian for QCD on the light cone. The propagators with horizontal bars represent "instantaneous" gluon and quark exchange which arise from reduction of the dependent fields in $A^+ = 0$ gauge. (a) Basic interaction vertices in QCD. (b) "Instantaneous" contributions.

The first three terms in V correspond to the familiar three and four gluon vertices, and the gluon-quark vertex [Fig. 12(a)]. The remaining terms represent new four-quanta interactions containing instantaneous fermion and gluon propagators [Fig. 12(b)]. All terms conserve total three-momentum $\underline{k} = (k^+, \vec{k}_\perp)$, because of the integral over $\underline{x}$ in V.

The matrix elements of the light-cone Hamiltonian for the continuum case can be found in Refs. [19, 28, 27]. For the sake of completeness, the explicit expressions are compiled in Tables 2a-d for the vertex V, the contraction C, and the seagull interaction S, respectively, to the extent they are needed in the present context. The light cone Hamiltonian $H_{\mathrm{LC}} = T + V + S + C$ is the sum of these three interactions and of the free or 'kinetic' energy

$$T = \sum_q \left(\frac{m_{\mathrm{F}}^2 + \vec{k}_\perp^{\,2}}{x}\right)_q (b_q^\dagger b_q + d_q^\dagger d_q) \; + \; \sum_q \left(\frac{\vec{k}_\perp^{\,2}}{x}\right)_1 a_q^\dagger a_q .$$

The creation operators $b_q^\dagger$, $d_q^\dagger$ and $a_q^\dagger$ create plane wave states for the electrons, positrons, and photons, respectively, characterized by the four kinematical quantum numbers $q \equiv (x, \vec{k}_\perp, \lambda)$, and the destruction operators b_q, d_q and a_q destroy them correspondingly. They obey the usual (anti-)commutation relations. Each single particle carries thus a longitudinal momentum fraction x, transverse momentum $\vec{k}_\perp$, and helicity λ. The fermions have mass m_{F} and kinetic energy $\frac{m_{\mathrm{F}}^2 + \vec{k}_\perp^{\,2}}{x}$, the photons are massless. The symbol $\sum_q$ denotes summation over the entire range of the quantum numbers. In the continuum limit sums are replaced by integrals, *i.e.* $\sum_q \longrightarrow C_L \int dq$, where

$$C_L \equiv \frac{\Omega P^+}{16\pi^3} \quad \text{and} \quad \int dq \equiv \sum_{\lambda=\pm 1} \int_0^1 dx \int_{-\infty}^{+\infty} d(\vec{k}_\perp)_x \int_{-\infty}^{+\infty} d(\vec{k}_\perp)_y \, .$$

The normalization volume is denoted by $\Omega \equiv 2L_\parallel (2L_\perp)^2$, and the total longitudinal momentum by P^+.

Table 2a: The matrix elements of the vertex interaction V. — The transversal polarisation vector is defined as $\vec{\epsilon}_\perp(\lambda) = (-\lambda\hat{x} - i\hat{y})/\sqrt{2}$. The coupling constant g is hidden in $\widetilde{g}$, with $\widetilde{g}^2 = g^2 \frac{2}{P^+\Omega}$. In the continuum limit one replaces sums by integrals and $\beta = \widetilde{g}^2$ by $\widetilde{\beta} \equiv C_L\beta = \frac{\alpha}{2\pi^2}$, since $g^2 \equiv 4\pi\alpha$ in our units. — The Gell-Mann matrices are denoted by $T^a_{c_1c_2}$, and the totally anti-symmetric structure constants of SU(N_c) by $C^a_{bc} \equiv c^{abc}$. The are related by $[T^a, T^b] = ic^{abc}T^c$.

Graph	Matrix Element = Momentum×Helicity×Flavor×Color Factor			
$V_{q\to qg}(1;2,3)$	$= \quad +\widetilde{g}\sqrt{\frac{1}{x_3}}\, m_F \left[\frac{1}{x_1} - \frac{1}{x_2}\right]$	$\delta^{\lambda_2}_{-\lambda_1}\,\delta^{\lambda_3}_{\lambda_1}$	$\delta^{f_2}_{f_1}$	$T^{a_3}_{c_1c_2}$
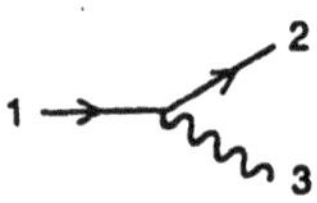	$+\widetilde{g}\sqrt{\frac{2}{x_3}}\,\vec{\epsilon}_\perp(\lambda_3)\cdot\left[\left(\frac{\vec{k}_\perp}{x}\right)_3 - \left(\frac{\vec{k}_\perp}{x}\right)_2\right]$	$\delta^{\lambda_2}_{+\lambda_1}\,\delta^{\lambda_3}_{\lambda_1}$	$\delta^{f_2}_{f_1}$	$T^{a_3}_{c_1c_2}$
	$+\widetilde{g}\sqrt{\frac{2}{x_3}}\,\vec{\epsilon}_\perp(\lambda_3)\cdot\left[\left(\frac{\vec{k}_\perp}{x}\right)_3 - \left(\frac{\vec{k}_\perp}{x}\right)_1\right]$	$\delta^{\lambda_2}_{+\lambda_1}\,\delta^{\lambda_3}_{-\lambda_1}$	$\delta^{f_2}_{f_1}$	$T^{a_3}_{c_1c_2}$
$V_{g\to q\bar{q}}(1;2,3)$	$= \quad +\widetilde{g}\frac{1}{\sqrt{x_1}}\, m_F \left[\frac{1}{x_2} + \frac{1}{x_3}\right]$	$\delta^{\lambda_3}_{+\lambda_2}\,\delta^{\lambda_3}_{+\lambda_1}$	$\delta^{f_3}_{f_2}$	$T^{a_1}_{c_2c_3}$
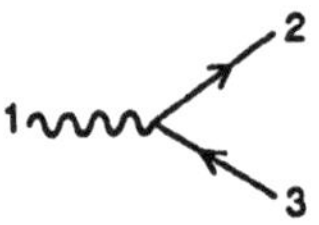	$+\widetilde{g}\sqrt{\frac{2}{x_1}}\,\vec{\epsilon}_\perp(\lambda_1)\cdot\left[\left(\frac{\vec{k}_\perp}{x}\right)_1 - \left(\frac{\vec{k}_\perp}{x}\right)_3\right]$	$\delta^{\lambda_3}_{-\lambda_2}\,\delta^{\lambda_3}_{-\lambda_1}$	$\delta^{f_3}_{f_2}$	$T^{a_1}_{c_2c_3}$
	$+\widetilde{g}\sqrt{\frac{2}{x_1}}\,\vec{\epsilon}_\perp(\lambda_1)\cdot\left[\left(\frac{\vec{k}_\perp}{x}\right)_1 - \left(\frac{\vec{k}_\perp}{x}\right)_2\right]$	$\delta^{\lambda_3}_{-\lambda_2}\,\delta^{\lambda_3}_{+\lambda_1}$	$\delta^{f_3}_{f_2}$	$T^{a_1}_{c_2c_3}$
$V_{g\to gg}(1;2,3)$	$= -\widetilde{g}\sqrt{\frac{x_3}{2x_1x_2}}\,\vec{\epsilon}^{\,\star}_\perp(\lambda_1)\cdot\left[\left(\frac{\vec{k}_\perp}{x}\right)_1 - \left(\frac{\vec{k}_\perp}{x}\right)_3\right]$	$\delta^{\lambda_3}_{-\lambda_2}$	1	$iC^{a_1}_{a_2a_3}$
	$-\widetilde{g}\sqrt{\frac{x_1}{2x_2x_3}}\,\vec{\epsilon}_\perp(\lambda_3)\cdot\left[\left(\frac{\vec{k}_\perp}{x}\right)_3 - \left(\frac{\vec{k}_\perp}{x}\right)_2\right]$	$\delta^{\lambda_2}_{\lambda_1}$	1	$iC^{a_1}_{a_2a_3}$
	$-\widetilde{g}\sqrt{\frac{x_2}{2x_3x_1}}\,\vec{\epsilon}_\perp(\lambda_3)\cdot\left[\left(\frac{\vec{k}_\perp}{x}\right)_3 - \left(\frac{\vec{k}_\perp}{x}\right)_1\right]$	$\delta^{\lambda_2}_{\lambda_1}$	1	$iC^{a_1}_{a_2a_3}$

$$
\begin{aligned}
V = & \textstyle\sum_{q_1,q_2,q_3} (b_1^\dagger b_2 a_3 - d_1^\dagger d_2 a_3)\; V_{q\to qg}(1;2,3) \\
& + \textstyle\sum_{q_1,q_2,q_3} (a_3^\dagger b_2^\dagger b_1 - a_3^\dagger d_2^\dagger d_1)\; V^\star_{q\to qg}(1;2,3) \\
& + \textstyle\sum_{q_1,q_2,q_3} (a_1^\dagger d_2 b_3\; V_{g\to q\bar{q}}(1;2,3) + b_3^\dagger d_2^\dagger a_1\; V^\star_{g\to q\bar{q}}(1;2,3)) \\
& + \textstyle\sum_{q_1,q_2,q_3} (a_1^\dagger a_2 a_3\; V_{g\to gg}(1;2,3) + a_3^\dagger a_2^\dagger a_1\; V^\star_{g\to gg}(1;2,3))
\end{aligned}
$$

Table 2b: The matrix elements of the contraction energy C. — The effective coupling constant g is given by $\tilde{g}^2 = g^2 \frac{2}{P^+\Omega}$. The color coefficient for the quarks are given by $C_F \equiv \sum_{a,c}(\mathrm{T}^a\mathrm{T}^a)_{cc'} = (N_c^2-1)/2N_c$, and for the gluons by $C_G \equiv \sum_{a,n_f} \mathrm{Tr}(\mathrm{T}^a\mathrm{T}^{a'}) = N_f/2$, respectively.

Graph	Matrix Element as Infinite Sums	as Finite Sums
	$C_q^{(g)} = \tilde{g}^2 C_F \sum_{x,\vec{k}_\perp}\left[\frac{1}{(x_1-x)^2} - \frac{1}{(x_1+x)^2}\right]$	$= 2\tilde{g}^2 C_F \sum_{\vec{n}_\perp,n=1}^{n_1}\left(\frac{LP^+}{2\pi n}\right)^2$
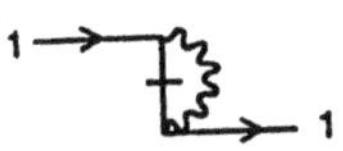	$C_q^{(q)} = \tilde{g}^2 \frac{C_F}{2} \sum_{x,\vec{k}_\perp}\left[\frac{1}{x(x_1+x)} + \frac{1}{x(x_1-x)}\right]$	$= \tilde{g}^2 \frac{C_F}{x_1} \sum_{\vec{n}_\perp,n=1}^{n_1} \frac{LP^+}{2\pi n}$
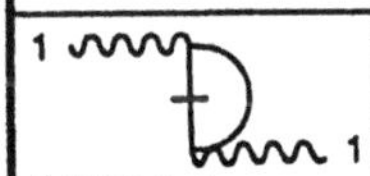	$C_g^{(g)} = \tilde{g}^2 C_G \sum_{x,\vec{k}_\perp}\left[\frac{-1}{x_1(x+x_1)} + \frac{-1}{x_1(x-x_1)}\right]$	$= -2\tilde{g}^2 \frac{C_G}{x_1} \sum_{\vec{n}_\perp,n=1}^{n_1} \frac{LP^+}{2\pi n}$
	$C_g^{(g)} = \tilde{g}^2 \frac{N_c}{4} \sum_{x,\vec{k}_\perp}\left[\frac{2(x-x_1)}{x(x+x_1)^2} - \frac{(x+x_1)^2}{x x_1 (x-x_1)^2}\right]$	
	$C_g^{(n)} = \quad 0$	
$C = \sum_{q_1} a_1^\dagger a_1$	$\left[C_g^{(g)}(1) + C_g^{(q)}(1) + C_g^{(n)}(1)\right] + (b_1^\dagger b_1 + d_1^\dagger d_1)$	$\left[C_q^{(g)}(1) + C_q^{(q)}(1)\right]$

Table 2c: The matrix elements of the seagull interaction S. — The coupling constant g is hidden in $\widetilde{g}^2 = g^2 \frac{2}{P^+\Omega}$. In the continuum limit one replaces sums by integrals and $\widetilde{g}^2$ by $\widetilde{g}^2 C_L = \frac{\alpha}{2\pi^2}$, since $g^2 \equiv 4\pi\alpha$ in our units.

Type	Graph	Element = Momentum×Helicity×Flavor×Color Factor			
S_1	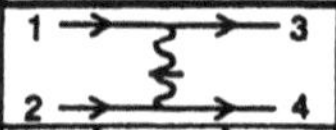	$S_{qq\to qq} = -\widetilde{g}^2 \frac{1}{(x_1-x_3)^2}$	$\delta^{\lambda_3}_{\lambda_1}\, \delta^{\lambda_4}_{\lambda_2}$	$\delta^{f_3}_{f_1}\, \delta^{f_4}_{f_2}$	$T^a_{c_1c_3}T^a_{c_2c_4}$
S_3	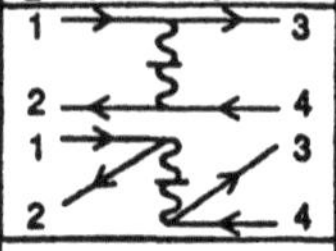	$S^{(s)}_{q\bar{q}\to q\bar{q}} = \widetilde{g}^2 \frac{2}{(x_1-x_3)^2}$	$\delta^{\lambda_3}_{\lambda_1}\, \delta^{\lambda_4}_{\lambda_2}$	$\delta^{f_3}_{f_1}\, \delta^{f_4}_{f_2}$	$T^a_{c_1c_3}T^a_{c_4c_2}$
		$S^{(a)}_{q\bar{q}\to q\bar{q}} = -\widetilde{g}^2 \frac{2}{(x_1+x_2)^2}$	$\delta^{\lambda_2}_{-\lambda_1}\, \delta^{\lambda_4}_{-\lambda_3}$	$\delta^{f_2}_{f_1}\, \delta^{f_4}_{f_3}$	$T^a_{c_1c_2}T^a_{c_4c_3}$
S_5	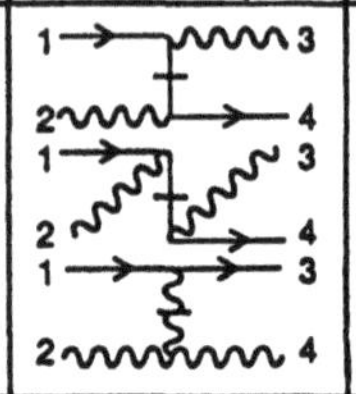	$S^{(s)}_{qg\to qg} =^{*} \widetilde{g}^2 \frac{1}{x_1-x_4}\frac{1}{\sqrt{x_2x_4}}$	$\delta^{\lambda_3}_{\lambda_1}\, \delta^{\lambda_4}_{\lambda_2}\, \delta^{\lambda_2}_{\lambda_1}$	$\delta^{f_3}_{f_1}$	$T^{a_4}_{c_1c}T^{a_2}_{cc_3}$
		$S^{(a)}_{qg\to qg} = \widetilde{g}^2 \frac{1}{x_1+x_2}\frac{1}{\sqrt{x_2x_4}}$	$\delta^{\lambda_3}_{\lambda_1}\, \delta^{\lambda_4}_{\lambda_2}\, \delta^{\lambda_2}_{-\lambda_1}$	$\delta^{f_3}_{f_1}$	$T^{a_2}_{c_1c}T^{a_4}_{cc_3}$
		$S^{(n)}_{qg\to qg} = \widetilde{g}^2 \frac{1}{(x_1-x_3)^2}\frac{(x_2+x_4)}{\sqrt{x_2x_4}}$	$\delta^{\lambda_3}_{\lambda_1}\, \delta^{\lambda_4}_{\lambda_2}$	$\delta^{f_3}_{f_1}$	$iT^a_{c_1c_3}C^a_{a_2a_4}$
S_7	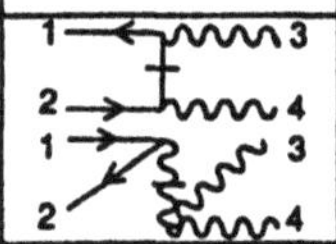	$S_{q\bar{q}\to gg} = \widetilde{g}^2 \frac{1}{x_1-x_3}\frac{1}{\sqrt{x_3x_4}}$	$\delta^{\lambda_2}_{-\lambda_1}\, \delta^{\lambda_4}_{-\lambda_3}\, \delta^{\lambda_3}_{\lambda_1}$	$\delta^{f_2}_{f_1}$	$T^{a_3}_{c_1c}T^{a_4}_{cc_2}$
		$S^{(n)}_{q\bar{q}\to gg} = \widetilde{g}^2 \frac{1}{(x_1+x_2)^2}\frac{(x_3-x_4)}{\sqrt{x_3x_4}}$	$\delta^{\lambda_2}_{-\lambda_1}\, \delta^{\lambda_4}_{-\lambda_3}$	$\delta^{f_2}_{f_1}$	$iT^a_{c_1c_2}C^a_{a_3a_4}$
S_9	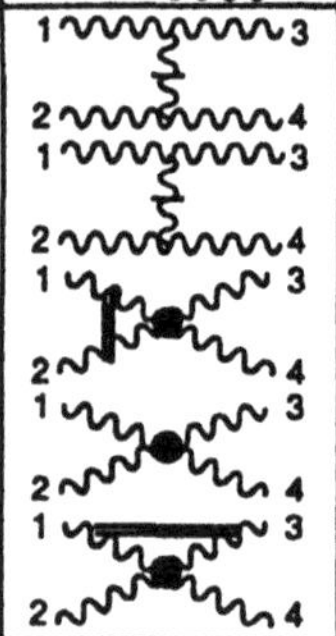	$S^{(s)}_{gg\to gg} = -\widetilde{g}^2 \frac{(x_1+x_3)}{4(x_1-x_3)^2}\frac{(x_2+x_4)}{\sqrt{x_1x_2x_3x_4}}$	$\delta^{\lambda_3}_{\lambda_1}\, \delta^{\lambda_4}_{\lambda_2}$	1	$C^a_{a_1a_3}C^a_{a_2a_4}$
		$S^{(a)}_{gg\to gg} = \widetilde{g}^2 \frac{1}{2(x_1+x_2)^2}\sqrt{\frac{x_1x_3}{x_2x_4}}$	$\delta^{\lambda_2}_{-\lambda_1}\, \delta^{\lambda_4}_{-\lambda_3}$	1	$C^a_{a_1a_2}C^a_{a_3a_4}$
		$S^{(na)}_{gg\to gg} = \widetilde{g}^2 \frac{1}{4\sqrt{x_1x_2x_3x_4}}$	$\delta^{\lambda_3}_{\lambda_1}\, \delta^{\lambda_4}_{\lambda_2}$	1	$C^a_{a_1a_2}C^a_{a_3a_4}$
		$S^{(nb)}_{gg\to gg} = \widetilde{g}^2 \frac{1}{4\sqrt{x_1x_2x_3x_4}}$	$\delta^{\lambda_3}_{\lambda_1}\, \delta^{\lambda_4}_{\lambda_2}$	1	$C^a_{a_1a_4}C^a_{a_3a_2}$
		$S^{(nc)}_{gg\to gg} = \widetilde{g}^2 \frac{1}{4\sqrt{x_1x_2x_3x_4}}$	$\delta^{\lambda_2}_{-\lambda_1}\, \delta^{\lambda_4}_{-\lambda_3}$	1	$C^a_{a_1a_3}C^a_{a_2a_4}$

$$
\begin{aligned}
S = &\textstyle\sum_{q_1,q_2,q_3,q_4} (b^\dagger_1 b^\dagger_2 b_3 b_4 + d^\dagger_1 d^\dagger_2 d_3 d_4)\; S_1(1,2;3,4) \\
+ &\textstyle\sum_{q_1,q_2,q_3,q_4} b^\dagger_1 d^\dagger_2 b_3 d_4 \; S_3(1,2;3,4) \\
+ &\textstyle\sum_{q_1,q_2,q_3,q_4} (b^\dagger_1 a^\dagger_2 b_3 a_4 + d^\dagger_1 a^\dagger_2 d_3 a_4)\; S_5(1,2;3,4) \\
+ &\textstyle\sum_{q_1,q_2,q_3,q_4} (b^\dagger_1 d^\dagger_2 a_3 a_4 + a^\dagger_4 a^\dagger_3 d_2 b_1)\; S_7(1,2;3,4) \\
+ &\textstyle\sum_{q_1,q_2,q_3,q_4} a^\dagger_1 a^\dagger_2 a_3 a_4 \; S_9(1,2;3,4)
\end{aligned}
$$

Table 2d: The matrix elements of the fork interaction F. — The coupling constant g is hidden in $\tilde{g}^2 = g^2 \frac{2}{P^+\Omega}$.

Type	Graph	Momentum × Helicity × Flavor × Color Factor			
F_3		$F_{q\to qq\bar{q}} = \tilde{g}^2 \frac{2}{(x_1-x_2)^2}$	$\delta^{\lambda_2}_{\lambda_1}\, \delta^{\lambda_4}_{-\lambda_3}$	$\delta^{f_2}_{f_1}\, \delta^{f_4}_{f_3}$	$T^a_{c_1c_2} T^a_{c_3c_4}$
F_5		$F^{(q)}_{q\to qgg} = \tilde{g}^2 \frac{1}{(x_1-x_4)} \frac{1}{\sqrt{x_3x_4}}$	$\delta^{\lambda_2}_{\lambda_1}\, \delta^{\lambda_4}_{-\lambda_3}\, \delta^{\lambda_4}_{+\lambda_1}$	$\delta^{f_2}_{f_1}$	$T^{a_3}_{c_1c} T^{a_4}_{cc_2}$
	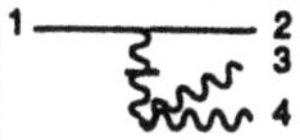	$F^{(g)}_{q\to qgg} = \tilde{g}^2 \frac{1}{(x_1-x_2)^2} \sqrt{\frac{x_3}{x_4}}$	$\delta^{\lambda_2}_{\lambda_1}\, \delta^{\lambda_4}_{-\lambda_3}$	$\delta^{f_2}_{f_1}$	$iT^a_{c_1c_2} C^a_{a_3a_4}$
F_7	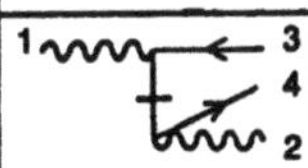	$F^{(qd)}_{g\to gq\bar{q}} = \tilde{g}^2 \frac{1}{(x_1-x_3)} \frac{1}{\sqrt{x_1x_2}}$	$\delta^{\lambda_2}_{\lambda_1}\, \delta^{\lambda_4}_{-\lambda_3}\, \delta^{\lambda_4}_{-\lambda_1}$	$\delta^{f_4}_{f_3}$	$T^{a_1}_{c_3c} T^{a_2}_{cc_4}$
		$F^{(qe)}_{g\to gq\bar{q}} = -\tilde{g}^2 \frac{1}{(x_1-x_4)} \frac{1}{\sqrt{x_1x_2}}$	$\delta^{\lambda_2}_{\lambda_1}\, \delta^{\lambda_4}_{-\lambda_3}\, \delta^{\lambda_4}_{+\lambda_1}$	$\delta^{f_4}_{f_3}$	$T^{a_2}_{c_3c} T^{a_1}_{cc_4}$
		$F^{(g)}_{g\to gq\bar{q}} = \tilde{g}^2 \frac{1}{(x_1-x_2)^2} \frac{(x_1+x_2)}{\sqrt{x_1x_2}}$	$\delta^{\lambda_2}_{\lambda_1}\, \delta^{\lambda_4}_{-\lambda_3}$	$\delta^{f_4}_{f_3}$	$iC^a_{a_1a_2} T^a_{c_3c_4}$
F_9	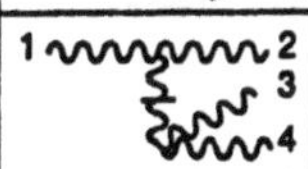	$F^{(g)}_{g\to ggg} = \tilde{g}^2 \frac{1}{2(x_1-x_2)^2} \frac{(x_1+x_2)x_3}{\sqrt{x_1x_2x_3x_4}}$	$\delta^{\lambda_2}_{\lambda_1}\, \delta^{\lambda_4}_{-\lambda_3}$	1	$C^a_{a_1a_2} C^a_{a_3a_4}$
	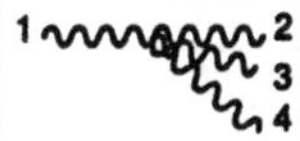	$F^{(n)}_{g\to ggg} = \tilde{g}^2 \frac{1}{2\sqrt{x_1x_2x_3x_4}}$	$\delta^{\lambda_3}_{\lambda_1}\, \delta^{\lambda_4}_{-\lambda_2}$	1	$C^a_{a_1a_2} C^a_{a_3a_4}$

$$
\begin{aligned}
F = & \textstyle\sum_{q_1,q_2,q_3,q_4} (b_1^\dagger b_2 d_3 b_4 + d_1^\dagger d_2 b_3 d_4)\ F_3(1;2,3,4) + \text{h.c.} \\
& + \textstyle\sum_{q_1,q_2,q_3,q_4} (b_1^\dagger b_2 a_3 a_4 + d_1^\dagger d_2 a_3 a_4)\ F_5(1;2,3,4) + \text{h.c.} \\
& + \textstyle\sum_{q_1,q_2,q_3,q_4} \qquad a_1^\dagger a_2 d_3 b_4\ F_7(1;2,3,4) + \text{h.c.} \\
& + \textstyle\sum_{q_1,q_2,q_3,q_4} \qquad a_1^\dagger a_2 a_3 a_4\ F_9(1;2,3,4) + \text{h.c.}
\end{aligned}
$$

4.1 Light-Cone Perturbation Theory for Gauge Theory

The light-cone Green's functions are the probability amplitudes that a state starting in Fock state $|i\rangle$ ends up in Fock state $|f\rangle$ a (light-cone) time τ later

$$\begin{aligned}\langle f|i\rangle\ G(f,i;\tau) &\equiv \langle f|e^{-iH_{LC}\tau/2}|i\rangle \\ &= i\int\frac{d\epsilon}{2\pi}\ e^{-i\epsilon\tau/2}\ G(f,i;\epsilon)\,\langle f|i\rangle\ ,\end{aligned}$$

where Fourier transform $G(f,i;\epsilon)$ can be written

$$\begin{aligned}\langle f|i\rangle\ G(f,i;\epsilon) &= \left\langle f\left|\frac{1}{\epsilon-H_{LC}+i0_+}\right|i\right\rangle \\ &= \left\langle f\left|\frac{1}{\epsilon-H_{LC}+i0_+}+\frac{1}{\epsilon-H_0+i0_+}V\frac{1}{\epsilon-H_0+i0_+}\right.\right. \\ &\quad \left.\left.+\frac{1}{\epsilon-H_0+i0_+}V\frac{1}{\epsilon-H_0+i0_+}V\frac{1}{\epsilon-H_0+i0_+}+\ldots\right|i\right\rangle .\end{aligned}$$

The rules for τ-ordered perturbation theory follow immediately when $(\epsilon-H_0)^{-1}$ is replaced by its spectral decomposition.

$$\frac{1}{\epsilon-H_0+i0_+}=\sum_{n,\lambda_i}\int\widetilde{\prod}\frac{dk_i^+\,d^2k_{\perp i}}{16\pi^3\,k_i^+}\frac{|n:\underline{k}_i,\lambda_i\rangle\ \langle n:\underline{k}_i,\lambda_i|}{\epsilon-\sum\limits_i(k^2+m^2)_i/k_i^++i0_+}$$

The sum becomes a sum over all states n intermediate between two interactions.

To calculate $G(f,i;\epsilon)$ perturbatively then, all τ-ordered diagrams must be considered, the contribution from each graph computed according to the following rules:

1. Assign a momentum k^μ to each line such that the total $k^+, k_\perp$ are conserved at each vertex, and such that $k^2=m^2$, *i.e.* $k^-=(k^2+m^2)/k^+$. With fermions associate an on-shell spinor.

$$u(\underline{k},\lambda)=\frac{1}{\sqrt{k^+}}\left(k^++\beta m+\vec{\alpha}_\perp\cdot\vec{k}_\perp\right)\begin{cases}\chi(\uparrow) & \lambda=\uparrow\\ \chi(\downarrow) & \lambda=\downarrow\end{cases}$$

or

$$v(\underline{k},\lambda)=\frac{1}{\sqrt{k^+}}\left(k^+-\beta m+\vec{\alpha}_\perp\cdot\vec{k}_\perp\right)\begin{cases}\chi(\downarrow) & \lambda=\uparrow\\ \chi(\uparrow) & \lambda=\downarrow\end{cases}$$

where $\chi(\uparrow)=1/\sqrt{2}\,(1,0,1,0)$ and $\chi(\downarrow)=1/\sqrt{2}\,(0,1,0,-1)^T$. For gluon lines, assign a polarization vector $\epsilon^\mu=(0,\ 2\vec{\epsilon}_\perp\cdot\vec{k}_\perp/k^+,\ \vec{\epsilon}_\perp)$ where $\vec{\epsilon}_\perp(\uparrow)=-1/\sqrt{2}\,(1,i)$ and $\vec{\epsilon}_\perp(\downarrow)=1/\sqrt{2}\,(1,-i)$.
2. Include a factor $\theta(k^+)/k^+$ for each internal line.
3. For each vertex include factors as illustrated in Fig. 13. To convert incoming into outgoing lines or vice versa replace

$$u\leftrightarrow v\ ,\qquad \overline{u}\leftrightarrow-\overline{v}\ ,\qquad \epsilon\leftrightarrow\epsilon^*$$

in any of these vertices.

	Vertex Factor	Color Factor
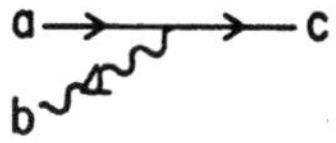	$g\bar{u}(c)\, \not{\epsilon}_b\, u(a)$	T^b
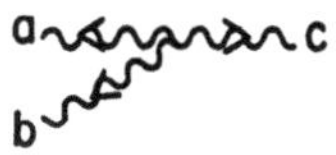	$g\{(p_a - p_b)\cdot \epsilon_c^* \epsilon_a \cdot \epsilon_b$ + cyclic permutations$\}$	iC^{abc}
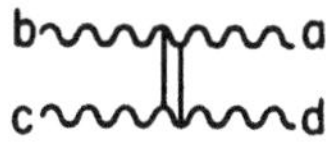	$g^2 \left\{\epsilon_b \cdot \epsilon_c \epsilon_a^* \cdot \epsilon_d^* + \epsilon_a^* \cdot \epsilon_c \epsilon_b \cdot \epsilon_d^*\right\}$	$iC^{abe}\, iC^{cde}$
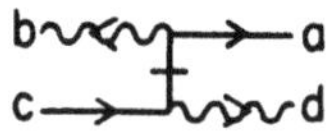	$g^2\, \bar{u}(a)\, \not{\epsilon}_b \dfrac{\gamma^+}{2(p_c^+ - p_d^+)} \not{\epsilon}_c^*\, u(c)$	$T^b\, T^d$
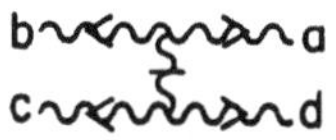	$g^2 \epsilon_a^* \cdot \epsilon_b \dfrac{(p_a^+ - p_b^+)(p_c^+ - p_d^+)}{(p_c^+ + p_b^+)} \epsilon_d^* \cdot \epsilon_c$	$iC^{abe}\, iC^{cde}$
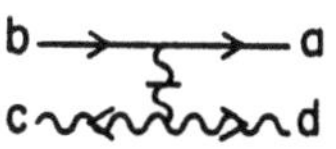	$g^2\, \bar{u}(a)\, \gamma^+ u(b) \dfrac{(p_c^+ - p_d^+)}{(p_c^+ + p_d^+)^2} \epsilon_d^* \cdot \epsilon_c$	$iC^{cde}\, T^e$
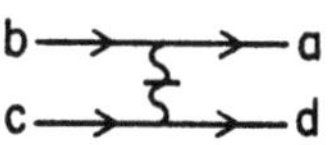	$g^2 \dfrac{\bar{u}(a)\, \gamma^+ u(b)\, \bar{u}(d)\, \gamma^+ u(c)}{(p_c^+ - p_d^+)^2}$	$T^e\, T^e$

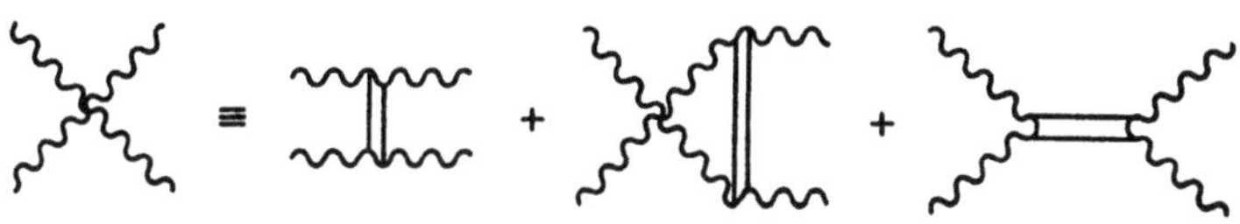

Fig. 13. Graphical rules for QCD in light-cone perturbation theory.

4. For each intermediate state there is a factor

$$\frac{1}{\epsilon - \sum\limits_{\text{interm}} k^- + i0_+}$$

   where ϵ is the incident P^-, and the sum is over all particles in the intermediate state.
5. Integrate $\int dk^+ d^2k_\perp/16\pi^3$ over each independent k, and sum over internal helicities and colors.
6. Include a factor -1 for each closed fermion loop, for each fermion line that both begins and ends in the initial state (*i.e.* $\overline{v} \ldots u$), and for each diagram in which fermion lines are interchanged in either of the initial or final states.

As an illustration, the second diagram in Fig. 13 contributes

$$\frac{1}{\epsilon - \sum\limits_{i=b,d}\left(\frac{k_\perp^2+m^2}{k^+}\right)_i} \cdot \frac{\theta(k_a^+ - k_b^+)}{k_a^+ - k_b^+}$$

$$\times \frac{g^2 \sum\limits_{\lambda} \overline{u}(b)\,\epsilon^*(\underline{k}_a - \underline{k}_b, \lambda)\, u(a)\,\overline{u}(d)\, \not\epsilon(\underline{k}_a - \underline{k}_b, \lambda)\, u(c)}{\epsilon - \sum\limits_{i=b,c}\left(\frac{k_\perp^2+m^2}{k^+}\right)_i - \frac{(k_{\perp a}-k_{\perp b})^2}{k_a^+ - k_b^+}} \cdot \frac{1}{\epsilon - \sum\limits_{i=a,c}\left(\frac{k_\perp^2+m^2}{k^+}\right)_i}$$

(times a color factor) to the $q\overline{q} \to q\overline{q}$ Green's function. (The vertices for quarks and gluons of definite helicity have very simple expressions in terms of the momenta of the particles.) The same rules apply for scattering amplitudes, but with propagators omitted for external lines, and with $\epsilon = P^-$ of the initial (and final) states.

The light-cone Fock state representation can thus be used advantageously in perturbation theory. The sum over intermediate Fock states is equivalent to summing all τ–ordered diagrams and integrating over the transverse momentum and light-cone fractions x. Because of the restriction to positive x, diagrams corresponding to vacuum fluctuations or those containing backward-moving lines are eliminated. For example, such methods can be used to compute perturbative contributions to the annihilation ratio $R_{e\bar{e}} = \sigma(e\bar{e} \to \text{hadrons})/\sigma(e\bar{e} \to \mu^+\mu^-)$ as well as the quark and gluon jet distribution. The computed distributions are functions of the light-cone variables, x, $k_\perp$, λ, which are the natural covariant variables for this problem. Since there are no Faddeev-Popov or Gupta-Bleuler ghost fields in the light-cone gauge $A^+ = 0$, the calculations are explicitly unitary. It is hoped that one can in this way check the three-loop calculation of Gorishny, *et al.*[69]

5. The Lorentz Symmetries of Light-Cone Quantization

It is important to notice that the light-cone quantization procedure and all amplitudes obtained in light-cone perturbation theory (graph by graph!) are manifestly invariant under a large class of Lorentz transformations:

1. boosts along the 3-direction — *i.e.* $p^+ \to Kp^+$, $p^- \to K^{-1}p^-$, $p_\perp \to p_\perp$ for each momentum;
2. transverse boosts — *i.e.* $p^+ \to p^+$, $p^- \to p^- + 2p_\perp \cdot Q_\perp + p^+ Q_\perp^2$, $p_\perp \to p_\perp + p^+ Q_\perp$ for each momentum ($Q_\perp$ like K is dimensionless);
3. rotations about the 3-direction. It is these invariances which also lead to the frame independence of the Fock state wave functions.

More generally, we can understand these properties from the fact that the maximum number (seven) of the ten Poincare generators are kinematic in light-cone and thus leave the state unchanged at $\tau = 0$. [9,70]

5.1 Light-Cone Poincare Generators

The seven generators that commute with the light-cone energy

$$P^- = P^0 - P^3 ,$$

are the three momenta,

$$P^+ = P^0 + P^3, \quad P_\perp = (P_1, P_2) ,$$

the longitudinal rotation and boost operators,

$$J_3, \quad K_3 ,$$

and the light-cone boost operators,

$$B_{\perp 1} = \frac{(K_1 + J_2)}{\sqrt{2}} , \qquad B_{\perp 2} = \frac{(K_2 - J_1)}{\sqrt{2}} .$$

Thus one can diagonalize the light cone energy P^- within a Fock basis where the constituents have fixed total $P^+, P_\perp$, and J_3. For convenience we shall define the light cone Hamiltonian as the operator

$$H_{LC} = P^- P^+ - P_\perp^2$$

so that the eigenvalues of H_{LC} correspond to the invariant spectrum $\mathcal{M}^2$ of the theory.

The boost invariance of the eigensolutions of H_{LC} reflects the fact that the boost operators $K_3, B_{\perp 1}$ and $B_{\perp 2}$ are kinematical. The remaining Poincare generators, the light-cone angular momentum operators,

$$S_{\perp 1} = \frac{(K_1 - J_2)}{\sqrt{2}}, \quad \text{and} \quad S_{\perp 2} = \frac{(K_2 + J_1)}{\sqrt{2}}$$

are dynamical and do not commute with P^- or H_{LC}.

In order to understand these features better, we shall discuss the construction of the LC Fock basis for mesons in QCD in some detail. It is easiest to start in a "standard frame" with total momentum $P^+_{\rm std} = 1, \vec{P}_{\perp \rm std} = \vec{0}_\perp$ (in any units!) and then boost to a general frame.[70] To simplify the notation we shall write the conserved three momenta in the form $\underline{k_i} \equiv (k_i^+, \vec{k}_{\perp i})$ which becomes $(x_a, \vec{k}_{\perp i})$ in the standard frame. We can then build the light-cone Fock states by applying the free quark, anti-quark, and gluon field operators to the free vacuum:

$$|n\rangle = b^\dagger(\underline{k_a}, \lambda_a) d^\dagger(\underline{k_b}, \lambda_b) a^\dagger(\underline{k_c}, \lambda_c) \, |0\rangle ,$$

where $\sum x_i = 1, \sum k_{\perp i} = 0_\perp$, and $\sum \lambda_i = \lambda$, since

$$J_3 \left| P^+ = 1, \vec{P}_\perp = \vec{0}_\perp, \lambda \right\rangle = \lambda \left| P^+ = 1, \vec{P}_\perp = 0_\perp, \lambda \right\rangle .$$

In addition, in each Fock state the color indices of the quark and gluon quanta can be combined to form $SU(3)_C$ color-singlet representations. (A general group theory procedure for finding all such irreducible representations is given by Kaluža.[30]) Since the Fock basis is complete, we can write the eigensolution to the pion wavefunction in the standard frame in the form

$$|\Psi_\pi(P^+ = 1, P_\perp = 0_\perp)\rangle = \sum_{\lambda_i} \int_0^1 \frac{1}{\sqrt{x}\sqrt{1-x}} \int \frac{d^2k_\perp}{2(2\pi)^3} \psi_{q\bar{q}}(x, k_\perp, \lambda_1)$$

$$\times\, b^\dagger(x, k_\perp, \lambda_1) d^\dagger(1-x, -k_\perp, -\lambda_1)\, |0\rangle$$

$$+ \sum \int \psi_{q\bar{q}g} b^\dagger d^\dagger a^\dagger\, |0\rangle + \ \dots .$$

Thus with this construction $\overrightarrow{P}_\perp |\Psi_\pi\rangle = \overrightarrow{0}_\perp$, and $P^+ |\Psi_\pi\rangle = 1\, |\Psi_\pi\rangle$. The eigenvalue problem for the pion in QCD is then

$$P^- |\Psi_\pi\rangle = m_\pi^2\, |\Psi_\pi\rangle \ .$$

which in the Fock basis reduces to the problem of diagonalizing the Heisenberg matrix:

$$\sum_n \langle m|H_{LC}|n\rangle \langle n|\Psi_\pi\rangle = m_\pi^2 \langle m|\Psi_\pi\rangle \ .$$

The eigensolutions then determine the complete Fock representation of the pion light-cone wavefunction.

Given the pion eigensolution in the standard frame, we can immediately construct the pion wavefunction at any total three-momentum $\underline{Q} = (Q^+, Q_\perp)$ since the boost operators $K_3, B_{\perp 1}$, and $B_{\perp 2}$ can all be constructed from the free quark and gluon fields. The boost operators have the action

$$e^{-i\omega K_3} P^+ e^{i\omega K_3} = e^\omega P^+ \ ,$$

and

$$e^{-i\overrightarrow{V}_\perp \cdot \overrightarrow{B}_\perp} \overrightarrow{P}_\perp = \overrightarrow{P}_\perp + P^+ \overrightarrow{V}_\perp \, .$$

Thus we define the boost operator

$$U(Q^+, \overrightarrow{Q}_\perp) = e^{-i\frac{\overrightarrow{Q}_\perp}{Q^+} \cdot \overrightarrow{B}_\perp} e^{-\ell n\, Q^+ K_3} \ ,$$

so that

$$U(Q^+, \overrightarrow{Q}_\perp) b^\dagger(x, \vec{k}_\perp, \lambda) U^{-1}(Q^+, \overrightarrow{Q}_\perp) = b^\dagger(xQ, \vec{k}_\perp + x\overrightarrow{Q}_\perp, \lambda) \ ,$$

etc. Thus the pion wavefunction in a general frame is

$$\Psi_\pi(Q^+, \overrightarrow{Q}_\perp) = U(Q^+, \overrightarrow{Q}_\perp) \Psi_\pi(Q^+ = 1, \overrightarrow{Q}_\perp = \overrightarrow{0}_\perp),$$

since $\sum_i(\vec{k}_{\perp i}+x_i\vec{Q}_\perp) = \vec{Q}_\perp$, and $\sum_i x_i Q^+ = Q^+$. Since U is only a function of the free fields, the result is the Fock expansion derived in Section 2.3. Thus, as emphasized above, the light-cone wavefunctions $\psi_n(x_i, \vec{k}_{\perp i}, \lambda_i)$ and its relative coordinates x_i and $\vec{k}_{\perp i}$ are independent of the total momenta $Q^+, \vec{Q}_\perp$. The actual particle momenta are with plus momentum $k^+ = x_i Q^+$, transverse momentum $\vec{k}_{\perp i} + x_i \vec{Q}_\perp$ and spin projection $J_3 = \lambda_i$. The spinors and polarization vectors for such particles are given explicitly in the proceeding section.

5.2 Spin on the Light-Cone [70]

If a theory is rotational invariant, then each eigenstate of the Hamiltonian which describes a state of nonzero mass can be classified in its rest frame by its spin eigenvalues

$$J^2 \left| P^0 = M, \vec{P} = \vec{0} \right\rangle = s(s+1) \left| P^0 = M, \vec{P} = \vec{0} \right\rangle ,$$

and

$$J_z \left| P^0 = M, \vec{P} = \vec{0} \right\rangle = s_z \left| P^0 = M, \vec{P} = \vec{0} \right\rangle .$$

This procedure is more complicated in the light-cone scheme since the angular momentum operator does not commute with H_{LC}. Nevertheless, one can construct light-cone operators [70] $\mathcal{J}^2 = \mathcal{J}_3^2 + \mathcal{J}_\perp^2$ and $\mathcal{J}_3$ where

$$\mathcal{J}_3 = J_3 + \epsilon_{ij} B_{\perp i} P_{\perp j}/P^+ ,$$

and

$$\mathcal{J}_\perp = \frac{1}{M}\epsilon_{k\ell}(S_{\perp\ell}P^+ - B_{\perp\ell}P^- - K_3 P_{\perp\ell} + \mathcal{J}_3 \epsilon_{\ell m} P_{\perp m}) ,$$

which, in principle, could be applied to an eigenstate $\left| P^+, \vec{P}_\perp \right\rangle$ to obtain the rest frame spin quantum numbers. This is straightforward for $\mathcal{J}_3$ since it is kinematical; in fact, $\mathcal{J}_3 = J_3$ in a frame with $\vec{P}_\perp = \vec{0}_\perp$. However, $\mathcal{J}_\perp$ is dynamical and depends on the interactions. Thus it is generally difficult to explicitly compute the total spin of a state using light-cone quantization. Nevertheless, this is not a problem in practice since, given the spectrum of the light-cone Hamiltonian, one can identify the rest-frame spin of each eigenstate simply by counting the number of degenerate levels appearing at each value of J_3.

5.3 Discrete Light-Cone Symmetries

The QCD Hamiltonian has a number of global symmetries which are also characteristic of its eigensolutions. It is thus useful to pre-diagonalize the Light-Cone Fock state basis with respect to all of the operators which commute with

H_{LC} and then diagonalize H_{LC} within each super–selection sector. The most important global symmetries are [70]
Light-Cone Parity,

$$I_P^{LC} = e^{-i\pi J_1} I_P ,$$

where

$$I_P^{LC} \psi_{q\bar{q}}(x, k_{\perp 1}, k_{\perp 2}, \lambda_i) = \psi_{q\bar{q}}(x, -k_{\perp 1}, k_{\perp 2}, -\lambda_i) ;$$

Light-Cone Time-Reversal,

$$I_T^{LC} = e^{-i\pi J_3} I_P I_T ,$$

where

$$I_P^{LC} \psi_{q\bar{q}}(x, \vec{k}_\perp \lambda_i) = \psi^*_{q\bar{q}}(x, -\vec{k}_\perp, \lambda_i) ;$$

and *Light-Cone Charge-Conjugation:*

$$I_P^{LC} \psi_{q\bar{q}}(x, \vec{k}_\perp, \lambda_1, \lambda_2) = -\psi_{q\bar{q}}(1-x, -\vec{k}_\perp, \lambda_2, \lambda_1) .$$

By pre-diagonalizing in the eigensectors of these symmetries, one reduces the matrix size of the representations of H_{LC} by a factor of two for each symmetry.

6. Renormalization and Ultra-Violet Regulation of Light-Cone-Quantized Gauge Theory

An important element in the light-cone Hamiltonian formulation of quantum field theories is the regulation of the ultraviolet region. In order to define a renormalizable theory, a covariant and gauge invariant procedure is required to eliminate states of high virtuality. The physics beyond the scale Λ is contained in the normalization of the mass $m(\Lambda)$ and coupling constant $g(\Lambda)$ parameters of the theory, modulo negligible corrections of order $1/\Lambda^n$ from the effective Lagrangian. The logarithmic dependence of these input parameters is determined by the renormalization group equations. In Lagrangian field theories the ultraviolet cut-off is usually introduced via a spectrum of Pauli-Villars particles or dimensional regulation. Another interesting possibility is to work with a super-symmetric extension of theory which is finite, and then introduce soft symmetry breaking to give the super-partners large mass of order Λ.

An analogous ultraviolet regularization must also can also be followed in the case of quantization in the light-cone Hamiltonian framework. For example, one can construct the ultraviolet regulated Hamiltonian H_{LC}^{Λ} for QED(3+1) directly from the Lagrangian using Pauli-Villars regulation for both the ultraviolet and infrared regions. The Pauli-Villars spectral conditions must be chosen to eliminate both logarithmic and potentially quadratic divergences.

As an example of this procedure, we have shown in Ref. 28 that the lepton mass renormalization counterterms computed in LCPTH using discretization is identical to that of the Lagrangian perturbation theory in the continuum limit.

It was also verified numerically (to 12 significant figures) that this procedure is also consistent within the context of the non-perturbative diagonalization of the light-cone Hamiltonian for the electron state within a truncated Fock space basis $|e\rangle$, $|e\gamma\rangle$.

The Pauli-Villars regulation allows a complete implementation of time-ordered Hamiltonian perturbation theory at $P \to \infty$, in a form which is essentially equivalent to LCPTH.[11] The renormalized amplitudes can be explicitly constructed in each order in perturbation theory simply by subtracting local mass vertex and wavefunction renormalization counterterms defined using the "alternating denominators" method. (See Fig. 4). In addition, it is shown in Ref. 11 that Z-graphs or instantaneous fermion exchange contributions can be automatically included leading to a numerator factor from each time ordering identical to the numerator of the corresponding Feynman amplitude. These methods have been successfully applied to the calculation of the electron magnetic moment to two and three loop order. More recently, Langnau [71] has extended the $g-2$ calculations in LCPTH using dimensional regulation for the transverse momentum integrations in both Feynman and light-cone gauge.

The above method for ultraviolet regulation is not sufficient for non-perturbative problems, such as the diagonalization of the light-cone Hamiltonian. In the previous sections we have discussed the discretization of the light-cone Fock basis using DLCQ. In such methods, one needs to provide *a priori* some type of truncation of the Fock state basis. Since wavefunctions and Green's functions decrease with virtuality, one expects that states very far off the light-cone energy shell will have no physical effect on a system, except for renormalization of the coupling constant and mass parameters. Thus it is natural to introduce a "global" cut-off such that a Fock state $|n\rangle$ is retained only if

$$\sum_{i\epsilon n} \frac{\vec{k}_{\perp i}^2 + m_i^2}{x_i} - M^2 < \Lambda^2 .$$

Here M is the mass of the system in the case of the bound state problem, or the total invariant mass $\sqrt{s}$ of the initial state in scattering theory. One can also limit the growth of the Fock state basis by introducing a "local" cutoff on each matrix element $\langle n|H_{LC}|m\rangle$ by requiring that the change in invariant mass squared

$$\left| \sum_{i\epsilon n} \frac{\vec{k}_{\perp i}^2 + m_i^2}{x_i} - \sum_{i\epsilon m} \frac{\vec{k}_{\perp i}^2 + m_i^2}{x_i} \right| < \Lambda^2 .$$

Similarly, one can use a lower cutoff on the invariant mass difference to regulate the infrared region.[72]

The global and local cutoff methods were used in Ref. 4 to derive factorization theorems for exclusive and inclusive processes at large momentum transfer in QCD. In particular, the global cut-off defines the Fock-state wavefunctions $\psi^{\Lambda}(x, \vec{k}_{\perp}, \lambda)$ and distribution amplitude $\phi(x, \Lambda)$, the non-perturbative input for computing hadronic scattering amplitudes. The renormalization group properties of the light-cone wavefunctions and the resulting evolution equations for the structure functions and distribution amplitudes are also discussed in Ref. 4. The calculated anomalous dimensions γ_n for the moments of these quantities agree with results obtained using the operator product expansion.[73]

The global cut-off conveniently truncates the ultraviolet and infrared regions of the Fock space basis, and it is easily implemented in practice. However, there several complicating features if this method is used as the sole ultraviolet cut-off of the field theory:

- Gauge-invariance is obviously destroyed by the implementation of a strict cut-off in momentum space. In fact, this problem can be largely avoided by using the following "gauge principle": [28] the matrix element of an instantaneous gluon exchange four-point interaction is nonzero only if the corresponding three-point gluon exchange interactions are allowed by the Fock space cutoff. Similarly, instantaneous fermion exchange matrix elements are non-zero only if the corresponding propagating fermion interactions are permitted. One can easily check that this principle retains gauge invariance in tree diagrams in gauge theory and preserves the boost invariance of the light-cone Hamiltonian theory.
- Use of the global cut-off alone implies that the cut-off of a self-energy insertion counterterm for any particle depends on the invariant mass of the entire Fock state. This implies that the renormalization counterterms for a given particle depends on the kinematics of the spectators appearing in that Fock state.[28] Formally, this dependence is power-law suppressed by at least a power of $1/\Lambda^2$, but in practice, it is advantageous to keep Λ^2 of reasonable size. The spectator problem is avoided if one uses the local cut-off.
- In general, light-cone quantization using the global or local cutoff can lead to terms in H_{LC}^{Λ} of the form $\delta m \bar{\psi} \frac{\gamma^+}{i\partial^+} \psi$. Although such a term is invariant under the large class of light-cone Lorentz transformations, it is not totally invariant. For example, such terms arise in order g^2 as a result of normal-ordering of the four-point interaction terms. (Note that this complication does not occur in a strictly covariant regulation procedure such as Pauli-Villars.) Thus in this cut-off procedure one has to allow for an extra mass counterterm insertions in the numerator matrix elements of the light-cone interaction Hamiltonian. Burkardt and Langnau [12] have suggested that the extra counterterms can be fixed by *a posteriori* imposing rotational symmetry on the bound state solutions, so that all Lorentz symmetries are restored.

Each of the proposed cut-offs thus have advantages and disadvantages for the DLCQ program. A global cut-off is necessary to limit the size of the Fock space for the numerical diagonalization of the light-cone Hamiltonian or to truncate it to a finite set of equations of motion. However, for the purpose of renormalization, it is possibly advantageous to simultaneously implement other regulators, such as the local cut-off, a Pauli-Villars spectrum, supersymmetric partners, etc.

Ideally, one can apply all of this to QCD(3+1). Once one has defined the regulated light-cone Hamiltonian, solved for its spectrum, as in the DLCQ procedure, the mass $m(\Lambda)$ and coupling constant $g(\Lambda)$ parameters can be fitted by normalizing the output mass and charge radius of the proton state, say, to experiment. Non-perturbative QCD would then be tested by comparison with the remaining hadron and nuclear spectrum and wavefunctions. We discuss the beginning of the application of this program to three-space one-time theories in the following sections.

7. Discretized Light-Cone Quantization: Applications to QCD(1+1)

As we have seen in the proceeding sections, QCD dynamics takes a rather simple form when quantized at equal light-cone "time" $\tau = t + z/c$. In light-cone gauge $A^+ = A^0 + A^z = 0$, the QCD light-cone Hamiltonian

$$H_{\mathrm{QCD}} = H_0 + gH_1 + g^2 H_2$$

contains the usual 3-point and 4-point interactions plus induced terms from instantaneous gluon exchange and instantaneous quark exchange diagrams. The perturbative vacuum serves as the lowest state in constructing a complete basis set of color-singlet Fock states of H_0 in momentum space. Solving QCD is then equivalent to solving the eigenvalue problem:

$$H_{\mathrm{QCD}}|\Psi >= M^2|\Psi >$$

as a matrix equation on the free Fock basis. The set of eigenvalues $\{M^2\}$ represents the spectrum of the color-singlet states in QCD. The Fock projections of the eigenfunction corresponding to each hadron eigenvalue gives the quark and gluon Fock state wavefunctions $\psi_n(x_i, k_{\perp i}, \lambda_i)$ required to compute structure functions, distribution amplitudes, decay amplitudes, etc. For example, the e^+e^- annihilation cross section into a given $J = 1$ hadronic channel can be computed directly from its $\psi_{q\overline{q}}$ Fock state wavefunction.

The basic question is whether one can actually solve the light-cone Hamiltonian eigenvalue problem, even numerically. This is the goal of the DLCQ method.[21] We first observe that the light-cone momentum space Fock basis becomes discrete and amenable to computer representation if one chooses (anti-)periodic boundary conditions for the quark and gluon fields along the $z^- = z - ct$ and $z_\perp$ directions. In the case of renormalizable theories, a covariant ultraviolet cutoff Λ is introduced which limits the maximum invariant mass of the particles in any Fock state. One thus obtains a finite matrix representation of $H_{\mathrm{QCD}}^{(\Lambda)}$ which has a straightforward continuum limit. The entire analysis is frame independent, and fermions present no special difficulties.

7.1 Construction of the Discrete LC Fock Basis

The key step in obtaining a discrete representation of the light-cone Hamiltonian in a form amenable to numerical diagonalization, is the construction of a complete, countable, Fock state basis,

$$\sum_n |n\rangle \langle n| = I .$$

This can be explicitly done in QCD by constructing a complete set of color-singlet eigenstates of the free Hamiltonian as products of representations of free quark and gluon fields. The states are chosen as eigenstates of the constants of the motion, P^+, $\vec{P}_\perp$, J_z, and the conserved charges. In addition, one can pre-diagonalize the Fock representation by classifying the states according to their discrete symmetries, as described in the previous section. This step alone reduces the size of the matrix representations by as much as a factor of 16.

The light-cone Fock representation can be made discrete by choosing periodic (or, in the case of fermions, anti-periodic) boundary conditions on the fields:

$$\psi(z^-) = \pm\psi(z^- - L) ,$$

$$\psi(x_\perp) = \psi(x_\perp - L_\perp)$$

Thus in each Fock state,

$$P^+ = \frac{2\pi}{L} K ,$$

and each constituent

$$k_i^+ = \frac{2\pi}{L} n_i ,$$

where the positive integers n_i satisfy

$$\sum_i n_i = K .$$

Similarly

$$\vec{k}_{\perp i} = \frac{\pi}{L_\perp} \vec{n}_{\perp i} ,$$

where the vector integers sum to $\vec{0}_\perp$ in the standard frame.

The positive integer K is called the "harmonic resolution." Notice that for any choice of K, there are only a finite number of partitions of the plus momenta, and thus only a finite set of rational values of $x_i = k_i^+/P^+ = n_i/K$ appear:

$$x_i = \frac{1}{K} , \quad \frac{2}{K} , \quad \ldots \frac{K-1}{K} .$$

Thus eigensolutions obtained by diagonalizing H_{LC} on this basis determine the deep inelastic structure functions $F_I(x)$ only at the set of rational discrete points x_i. The continuum limit thus requires extrapolation to $K \to \infty$. Note that the value of L is irrelevant, since it can always be scaled away by a Lorentz boost. Since H_{LC}, P^+, $\vec{P}_\perp$, and the conserved charges all commute, H_{LC} is block diagonal.

The DLCQ program becomes especially simple for gauge theory in one-space one-time dimensions because of the absence of transverse momenta but also because there are no gluon degrees of freedom. In addition, for a given value of the harmonic resolution K the Fock basis becomes restricted to finite dimensional representations. The dimension of the representation corresponds to the number of partitions of the integer K as a sum of positive integers n. The eigenvalue problem thus reduces to the diagonalization of a finite Hermitian matrix. The continuum limit is clearly $K \to \infty$.

Since continuum scattering states as well as single hadron color-singlet hadronic wavefunctions are obtained by the diagonalization of H_{LC}, one can also calculate scattering amplitudes as well as decay rates from overlap matrix elements of the interaction Hamiltonian for the weak or electromagnetic interactions. In principle, all higher Fock amplitudes, including spectator gluons, can be kept in the light-cone quantization approach; such contributions cannot generally be neglected in decay amplitudes involving light quarks.

One of the first applications [22] of DLCQ to local gauge theory was to QED in one-space and one-time dimensions. Since $A^+ = 0$ is a physical gauge, there are no photon degrees of freedom. Explicit forms for the matrix representation of H_{QED} are given in Ref. 22, The QED results agree with the Schwinger solution at zero fermion mass, or equivalently, infinite coupling strength.

More recently DLCQ [18] has been used to obtain the complete color-singlet spectrum of QCD in one space and one time dimension for $N_C = 2, 3, 4$. [74] The hadronic spectra are obtained as a function of quark mass and QCD coupling constant (see Fig. 14). Where they are available, the spectra agree with results obtained earlier; in particular, the lowest meson mass in SU(2) agrees within errors with lattice Hamiltonian results.[75] The meson mass at $N_C = 4$ is close to the value predicted by 't Hooft [74] in the large N_C limit. The DLCQ method also provides the first results for the baryon spectrum in a non–Abelian gauge theory. The lowest baryon mass is shown in Fig. 14 as a function of coupling constant. The ratio of meson to baryon mass as a function of N_C also agrees at strong coupling with results obtained by Frishman and Sonnenschein.[76] Precise values for the mass eigenvalue can be obtained by extrapolation to large K by fitting to forms with the correct functional dependence in $1/K$.

7.2 QCD(1+1) in the $m/g \to 0$ Limit

It is interesting to see how one QCD(1+1) and QED(1+1) become equivalent to theories of non-interacting hadrons theories in the Schwinger zero quark mass limit. The emergence of massless hadrons at zero quark mass in the non-Abelian theory may be understood [18] by studying the momentum space transforms of the $SU(N)$ currents (at $x^+ = 0$)

$$V_k^a = \tfrac{1}{2} \int_{-L}^{L} dx^- e^{-i \frac{k\pi}{L} x^-} j^{+a}(x^-)$$

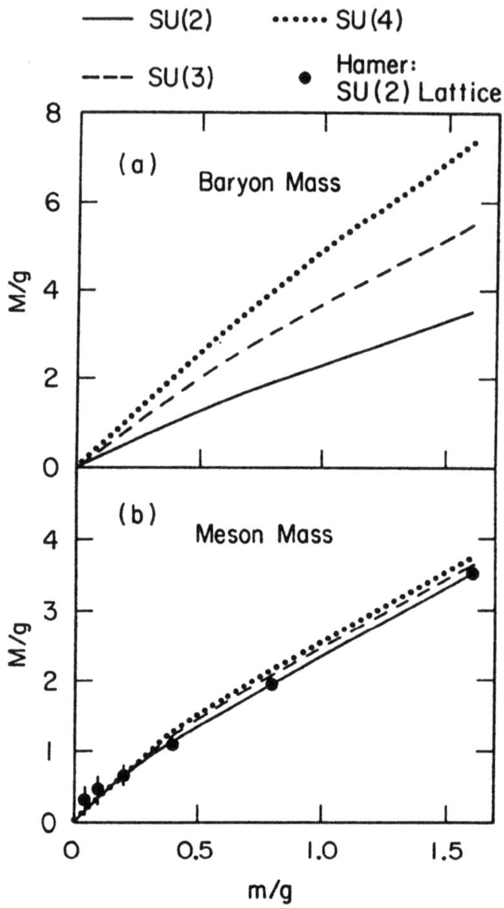

Fig. 14. The baryon and meson spectrum in QCD(1+1) computed in DLCQ for $N_C = 2, 3, 4$ as a function of quark mass and coupling constant. [18]

which satisfy $[V_k^a, V_l^b] = if^{abc}V_{k+l}^c + \frac{1}{2}\, l\delta^{ab}\delta_{k+l,0}$. The currents j^{+a} are defined by point splitting along x^-; however for $A^+ = 0$, the path- ordered exponential included to ensure gauge invariance reduces to one. The algebra may be extended to include the $U(1)$ current $j^+ = (\frac{2}{N})^{\frac{1}{2}} : \psi_R^\dagger \psi_R :$. The transformed operator V_k commutes with the other $SU(N)$ elements, and the related operator $a_k^\dagger \equiv (\frac{2}{k})^{\frac{1}{2}} \epsilon(k) V_k$ satisfies the free boson commutation relations $[a_k, a_k^\dagger] = \delta_{k,l}$.

The interacting part of the Hamiltonian is greatly simplified when expressed in terms of these operators:

$$P_I^- = -\frac{g^2}{16}\int_{-L}^{L} dx^- dy^- |x^- - y^-| j^{+a}(x^-) j^{+a}(y^-)$$

becomes

$$\frac{L}{2\pi}\frac{g^2}{2\pi}\sum_{k=-\infty}^{\infty}\frac{1}{k^2}V_k^a V_{-k}^a.$$

Because $V_0^a = Q^a$, the contribution at $k = 0$ is proportional to the total charge $Q^a Q^a$ and so may be discarded.

The V_k are color-singlet bi-linear operators in ψ_R, and so may be used to create mesonic-like states with momentum $P^+ = \frac{2\pi k}{L}$. In the limit where m/g is zero, the entire Hamiltonian is given by the above equation. Because the V_k commute with the V_k^a which appear in P_I^-,

$$M^2 V_k |0\rangle = \frac{2\pi k}{L}[P^-, V_k] |0\rangle = 0.$$

Not only is the state created by acting with V_k on the vacuum an exactly massless eigenstate in this limit, but states formed by repeated applications are also exactly massless. Furthermore, acting with V_k on an eigenstate of non-zero mass produces a degenerate state of opposite parity. This argument is independent of the value of the numerical momentum K and so gives an exact continuum result.

If the gauge group is $U(N)$ rather than $SU(N)$, the additional term associated with the extra $U(1)$,

$$\frac{L}{2\pi}\frac{g^2}{2\pi}\sum_{k=1}^{\infty}\frac{1}{k}a_k^\dagger a_k,$$

appears in P^-. The a_k satisfy free bosonic commutation relations, and this additional interaction is therefore the discrete light-cone Hamiltonian for free bosons of mass squared $g^2/2\pi$. These formerly massless states created by the $a_k^\dagger$ are promoted to the free massive bosons found in the Schwinger model and are discussed in Refs. 77 and 22. The quark wavefunctions for these states at infinite coupling or zero fermion mass are constant in x, reflecting their point-like character.

7.3 Structure Functions for QCD(1+1) [18]

As we have emphasized, when the light-cone Hamiltonian is diagonalized at a finite resolution K, one gets a complete set of eigenvalues corresponding to the total dimension of the Fock state basis. A representative example of the spectrum is shown in Fig. 15 for baryon states ($B = 1$) as a function of the dimensionless variable $\lambda = 1/(1+\pi m^2/g^2)$. Notice that spectrum automatically includes continuum states with $B = 1$.

The structure functions for the lowest meson and baryon states in SU(3) at two different coupling strengths $m/g = 1.6$ and $m/g = 0.1$ are shown in Figs. 16 and 17. Higher Fock states have a very small probability; representative contributions to the baryon structure functions are shown in Figs. 18 and 19. For comparison, the valence wavefunction of a higher mass state which can be identified as a composite of meson pairs (analogous to a nucleus) is shown in Fig. 20. The interactions of the quarks in the pair state produce Fermi motion beyond $x = 0.5$. Although these results are for one-time one-space theory they

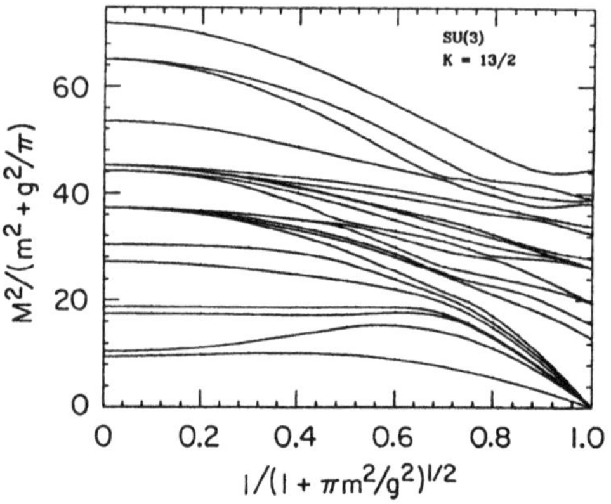

Fig. 15. Representative baryon spectrum for QCD in one-space and one-time dimension.[18]

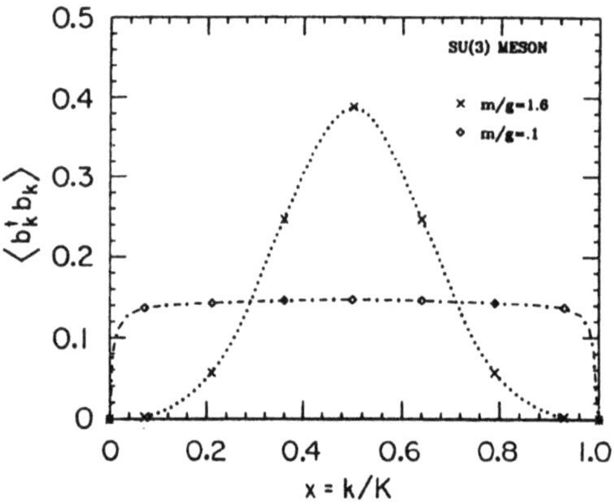

Fig. 16. The meson quark momentum distribution in QCD[1+1] computed using DLCQ.[18]

do suggest that the sea quark distributions in physical hadrons may be highly structured. We will discuss this possibility further in the next section.

8. The Heavy Quark Content of the Proton

The DLCQ results for sea quark distributions in QCD(1+1) may have implications for the heavy quark content of physical hadrons. One of the most intriguing unknowns in nucleon structure is the strange and charm quark structure of the nucleon wavefunction.[78] The EMC spin crisis measurements indicate a significant $s\bar{s}$ content of the proton, with the strange quark spin strongly anti-correlated with the proton spin. Just as striking, the EMC measurements [79] of the charm structure function of the Fe nucleus at large $x_{bj} \sim 0.4$ appear to be considerably larger than that predicted by the conventional photon-gluon

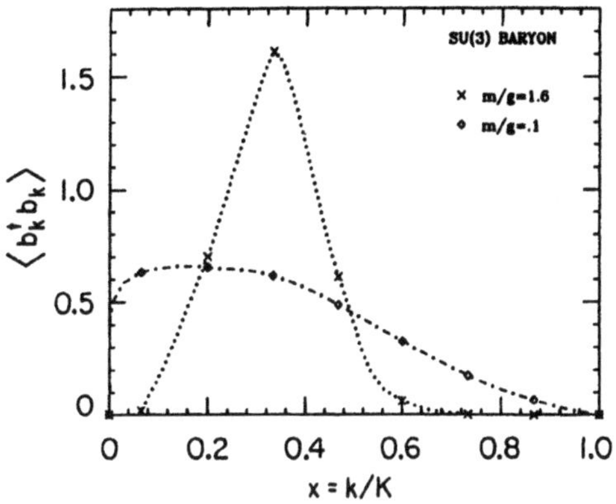

Fig. 17. The baryon quark momentum distribution in QCD[1+1] computed using DLCQ.[18]

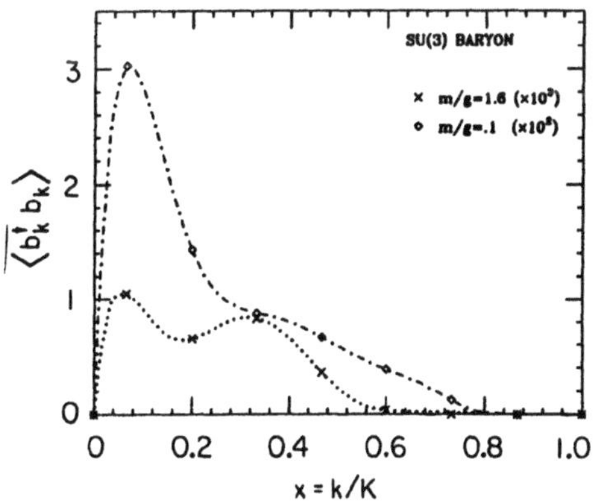

Fig. 18. Contribution to the baryon quark momentum distribution from $qqq\bar{q}\bar{q}$ states for QCD[1+1].[18]

fusion model, indicating an anomalous charm content of the nucleon at large values of x. The probability of intrinsic charm has been estimated [79] to be 0.3%.

As emphasized in the previous sections, the physical content of a hadron in terms of its quark and gluon constituents, including sea-quark distributions, is represented by its light-cone wavefunctions $\psi_n(x_i, p_{\perp i}, \lambda_i)$, which the projections of the hadron wavefunction on the complete set of Fock states defined at fixed light-cone time $\tau = t + z/c$.[19] Here $x_i = (E_i + p_{Li})/(E + p_L)$, with $\sum_i x_i = 1$, is the fractional (light-cone) momentum carried by parton i. The determination of the light-cone wavefunctions requires diagonalizing the light-cone Hamiltonian on the free Fock basis. As we have discussed, this has, in fact, been done for QCD in one-space and one-time dimension.[18]

In Fig. 21 we show recent results obtained by Hornbostel [80] for the structure functions of the lowest mass meson in QCD(1+1) wavefunctions for $N_C = 3$ and two quark flavors. As seen in the figure, the heavy quark distri-

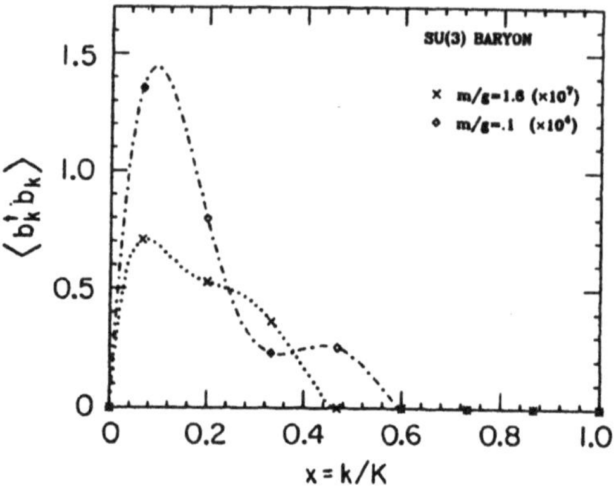

Fig. 19. Contribution to the baryon quark momentum distribution from $qqq\bar{q}\bar{q}\bar{q}$ states for QCD[1+1].[18]

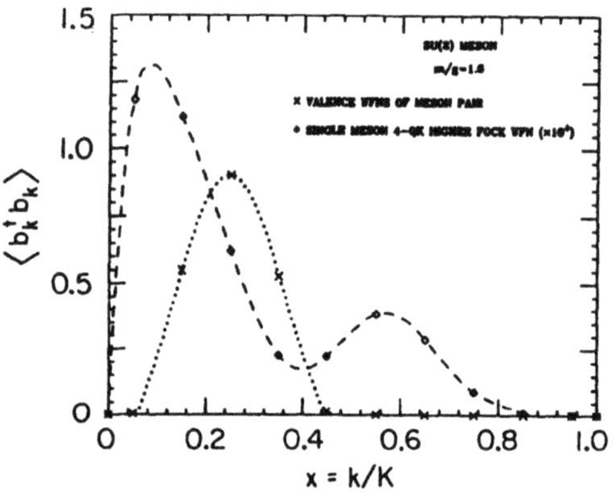

Fig. 20. Comparison of the meson quark distributions in the $qq\bar{q}\bar{q}$ Fock state with that of a continuum meson pair state. The structure in the former may be due to the fact that these four-particle wavefunctions are orthogonal.[18]

bution arising from the $q\bar{q}Q\bar{Q}$ Fock component has a two-hump character. The second maximum is expected since the constituents in a bound state tend to have equal velocities. The result is insensitive to the value of the Q^2 of the deep inelastic probe. Thus intrinsic charm is a feature of exact solutions to QCD(1+1). Note that the integrated probability for the Fock states containing heavy quarks falls nominally as g^2/m_Q^4 in this super-renormalizable theory, compared to g^2/m_Q^2 dependence expected in renormalizable theories.

In the case of QCD(3+1), we also expect a two-component structure for heavy-quark structure functions of the light hadrons. The low x_F enhancement reflects the fact that the gluon-splitting matrix elements of heavy quark production favor low x. On the other hand, the $Q\bar{Q}q\bar{q}$ wavefunction also favors equal velocity of the constituents in order to minimize the off-shell light-cone energy and the invariant mass of the Fock state constituents. In addition, the

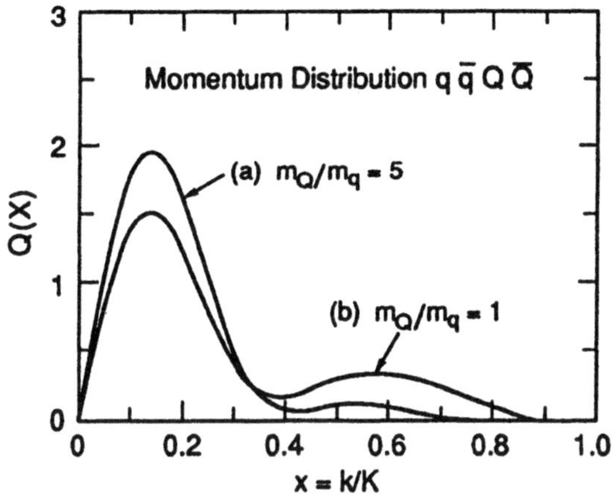

Fig. 21. The heavy quark structure function $Q(x) = G_{Q/M}(x)$ of the lightest meson in QCD(1+1) with $N_c = 3$ and $g/m_q = 10$. Two flavors are assumed with (a) $m_Q/m_q = 1.001$ and (b) $m_Q/m_q = 5$. The curves are normalized to unit area. The probability of the $q\bar{q}Q\bar{Q}$ state is 0.56×10^{-2} and 0.11×10^{-4}, respectively. The DLCQ method for diagonalizing the light-cone Hamiltonian is used with anti-periodic boundary conditions. The harmonic resolution is taken at $K = 10/2$. (From Ref. 80.)

non-Abelian effective Lagrangian analysis discussed above allows a heavy quark fluctuation in the bound state wavefunction to draw momentum from all of the hadron's valence quarks at order $1/m_Q^2$. This implies a significant contribution to heavy quark structure functions at medium to large momentum fraction x. The EMC measurements of the charm structure function of the nucleon appear to support this picture.[79]

It is thus useful to distinguish *extrinsic and intrinsic* contributions to structure functions. The extrinsic contributions are associated with the substructure of a single quark and gluon of the hadron. Such contributions lead to the logarithmic evolution of the structure functions and depend on the momentum transfer scale of the probe. The intrinsic contributions involve at least two constituents and are associated with the bound state dynamics independent of the probe. (See Fig. 22.) The intrinsic gluon distributions [17] are closely related to the retarded mass-dependent part of the bound-state potential of the valence quarks.[19] In addition, because of asymptotic freedom, the hadron wavefunction has only an inverse power $\mathcal{M}^{-2}$ suppression for high mass fluctuations, whether heavy quark pairs or light quark pairs at high invariant mass $\mathcal{M}$. This "intrinsic hardness" of QCD wavefunctions leads to a number of interesting phenomena, including a possible explanation for "cumulative production," high momentum components of the nuclear fragments in nuclear collisions. This is discussed in detail in Ref. 81.

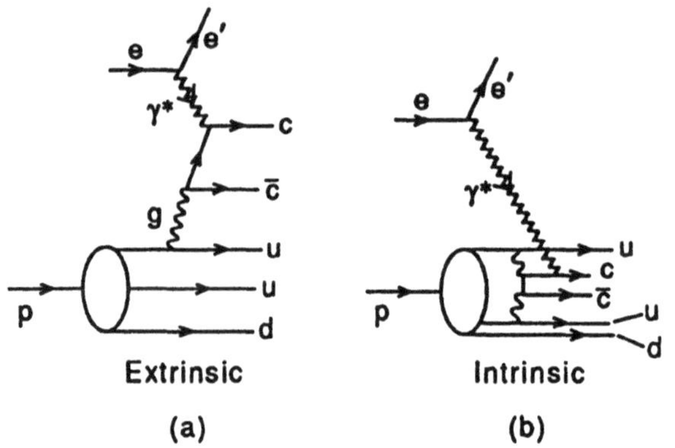

Fig. 22. Illustrations of (a) extrinsic (leading twist) and (b) intrinsic (higher twist $\mathcal{O}(\mu^2/m_c^2)$) QCD contributions to the charm structure function of the proton $G_{c/p}(x)$. The magnitude of the intrinsic contribution is controlled by the multi-gluon correlation parameter μ in the proton wavefunction. The intrinsic contribution dominates $G_{c/p}(x)$ at large x.

9. Calculation of the e^+e^- Annihilation Cross Section

An important advantage of the free LC Fock basis is that the electroweak currents have a simple representation. Thus once one diagonalizes the light-cone Hamiltonian, one can immediately compute current matrix elements, such as the proton–anti-proton time-like form factors $\langle 0|\, j^\mu(0)\, |\psi\rangle_{p\bar{p}}\,(s)$ or any given hadronic final state contribution to the total annihilation cross section $\sigma_{e^+e^-}(s)$. This program has recently been carried out explicitly using the DLCQ method and Lanczos tri-diagonalization by Hiller [13] for QED(1+1). A typical result is shown in Fig. 23. It would be interesting to repeat this non-perturbative calculation for a renormalizable theory like the Gross-Neveu model in (1+1) dimensions, and analyze how the channel-by-channel calculation merges into the asymptotic freedom result.

10. Applications of DLCQ to Gauge Theories in 3+1 Dimensions

The diagonalization of the light-cone Hamiltonian

$$H_{\rm LC}\,|\psi_i\rangle = M_i^2\,|\psi_i\rangle\,, \tag{10.1}$$

provides not only the eigenvalues M_i^2 but also the relativistic boost-invariant eigenfunctions $|\psi_i\rangle$. In the following sections we will discuss specific implementations of the DLCQ method for quantum electrodynamics and QCD in 3 + 1 dimensions. Although the QED spectrum is well understood from Bethe

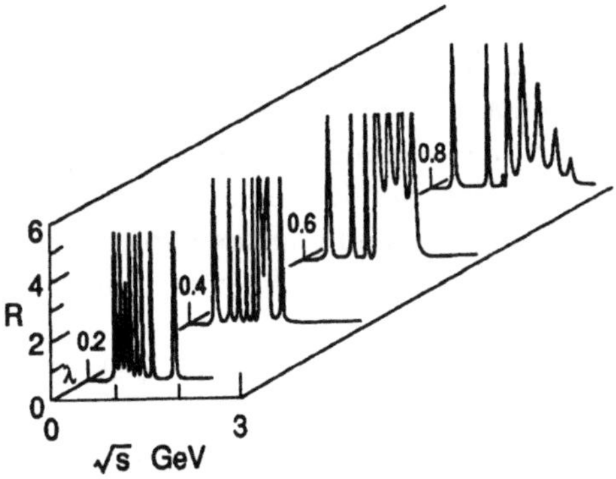

Fig. 23. Calculation of $R_{e^+e^-}(s)$ in QED(1+1) using the DLCQ method. The results are shown for different coupling constants. For display purposes, the plot is clipped at $R = 5$. In addition, in order to give finite widths to what would have been δ-functions, the infinitesimal ϵ was set to 0.01 (from Ref. 13).

Salpeter and other approaches, it is important and interesting to study this system at strong coupling strength for possible clues to confinement and hadronization mechanisms in QCD.[10] Furthermore at moderate α we can make contact with the precision QED results of Yennie, [82] Lepage, Bodwin, and others, and possibly understand the validity of standard perturbative expansions in QED. However, most important for our purposes, positronium serves as a crucial system to validate the DLCQ methods. In addition to the work discussed here, Kaluža [30] has recently used a DLCQ diagonalization approach to obtain the lepton structure function in positronium.

In the complete formulation of DLCQ, one constructs a complete discretized light-cone Fock basis in momentum space. The LC Hamiltonian can then be visualized as a matrix with a finite number of rows and columns assuming an invariant ultraviolet cut-off. Next, one formulates all necessary model assumptions, in accord with covariance and gauge-invariance, thus obtaining a discrete representation of the quantum filed theory. At any stage, one can go to the continuum limit, convert the matrix equation to an integral equation, and solve it with suitably optimalized numerical methods. One should emphasize, that the regularization scheme of DLCQ [28] explicitly allows for such a procedure, since the regularization scales are equal both for discretization and the continuum, contrary to lattice gauge theory, for example.

10.1 Testing Discretized Light Cone Quantization with Positronium

In the simplified DLCQ model [83] we shall discuss here, we will consider only the charge zero sector of QED(3+1) and include only the $J_z = 0$ *electron-positron* ($e\bar{e}$) and the *electron-positron-photon* ($e\bar{e}\gamma$) Fock states, denoted collectively by $|e\bar{e}\rangle$ and $|e\bar{e}\gamma\rangle$, respectively. In effect we have analyzed the muonium system μ^+e^- at equal lepton mass to avoid complications from the annihilation kernels. Even when one restricts the Fock states to one dynamical photon, one is considering a complex non-perturbative problem, similar to ladder approximation in the Bethe-Salpeter formalism. The light-cone approach has the advantage that one obtains the Dirac-Coulomb equation in the heavy muon limit. (In the Bethe-Salpeter approach, one must include all crossed graph irreducible kernels to derive the Dirac equation.) However, it should be emphasized that in any formalism the physics of the Lamb Shift and vertex corrections to the hyperfine interaction cannot occur until one includes the contributions of at least two dynamical photons "in flight."

It is convenient to introduce the projectors $P + Q = 1$, with $P = \sum_i |(e\bar{e})_i\rangle \langle(e\bar{e})_i|$ and $Q = \sum_i |(e\bar{e}\gamma)_i\rangle \langle(e\bar{e}\gamma)_i|$. The index i runs over all discrete light-cone momenta and helicities of the partons (electron e, positron $\bar{e}$ and photon γ) subject to fixed total momenta and to covariant regularization by a sharp momentum cut-off.[28] The light-cone Hamiltonian thus becomes a block matrix. There are a number of restrictions and simplifications due needed to maintain gauge invariance when the Fock space is limited in momentum space. The structure of these blocks in terms of matrix elements needs to be discussed in some detail. For example, the matrix elements as depicted in Fig. 24 for the Q-space contain either an instantaneous boson or an instantaneous fermion line. According to the general gauge principle [28,27] for DLCQ, one should include the instantaneous graphs only if the 'instantaneous parton' will be accompanied by a real 'dynamic parton' with the same space-like momentum and in the same Fock space configuration. Otherwise gauge invariance of the scattering amplitudes is violated already on the tree level. Thus, diagram (a) has to be excluded, since there are no $|e\bar{e}\gamma\gamma\rangle$ Fock states in the model, as well as diagram (b) since the two photon states are absent. Actually, only diagram (d) survives the gauge cut-off in the Q-space. Similarly in the P-space, only diagram (a) of Fig. 25 survives.

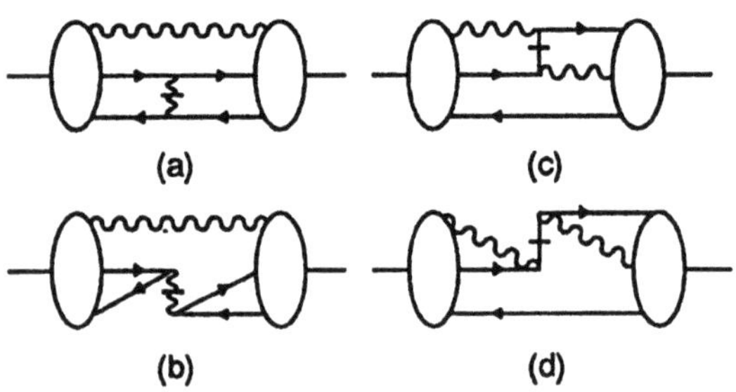

Fig. 24. The instantaneous interactions in the Q-space. – Graph (a) and (b): The instantaneous boson interactions $S^{(s)}_{q\bar{q}\to q\bar{q}}$ and $S^{(a)}_{q\bar{q}\to q\bar{q}}$, respectively. Graph (c) and (d): The instantaneous fermion interactions $S^{(s)}_{qg\to qg}$ and $S^{(a)}_{qg\to qg}$, respectively.

As part of the model, one additional simplification has been made, namely to omit diagram (d) of Fig. 24, *i.e.*

$$S^{(a)}_{qg\to qg} = 0 \text{ in } Q\text{−space}. \tag{10.2}$$

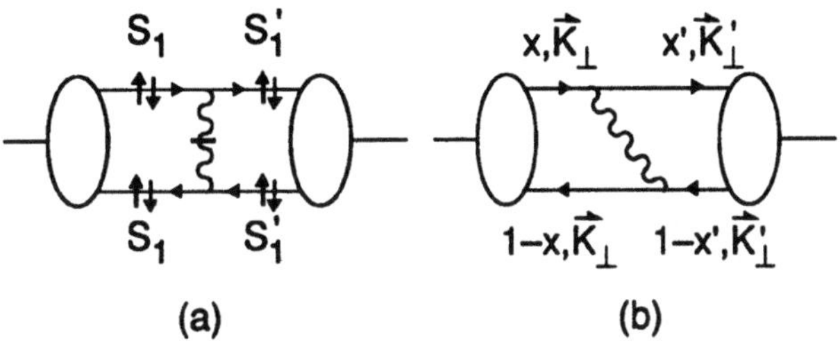

Fig. 25. The off-diagonal matrix elements in P-space. – (a): the instantaneous boson graph $S^{(s)}_{q\bar{q}\to q\bar{q}}$; (b): the iterated graph $W = VGV$. – The figure displays also the space-like momentum assignment of the fermions; those of the boson are fixed by momentum conservation. Graph (b) holds for $x > x'$, the corresponding one for $x < x'$ is not shown.

One has no good reason to do so except mathematical simplification, since the Q-space matrix is now rendered diagonal. An equivalent assumption has been made in all of the preceding work.[4,7,20]

By inverting the matrix $(M_i^2 - H_{\rm LC}) \equiv (\omega - H_{\rm LC})$ in the Q-space, *i.e.*

$$Q\,|\psi_i\rangle = Q\frac{1}{\omega - H_{\rm LC}}QH_{\rm LC}P\,|\psi_i\rangle\,, \tag{10.3}$$

Equation (10.1) can be identically rewritten as

$$H_{\rm eff}(\omega)\,|\psi_i(\omega)\rangle = M_i^2(\omega)\,|\psi_i(\omega)\rangle\,, \tag{10.4}$$

the 'effective Hamiltonian' acting only in P-space, *i.e.*

$$H_{\rm eff}(\omega) \equiv PH_{\rm LC}P + PH_{\rm LC}Q\,\frac{1}{\omega - H_{\rm LC}}\,QH_{\rm LC}P. \tag{10.5}$$

Once $|\psi_i(\omega)\rangle = P\,|\psi_i\rangle$ is known, one can calculate the Q-space wave function from Eq. (10.3) by quadrature.

Despite acting only in P-space, Eq. (10.4) is not simpler to solve than the full problem, Eq. (10.1). But it can be approximated easier. Since the Q-space matrix is diagonal by construction, $\omega - H_{\rm LC}$ can be inverted trivially. Characterizing the electron by its Lagrangian mass $m_{\rm F}$, its longitudinal momentum fraction x, its transverse momentum $\vec{k}_\perp$, and by its spin projection s_1, and correspondingly the positron as displayed in Fig. 25, the effective Hamiltonian

$$H_{\rm eff}(\omega) = \frac{m_{\rm F}^2 + \vec{k}_\perp^{\,2}}{x(1-x)} + C + S + W(\omega) = \frac{m_{\rm F}^2 + \vec{k}_\perp^{\,2}}{x(1-x)} + V_{\rm eff}(\omega^*) \tag{10.6}$$

contains thus the free part, the diagonal contraction terms C in the P-space, the seagull interaction $S = S^{(s)}_{q\bar{q}\to q\bar{q}}$ in the P-space, and the iterated vertex interaction $W(\omega) = V\dfrac{Q}{\omega - M^2_{e\bar{e}\gamma}}V$. The latter connects the P-space with the Q-space through the vertex interaction $V = V_{q\to qg}$ with the 'energy' denominator $\omega - M^2_{e\bar{e}\gamma}$. Note, that the effective potential $V_{\rm eff}$ is strictly proportional to α.

In general ω should be chosen as the eigenvalue M_i^2. To correct for the violation of gauge invariance by Eq. (10.2) approximately, however, one replaces the eigenvalue ω with a function of $(x, \vec{k}_\perp)$, *i.e.* with the symmetrized mass (squared)

$$\omega \Rightarrow \omega^* = \frac{1}{2}\left(\frac{m_F^2 + \vec{k}_\perp^2}{x(1-x)} + \frac{m_F^2 + \vec{k}_\perp'^2}{x'(1-x')}\right). \tag{10.7}$$

This (second ad-hoc) assumption restores the gauge-invariance of the $e\bar{e}$-scattering amplitude in the P-sector, at least. This completes the model.

The projection technique of deriving an effective Hamiltonian is fairly standard in many-body theory, [84] and has been applied to light-cone formulation before.[4,20] Since we have truncated the Fock states, the model can be regarded as a light-cone gauge theory analogue of the Tamm-Dancoff approach [7] used in equal-time theory. A similar approach was applied recently [85,86] to a scalar field model in light-cone coordinates.

10.2 The Light-Cone Tamm-Dancoff Equation

In the continuum, the matrix equation (10.5) becomes an integral equation

$$\left\{\frac{m_F^2 + \vec{k}_\perp^2}{x(1-x)} - M_i^2\right\}\psi_i(x, \vec{k}_\perp, s_1, s_2)$$
$$+ \sum_{s_1', s_2'} \int_D dx' d^2\vec{k}_\perp' \left\langle x, \vec{k}_\perp; s_1, s_2 \middle| V_{\text{eff}}(\omega) \middle| x', \vec{k}_\perp'; s_1', s_2' \right\rangle \psi_i(x', \vec{k}_\perp', s_1', s_2') = 0. \tag{10.8}$$

The finite domain of integration D is set by covariant Fock space regularization, [28]

$$\frac{m_F^2 + \vec{k}_\perp^2}{x(1-x)} \leq \Lambda^2 + 4m_F^2, \tag{10.9}$$

with given cut-off scale Λ. Combined with Eq. (10.7), we shall speak of this equation as the 'Light-cone Tamm-Dancoff' equation.

The effective interaction V_{eff}, which is also displayed diagrammatically in Figs. 25 and 26, appears to have two kinds of singularities, namely a 'Coulomb singularity' at ($x = x'$ and $\vec{k}_\perp = \vec{k}_\perp'$), and a 'colinear singularity' at ($x = x'$ and $\vec{k}_\perp \neq \vec{k}_\perp'$). The latter is caused by the instantaneous interaction

$$\left\langle x, \vec{k}_\perp; s_1, s_2 \middle| S \middle| x', \vec{k}_\perp'; s_1', s_2' \right\rangle = -\frac{\alpha}{2\pi^2}\frac{2}{(x-x')^2}\delta_{s_1,s_1'}\delta_{s_2,s_2'}, \tag{10.10}$$

but it will be cancelled by a corresponding term in the iterated vertex interaction $W(\omega)$. Strictly speaking, one should treat the ***diagonal*** interaction matrix element for $x = x'$ *and* $\vec{k}_\perp = \vec{k}_\perp'$ differently from the ***off-diagonal*** elements $(x, \vec{k}_\perp) \neq (x', \vec{k}_\perp')$, which is always possible going back to the discretized case. The diagonal matrix element $C + W(\omega)$ is displayed in Fig. 26 and vanishes

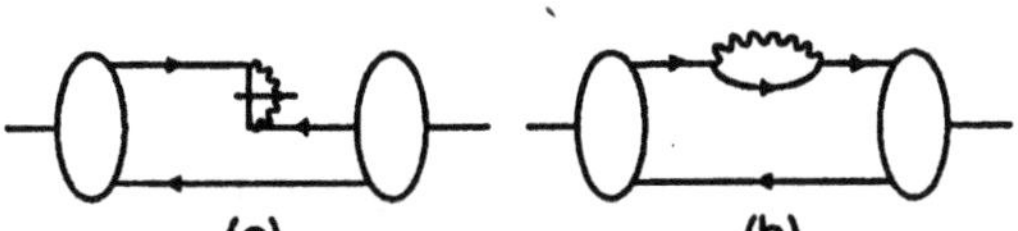

Fig. 26. **The diagonal matrix elements in P-space. – (a): the instantaneous contraction terms, (b): the iterated graph $W = VGV$ (self-mass diagram).**

strictly for $\omega = \omega^*$ due to mass renormalization.[83] One might expect a diagonal contribution from the instantaneous interaction S, but its matrix element vanishes for $x = x'$ in DLCQ.[28]

As an example, consider the off-diagonal matrix element of the iterated vertex interaction $\left\langle x, \vec{k}_\perp; \uparrow, \downarrow \middle| W(\omega) \middle| x', \vec{k}_\perp{}'; \uparrow, \downarrow \right\rangle$ for $x > x'$. Straightforward insertion of the matrix elements (See Tables 2a-d) gives

$$W(\omega) = \frac{2\beta}{\mathcal{D}(x, x'; \omega)}\, \vec{\epsilon}_\perp(\uparrow) \cdot \left(\frac{\vec{k}_\perp - \vec{k}_\perp{}'}{x - x'} - \frac{\vec{k}_\perp{}'}{x'}\right) \vec{\epsilon}_\perp{}^*(\uparrow) \cdot \left(\frac{\vec{k}_\perp - \vec{k}_\perp{}'}{x - x'} + \frac{\vec{k}_\perp{}'}{1 - x'}\right)$$

$$+ \frac{2\beta}{\mathcal{D}(x, x'; \omega)}\, \vec{\epsilon}_\perp(\downarrow) \cdot \left(\frac{\vec{k}_\perp - \vec{k}_\perp{}'}{x - x'} - \frac{\vec{k}_\perp}{x}\right) \vec{\epsilon}_\perp{}^*(\downarrow) \cdot \left(\frac{\vec{k}_\perp - \vec{k}_\perp{}'}{x - x'} + \frac{\vec{k}_\perp}{1 - x}\right). \tag{10.11}$$

The denominator $\mathcal{D}(x, x'; \omega) \equiv -(x - x')\,(\omega - M^2_{e\bar{e}\gamma})$ is introduced for convenience. The polarization sums can be expressed in terms of the transverse scalar and vector products, $\vec{k}_\perp \cdot \vec{k}_\perp{}'$ and $\vec{k}_\perp \wedge \vec{k}_\perp{}'$, respectively. One obtains straightforwardly $2(\vec{\epsilon}_\perp(\uparrow) \cdot \vec{k}_\perp)(\vec{\epsilon}_\perp{}^*(\downarrow) \cdot \vec{k}_\perp{}') = \vec{k}_\perp \cdot \vec{k}_\perp{}' - i\vec{k}_\perp \wedge \vec{k}_\perp{}'$. The colinear singularities in

$$W(\omega) = \frac{\alpha}{2\pi^2} \frac{1}{\mathcal{D}(x, x'; \omega)} \left[\frac{\vec{k}_\perp^2}{(x - x')^2}\left(1 + \frac{x'(1 - x')}{x(1 - x)}\right) + \frac{\vec{k}_\perp'^2}{(x - x')^2}\left(1 + \frac{x(1 - x)}{x'(1 - x')}\right)\right.$$

$$\left. - \frac{4\vec{k}_\perp \vec{k}_\perp{}'}{(x - x')^2} - \frac{\vec{k}_\perp \cdot \vec{k}_\perp{}' - i\vec{k}_\perp \wedge \vec{k}_\perp{}'}{x'x} - \frac{\vec{k}_\perp \cdot \vec{k}_\perp{}' + i\vec{k}_\perp \wedge \vec{k}_\perp{}'}{(1 - x')(1 - x)}\right] \tag{10.12}$$

cancel against those of S, and the effective interaction becomes finally for $\omega = \omega^*$

$$\left\langle x, \vec{k}_\perp; \uparrow, \downarrow \middle| V_{\text{eff}}(\omega^*) \middle| x', \vec{k}_\perp{}'; \uparrow, \downarrow \right\rangle = -\frac{\alpha}{2\pi^2} \frac{1}{\mathcal{D}(x, x')} \left[m_F^2 \left(\frac{1}{x'x} + \frac{1}{(1 - x')(1 - x)}\right)\right.$$

$$\left. + \frac{\vec{k}_\perp^2}{x(1 - x)} + \frac{\vec{k}_\perp'^2}{x'(1 - x')} + \frac{\vec{k}_\perp \cdot \vec{k}_\perp{}' - i\vec{k}_\perp \wedge \vec{k}_\perp{}'}{x'x} + \frac{\vec{k}_\perp \cdot \vec{k}_\perp{}' + i\vec{k}_\perp \wedge \vec{k}_\perp{}'}{(1 - x')(1 - x)}\right], \tag{10.13}$$

with $\mathcal{D}(x, x') \equiv \mathcal{D}(x, x'; \omega^*)$. The effective potential has no ultraviolet or infrared singularities. Only the usual integrable 'Coulomb singularity' in $\mathcal{D}(x, x')$ remains.

10.3 The Light-Cone Coulomb Schrödinger Equation

At this stage, the original matrix equation (10.1) has been approximated by the Tamm-Dancoff Equation, Eq. (10.8). For orientation, it is useful to consider the non-relativistic limit ($\vec{k}_\perp'^2 \ll m_F^2$ and $(x-\frac{1}{2})^2 \ll 1$). In this limit the TDE, and particularly Eq. (10.13) are easily converted [4] into the 'Light-Cone Schrödinger equation'

$$\frac{m_F^2+\vec{k}_\perp^2}{x(1-x)}\psi(x,\vec{k}_\perp)-\frac{\alpha}{2\pi^2}\int_D dx' d^2\vec{k}_\perp' \frac{8m_F^2\,\psi(x',\vec{k}_\perp')}{4m_F^2(x-x')^2+(\vec{k}_\perp'-\vec{k}_\perp)^2}=M^2\,\psi(x,\vec{k}_\perp), \tag{10.14}$$

One should note however, that this equation is kind of a hybrid since the non-relativistic limit is taken only in the potential energy. Therefore, it cannot and does not [83] precisely yield Bohr spectrum. When the non-relativistic limit is taken consistently by replacing the longitudinal momentum fraction with a 'parallel momentum' $k_\parallel \equiv 2m_F(x-\frac{1}{2})$, collecting the momenta into a 3-vector $\vec{k}=(k_\parallel,\vec{k}_\perp)$, substituting the kinetic energy $(m_F^2+\vec{k}_\perp^2)/x(1-x)$ by $4m_F^2+4\vec{k}^2$, and using the *definition* $M_i^2 \equiv 4m_F^2+4m_F E$, one arrives straightforwardly at the usual Coulomb Schrödinger equation in momentum space, *i.e.*

$$\frac{\vec{k}^2}{2m_r}\psi(\vec{k})-\frac{\alpha}{2\pi^2}\int_D d^3\vec{k}'\,\frac{1}{(\vec{k}-\vec{k}')^2}\psi(\vec{k}')=E\,\psi(\vec{k}), \tag{10.15}$$

including the correct reduced mass $m_r = m_F/2$. Fock space regularization converts itself into a 3-momentum cut-off, *i.e.* the domain of integration D is given by $\vec{k}^2 \le \Lambda^2/4$.

10.4 Solving the Integral Equations

As one sees from the above discussion, even at the level of only one dynamical photon, there are available a whole sequence of approximations: DLCQ-Matrix equation $\Longrightarrow$ Tamm-Dancoff Equation $\Longrightarrow$ Light-Cone Schrödinger equation $\Longrightarrow$ Coulomb Schrödinger equation. Each of these approximation equations have been recently investigated by numerical means.[30,83] the numerical effort turns out to be remarkably small, provided the numerical methods are optimized to the particular problem.

10.5 Implementing the Symmetries

A particularly important optimization for numerical solutions is the utilization of the light-cone symmetries. Some approaches to gauge field theory do not respect the elementary symmetries of the Lagrangian, by nature of their construction. However, the exact Lagrangian symmetries need not be violated by DLCQ or its approximations. For example, the Lagrangian is invariant under an arbitrary *rotation* of the coordinate system in the $x-y$-plane, corresponding to

$\frac{\pi}{\alpha}\langle x, k_\perp, J_z, s_1, s_2 \vert\, \widetilde{V}_{\rm eff}(\omega^*)\,\vert x', k'_\perp, J_z, s'_1, s'_2\rangle$	Helicity factors
$m_{\rm F}^2 \frac{(x'-x)^2}{(1-x')x'(1-x)x}\,{\rm A}$	$\delta_{-s_2,s'_2}\ \delta_{s_2,s'_1}\ \delta_{s_1,-s'_1}$
$-m_{\rm F}\frac{1}{x'x}s_1\left\{\frac{1}{k_\perp}\,{\rm B} + k_\perp\frac{1-x'}{1-x}\,{\rm A}\right\}$	$\delta_{s_2,s'_2}\ \delta_{s_1,s'_2}\ \delta_{-s_1,s'_1}$
$+m_{\rm F}\frac{1}{x'x}s_1\left\{\frac{1}{k'_\perp}\,{\rm B} + k'_\perp\frac{1-x}{1-x'}\,{\rm A}\right\}$	$\delta_{s_2,s'_2}\ \delta_{s_1,-s'_2}\ \delta_{-s_1,s'_1}$
$+m_{\rm F}\frac{1}{(1-x')(1-x)}s_1\left\{\frac{1}{k_\perp}\,{\rm B} + k_\perp\frac{x'}{x}\,{\rm A}\right\}$	$\delta_{s_2,-s'_2}\ \delta_{-s_1,s'_2}\ \delta_{s_1,s'_1}$
$+m_{\rm F}\frac{1}{(1-x')(1-x)}s_1\left\{\frac{1}{k'_\perp}\,{\rm B} + k'_\perp\frac{x}{x'}\,{\rm A}\right\}$	$\delta_{s_2,-s'_2}\ \delta_{s_1,s'_2}\ \delta_{s_1,s'_1}$
$-\,{\rm A}\left\{m_{\rm F}^2\left(\frac{1}{xx'}+\frac{1}{(1-x)(1-x')}\right)+\frac{k_\perp^{\,2}}{x(1-x)}+\frac{k_\perp'^{\,2}}{x'(1-x')}\right\}$	$\delta_{s_2,s'_2}\ \delta_{s_2,-s'_1}\ \delta_{s_1,s'_1}$
$-\,{\rm A}\,\frac{1}{xx'(1-x)(1-x')}k_\perp k'_\perp$	$\delta_{s_2,s'_2}\ \delta_{s_1,s'_2}\ \delta_{s_1,s'_1}$
$+\,{\rm B}\left(\frac{1}{xx'}+\frac{1}{(1-x)(1-x')}\right)$	$\delta_{s_2,s'_2}\ \delta_{s_2,-s'_1}\ \delta_{s_1,s'_1}$
$+\,{\rm B}\left(\frac{1}{xx'}+\frac{1}{(1-x)(1-x')}\right)\frac{1}{k_\perp k'_\perp}m_{\rm F}^2$	$\delta_{s_2,s'_2}\ \delta_{s_1,s'_2}\ \delta_{s_1,s'_1}$

Abbreviations:

$${\rm A} = \frac{1}{\sqrt{a^2-4k_\perp^2k_\perp'^2}}; \qquad {\rm B} = \frac{1}{2}\left(1+a\,{\rm A}\right);$$

$$a = -(x-x')^2\frac{m_{\rm F}^2}{2}\left(\frac{1}{x'x}+\frac{1}{(1-x')(1-x)}\right)-\left(k_\perp^2+k_\perp'^2\right)$$
$$+(x-x')\left\{\frac{k_\perp'^2}{2}\left(\frac{1}{1-x'}-\frac{1}{x'}\right)+\frac{k_\perp^2}{2}\left(\frac{1}{x}-\frac{1}{1-x}\right)\right\}$$

Table 3. The matrix elements of the effective interaction for $J_z = 0$ and $x > x'$.

conservation of the projection of the total angular momentum J_z. Introducing the coordinates $(\vec{k}_\perp)_x = k_\perp \cos\varphi$ and $(\vec{k}_\perp)_y = k_\perp \sin\varphi$, one can Fourier transform the continuum version of the Tamm-Dancoff Eq. (10.8), and in particular the effective potential $V_{\rm eff} = V_{\rm eff}(\omega^*)$ according to

$$\frac{1}{2\pi}\int_0^{2\pi} d\varphi e^{-iL_z\varphi}\int_0^{2\pi} d\varphi' e^{+iL'_z\varphi'}\,\langle x, k_\perp, \varphi; s_1, s_2\vert\, V_{\rm eff}\,\vert x', k'_\perp, \varphi'; s'_1, s'_2\rangle$$
$$= \langle x, k_\perp, L_z; s_1, s_2\vert\, \widetilde{V}_{\rm eff}\,\vert x', k'_\perp, L'_z; s'_1, s'_2\rangle\,. \tag{10.16}$$

In this way, one replaces the azimuthal angle φ by the projection of the orbital angular momentum $L_z = 0, \pm 1, \ldots$ as a variable, although neither L_z nor $S_z = s_1 + s_2$ are individually a good quantum number. The explicit expressions for the matrix elements of $\widetilde{V}_{\rm eff}$, are derived straightforwardly from those in Tables 2a-d. For the case $J_z = 0$, they are compiled in Tables 3 and 4.

The Lagrangian is also invariant under the operation of *charge conjugation* $\mathcal{C}$, *parity* $\mathcal{P}$, and *time reversal* $\mathcal{T}$. Neither $\mathcal{P}$ nor $\mathcal{T}$, however, is an explicit symmetry of the light-cone Hamiltonian, because $\mathcal{P}$ and $\mathcal{T}$ do not leave the $x^+ = 0$ plane invariant. (Parity interchanges x^+ and x^-.) However, $\exp(-i\pi J_3)\mathcal{P}\mathcal{T}$,

$\frac{\pi}{\alpha}\langle x, k_\perp, J_z, s_1, s_2 \vert \widetilde{V}_{\text{eff}}(\omega^*) \vert x', k'_\perp, J_z, s'_1, s'_2 \rangle$	Helicity factors		
$m_{\mathrm{F}}^2 \frac{(x'-x)^2}{(1-x')x'(1-x)x}\,\mathrm{A}$	δ_{-s_2,s'_2}	δ_{s_2,s'_1}	$\delta_{s_1,-s'_1}$
$+m_{\mathrm{F}} \frac{1}{(1-x')(1-x)} s_1 \left\{ \frac{1}{k_\perp}\,\mathrm{B} + k_\perp \frac{x'}{x}\,\mathrm{A} \right\}$	δ_{-s_2,s'_2}	δ_{s_2,s'_1}	δ_{s_1,s'_1}
$+m_{\mathrm{F}} \frac{1}{(1-x')(1-x)} s_1 \left\{ \frac{1}{k'_\perp}\,\mathrm{B} + k'_\perp \frac{x}{x'}\,\mathrm{A} \right\}$	δ_{-s_2,s'_2}	$\delta_{s_2,-s'_1}$	δ_{s_1,s'_1}
$-m_{\mathrm{F}} \frac{1}{x'x} s_1 \left\{ \frac{1}{k_\perp}\,\mathrm{B} + k_\perp \frac{1-x'}{1-x}\,\mathrm{A} \right\}$	δ_{s_2,s'_2}	δ_{-s_2,s'_1}	$\delta_{s_1,-s'_1}$
$+m_{\mathrm{F}} \frac{1}{x'x} s_1 \left\{ \frac{1}{k'_\perp}\,\mathrm{B} + k'_\perp \frac{1-x}{1-x'}\,\mathrm{A} \right\}$	δ_{s_2,s'_2}	δ_{s_2,s'_1}	$\delta_{s_1,-s'_1}$
$-\,\mathrm{A}\left\{ m_{\mathrm{F}}^2 \left(\frac{1}{xx'} + \frac{1}{(1-x)(1-x')} \right) + \frac{k_\perp^2}{x(1-x)} + \frac{k_\perp'^2}{x'(1-x')} \right\}$	δ_{s_2,s'_2}	$\delta_{s_2,-s'_1}$	δ_{s_1,s'_1}
$-\,\mathrm{A}\,\frac{1}{xx'(1-x)(1-x')} k_\perp k'_\perp$	δ_{s_2,s'_2}	δ_{s_1,s'_2}	δ_{s_1,s'_1}
$+\,\mathrm{B}\left(\frac{1}{xx'} + \frac{1}{(1-x)(1-x')} \right)$	δ_{s_2,s'_2}	$\delta_{s_2,-s'_1}$	δ_{s_1,s'_1}
$+\,\mathrm{B}\left(\frac{1}{xx'} + \frac{1}{(1-x)(1-x')} \right) m_{\mathrm{F}}^2 \frac{1}{k_\perp k'_\perp}$	δ_{s_2,s'_2}	δ_{s_1,s'_2}	δ_{s_1,s'_1}
Abbreviations:			
$\mathrm{A} = \frac{1}{\sqrt{a^2 - 4k_\perp^2 k_\perp'^2}}; \qquad \mathrm{B} = \frac{1}{2}\left(1 + a\,\mathrm{A}\right);$			
$a = -(x'-x)^2 \left(\frac{1}{x'x} + \frac{1}{(1-x')(1-x)} \right) \frac{m_{\mathrm{F}}^2}{2} - \left(k_\perp^2 + k_\perp'^2 \right) + (x'-x) \left\{ \frac{k_\perp^2}{2} \left(\frac{1}{1-x} - \frac{1}{x} \right) + \frac{k_\perp'^2}{2} \left(\frac{1}{x'} - \frac{1}{1-x'} \right) \right\}$			

Table 4. The matrix elements of the effective interaction for $J_z = 0$ and $x < x'$.

for example, or $\exp(-i\pi J_1)\mathcal{P}$ are exact symmetries.[18,30,87] In the numerical work [83] quoted below the combined symmetry $\mathcal{PT}$ with eigenvalues ± 1 will be used.

When including both the rotational and the combined $\mathcal{PT}$ symmetry, Eq. (10.8) can be rewritten in terms of $\Psi(x, k_\perp; J_z) \equiv \frac{1}{\sqrt{2\pi}} \int d\varphi e^{iL_z\varphi} \psi(x, \vec{k}_\perp; s_1, s_2)$, *i.e.* the Fourier-transformed wavefunctions. The "coordinate" J_z will be dropped in the sequel for notational simplicity. Since J_z is an exact symmetry, the matrix elements $\langle J_z \vert \widetilde{V}_{\text{eff}} \vert J'_z \rangle$ with different J_z vanish strictly in the present representation.

10.6 Solving the Light-Cone Tamm-Dancoff Equation

How does one solve an equation like (10.8) in practice? — As a rule, one can evaluate the integrals by Gaussian quadratures, mapping the integral equation onto another matrix equation. By converting the integration over the longitudinal momentum x into a Gaussian sum with weights $\omega_i^{(N)}$, the Tamm-Dancoff Equation (10.8) becomes

$$\left\{\frac{m_F^2+k_\perp^2}{x_i(1-x_i)}-M^2\right\}\Psi(x_i,k_\perp)-\sum_{j=1}^{N}\omega_j^{(N)}\int dk_\perp'^2\,\langle x_i,k_\perp|\,\widetilde{V}_{\text{eff}}\,|x_j,k_\perp'\rangle\Psi(x_i,k_\perp')$$
$$+C_{cc}(x_i,k_\perp)\Psi(x_i,k_\perp)+\sum_{j=1}^{N}\omega_j^{(N)}\int dk_\perp'^2\,\langle x_i,k_\perp|\,\widetilde{V}_{\text{eff}}\,|x_j,k_\perp'\rangle\Psi(x_j,k_\perp')=0.$$
$$(10.17)$$

(The domain of summation and integration is the same as in the Tamm-Dancoff Equation as given by Eq. (10.9).) In this expression two terms which sum to zero are included, *i.e.* $C_{cc}(x,\vec{k}_\perp)\psi(x,\vec{k}_\perp)-C_{cd}(x,\vec{k}_\perp)\psi(x,\vec{k}_\perp)$. With the continuum part defined by

$$C_{cc}(x,k_\perp)\equiv\int_D dx'dk_\perp'^2\,\langle x,k_\perp|\,\widetilde{V}_{\text{eff}}\,|x',k_\perp'\rangle\ , \qquad (10.18)$$

one easily identifies its discretized partner C_{cd} in Eq. (10.17). Their sum thus vanishes in the continuum limit.

The reason for introducing the 'Coulomb counter terms' is the following: The kernel of the Tamm-Dancoff Equation is singular, as can be seen explicitly in the approximate Eqs. (10.14) and (10.15). Despite being integrable, this singularity is a numerical nightmare, and is present whenever one deals with a Coulomb-like problem in a momentum-space representation. For example, it is close to impossible to get numerically stable solutions as function of the number of integration points (and resolution) N. However, when the Coulomb terms are added, the singularity in the kernel and in the (discrete) counter term tend to cancel in the vicinity of the singularity, since by construction, they have the same residue. What remains is a smooth function which can be approximated easier. The continuum part of the counterterm restores the original equation. Ideally it should be calculated analytically, or if this turns out too difficult, it can be evaluated numerically with ultra high resolution.

Instead of $x, k_\perp$ (and φ) one can use 'spherical momentum coordinates' r,θ as defined by $k_\perp=\sqrt{\Lambda^2+4m_F^2}\sin\theta$ and $x=\frac{1}{2}+r\cos\theta$. The variables r and $\cos\theta$ are discretized in the intervals $[0,\frac{\Lambda}{2\sqrt{\Lambda^2+4m_F^2}}]$ and $[-1,1]$ with ω_i as the Gaussian weights. In order to get an eigenvalue problem with a symmetric matrix the wavefunction Ψ is substituted by $\Psi(r_i,\cos\theta_j)=\Phi(r_i,\cos\theta_j)/r_i\sqrt{\omega_i\omega_j}$. The actual matrix elements are therefore $\langle r_i,\cos\theta_j|\,\widetilde{V}_{\text{eff}}\,|r_k',\theta_l'\rangle\,\Lambda^2\,r_ir_k'\,\sqrt{\omega_i\omega_j\omega_k\omega_l}$. For convenience the same number of integration points N in r and $\cos\theta$ will be used in the sequel. Further details are given in Ref. 83.

The spectrum of the Tamm-Dancoff Equation obtained using the above method is displayed in Fig. 27 as a function of the resolution. It is remarkable how fast the lowest two eigenvalues approach a limiting value. These two states are identified as the singlet and the triplet state of positronium, as verified by the fact that their wave function has the correct symmetries. It is not surprising to see the comparatively slow convergence of the higher excited states. Although their wave functions in momentum space are also localized near $x\approx\frac{1}{2}$ and $k_\perp\approx 0$, they have more nodal structures. Consequently, more integration points are needed to resolve their structure.

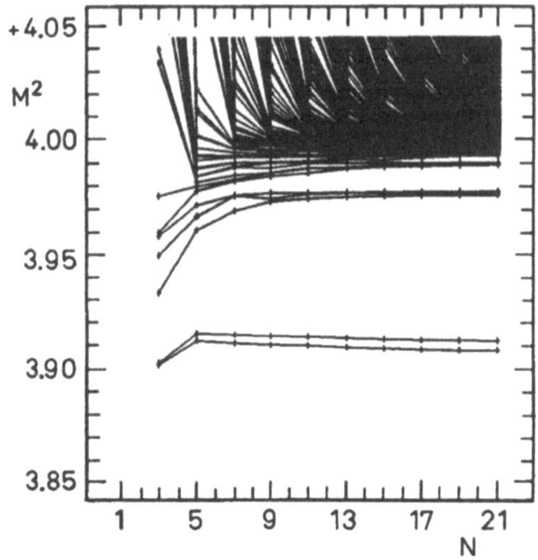

Fig. 27. The invariant mass squared eigenvalues of the Tamm-Dancoff equation *versus* the number of integration points N. – Note the good convergence with N, and the appearance of the hyperfine splitting. Calculations are done for $J_z = 0$, $\Lambda = m_F$, and $\alpha = 0.3$. The numerically integrated Coulomb counter term for the Tamm-Dancoff equation is included.

We should emphasize two points. First, the numerical methods are obviously very efficient. For example, only a 25×25 matrix (for $N = 5$) is needed to render the singlet and the triplet state reasonably stable as function of N. This corresponds to only two transverse momentum states. Second, one has established that the longitudinal and transverse continuum limit of DLCQ exists. One should emphasize that the light-cone approach is well-defined, covariant, and numerically very economical. Most of the results have been generated by diagonalizing matrices of dimension as small as 225×225. A particularly important role for achieving this result is played by the Coulomb counter terms. In general they are quite necessary for investigating numerically Coulomb-like problems in momentum representation. The methods applied are not only efficient but also precise. The calculation have been done for two vastly different values of the fine structure constant, namely $\alpha = 0.3$ and $\alpha = 1/137$. In order to extract the hyperfine shift the latter requires a numerical stability within ten significant figures (10^{-10}!).

On can also examine the convergence of the low lying spectrum as a function of scale Λ and make quantitative comparisons with analytical results. In Table 5 the binding coefficient of the singlet mass ($B_s \equiv (4(2 - M_s)/\alpha^2)$ and the singlet to triplet mass difference in the form of the hyperfine coefficient ($C_{\rm hf} \equiv (M_t - M_s)/\alpha^4$) are tabulated for five values of Λ and two values of α. The extrapolation to $\Lambda \to \infty$ is made by a Padé approximation. One should emphasize, that the deviation of the calculated mass squared eigenvalues from the free value is extremely small for $\alpha = \frac{1}{137}$. A reliable extraction of the data in Table 5 thus requires numerical accuracy to ten significant figures. The fact

$\alpha = \frac{3}{10}$			$\alpha = \frac{1}{137}$		
Λ	B_s	C_{hf}	$\Lambda\frac{1}{3\alpha}$	B_s	C_{hf}
1.0	1.0503	0.1348	1.0	0.9345	0.1023
2.0	1.1834	0.2888	2.0	0.9922	0.1955
3.0	1.2390	0.3857	3.0	1.0053	0.2366
4.0	1.2723	0.4533	4.0	1.0127	0.2581
5.0	1.2960	0.5037	5.0	1.0211	0.2667
∞	1.4025	0.8317	∞	1.0459	0.3140

Table 5. The singlet binding and the hyperfine coefficient for the Light-cone Tamm-Dancoff Equation as function of Λ and α. – Calculations are done $N = 15$ integration points, and extrapolated to $\Lambda \to \infty$ with a Padé-Approximand $f(\Lambda) = (c_1 + c_2/\Lambda)/(1 + c_3/\Lambda)$ using the values $\Lambda = 1.0$, 3.0, and 5.0.

that the calculations do not become numerically unstable as a function of Λ is taken as a final and overall indicator that one can master the numerical aspects of the problem.

What should one expect analytically? In an expansion up to order α^4, the singlet and the triplet mass of positronium (excluding annihilation) is given by [88]

$$M_s = 2 - \frac{1}{4}\alpha^2(1 + \frac{63}{48}\alpha^2) \quad \text{and} \quad M_t = 2 - \frac{1}{4}\alpha^2(1 - \frac{1}{48}\alpha^2) , \tag{10.19}$$

respectively, where here (and in the following) masses are given in units of the physical electron mass. The hyperfine coefficient is then the Fermi value $(C_{hf})_{\text{Fermi}} = \frac{1}{3}$. Bodwin *et al.*[82] have summarized the analytical work for the higher order corrections to the hyperfine shift in positronium:

$$C_{hf} = \frac{1}{2}\left[\frac{2}{3}\left(+\frac{1}{2}\right) - \frac{\alpha}{\pi}\left(\ln 2 + \frac{16}{9}\right) + \frac{5}{12}\alpha^2 \ln\frac{1}{\alpha} + K\alpha^2 + K'\alpha^3\right] . \tag{10.20}$$

The term $\frac{1}{2}$ is set in parentheses since it originates in the photon annihilation term. The impact of the coefficient K is small; its numerical value [82] is $K = +0.427$. A complete calculation is not yet available for K', except that is contains a $\ln\alpha$ term; it is set zero. Eq. (10.20) predicts therefore the values $C_{hf} = 0.333$ for $\alpha = \frac{1}{137}$ and $C_{hf} = 0.257$ for $\alpha = 0.3$. It should be noted that part of the higher order corrections come from contributions in which two dynamical photons are in flight; thus a strict correspondence with the spectrum of light-cone model is not expected.

The numbers in Table 5 agree with the analytical predictions only to first approximation. In particular, they are not as accurate as the recent results of Koniuk *et al.*[89,90] for a corresponding model using *equal time* quantization. Although the present hyperfine shift for $\alpha = \frac{1}{137}$ is reasonably in between the Fermi and the Bodwin *et al.*values, the singlet state is slightly over-bound by four percent — in view of the numerical accuracy a small but significant deviation. For $\alpha = \frac{3}{10}$ the discrepancy is even more accentuated. Instead of $B_s = 1.118$ one gets 1.403, which shows that the α^4 coefficient is overrated by about a factor 3. The hyperfine shift points to the same direction, it is larger than the analytical value by roughly a factor 3.

One must conclude that the Tamm-Dancoff Equation overdoes the relativistic effects. We conjecture that the right correction will come when the ad-hoc assumption, Eq. (10.2), will be relaxed and the restriction to one dynamical photon is removed.

10.7 Summary and Discussion of the Tamm-Dancoff Equation Results

The numerical tests of the light-cone Tamm Dancoff approach in positronium provide some confidence that one can use light-cone Fock methods to solve relativistic bound state problems in gauge theory. The Tamm-Dancoff Equation reproduces the expected Bohr spectrum $M_n \sim 2m(1 - B_n\frac{\alpha^2}{4})$ almost quantitatively, as well a the typical relativistic deviations like the hyperfine shift $\nu = \alpha^4 C_{\rm hf}(\alpha)$. The binding coefficients $B_n \sim \frac{1}{n^2}$ are reproduced with small but significant deviations, one percent by order of magnitude. They are much smaller for the physical value $\alpha = 1/137$ than for the very large value $\alpha = 0.3$. Similarly, the hyperfine coefficient for $\alpha = 1/137$ is close to the correct value $C_{\rm hf} \sim \frac{1}{3}$, but for $\alpha = 0.3$ it is almost twice as large.

But the Tamm-Dancoff Equation does not solve the original physical problem. It mistreats it by a so far uncertain approximation, stated in Eq. (10.2). It is however possible to relax this constraint Eq. (10.2) and generate the full resolvent without uncontrolled approximations: Quite in general, the full resolvent $\mathrm{G}(\omega)$ in the Q-space can be expanded in terms of the free resolvent $\mathrm{G}_0(\omega)$ and of the instantaneous annihilation interaction $S^{(a)}$, *i.e.* $\mathrm{G}(\omega) = \mathrm{G}_0(\omega) + \mathrm{G}_0(\omega)S^{(a)}\mathrm{G}(\omega)$. With $W(\omega) = V\mathrm{G}(\omega)V^\dagger$ the next-to-leading order becomes $W^{(2)}(\omega) = V\mathrm{G}_0(\omega)S^{(a)}\mathrm{G}_0(\omega)V^\dagger$. The superscript indicates the power of α. Figure 28 collects essentially all possible graphs classified according to whether the two vertices sit on the same or on a different fermion line. Considering the same case as in section 3, graphs 28(a) and 28(a′) turn out to vanish, strictly. Graphs 28(b) and 28(b′) must be absorbed into mass renormalization, only graphs 28(c) and 28(c′) need to be considered. As it turns out they can be re-summed explicitly *to all orders*, with the exact result

$$\widetilde{W}(\omega) = \sum_{n=2}^{\infty} W^{(n)}(\omega) = \frac{1}{1+a} W^{(2)}(\omega). \tag{10.21}$$

Both a and $W^{(2)}$ diverge logarithmically with the scale Λ. For sufficiently large values one gets by order of magnitude $a \sim \alpha \ln \Lambda/m_{\rm F}$ and $W^{(2)} \sim \alpha^2 \ln \Lambda/m_{\rm F}$. The re-summed interaction $\widetilde{W}$ therefore is proportional to α instead of to α^2,

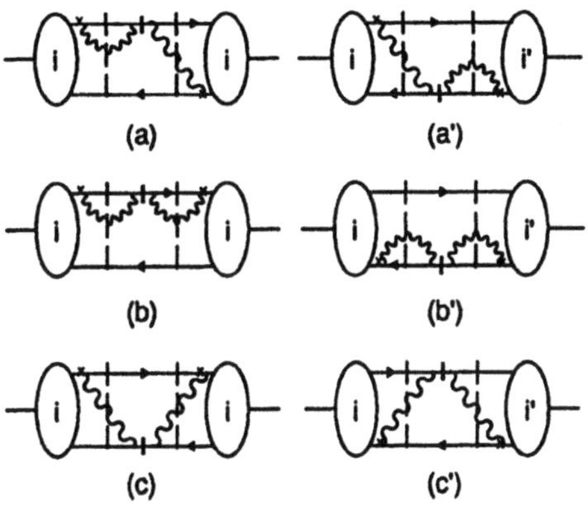

Fig. 28. Some typical graphs to order α^2 which begin and end with a vertex interaction $V_{q\to qg}(\times)$. The instantaneous interaction $S^{(a)}_{qg\to qg}$ is sandwiched in between them. In diagram (a) and (a') the vertices are one different, else they are on the same lines.

and independent of Λ it is probably able to account for the small but significant deviations in the binding coefficients and the hyperfine shift.

11. The Lanczos Method for DLCQ

The most serious practical difficulty for implementing DLCQ matrix diagonalization for physical theories in 3 + 1 dimensions, is the rapidly growing size of the matrix representation as one increases the size of the Fock basis. Fortunately, the matrix representations of H_{LC} in the free Fock basis are extremely sparse, and one can take advantage of efficient algorithms for diagonalizing such matrices. An important example of such an algorithm is the Lanczos method, which has been used for DLCQ problems in papers by Hiller,[13] Kaluža,[30] and Hollenberg, *et al.*[2] For example, let H be an $N \times N$ Hermetian matrix. Apply H to a test state vector $|v_1\rangle$. The result is

$$H\,|v_1\rangle = \alpha_1\,|v_1\rangle + \beta_1\,|v_2\rangle \ ,$$

where $|v_2\rangle$ is orthonormal to $|v_1\rangle$. Applying H to the new state vector gives:

$$H\,|v_2\rangle = \beta_1\,|v_1\rangle + \alpha_2\,|v_2\rangle + \beta_2\,|v_3\rangle \ .$$

where $|v_3\rangle$ is orthonormal to $|v_1\rangle$ and $|v_2\rangle$. However, if we apply H to $|v_3\rangle$ one only gets three non-zero terms:

$$H\,|v_3\rangle = 0\,|v_1\rangle + \beta_2\,|v_2\rangle + \alpha_3\,|v_3\rangle + \beta_3\,|v_4\rangle \ .$$

Thus, by construction, $\langle v_i|\,H\,|v_j\rangle$ is tri-diagonal, and the eigenvalues of its first $P \times P$ submatrix of this matrix converges rapidly to the lowest P eigenvalues

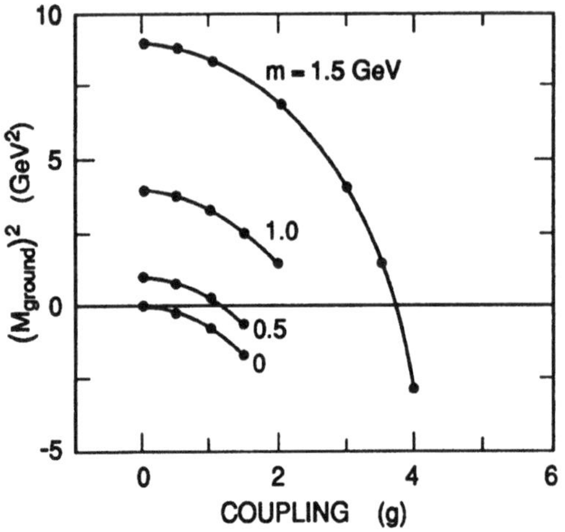

Fig. 29. First computation of the low-lying meson spectrum of $QCD(3+1)$ computed using Lanczos diagonalization. Only the lowest two Fock classes are retained. See Ref. 29.

of H. The computer time for obtaining these eigenvalues grows like fN^2 where f is a measure of the sparseness of H. This is much less than the time required for diagonalizing H itself, which grows like N^3. In the work of Hollenberg *et al.* one can handle matrices of sizes approaching $N = 10^6$.

12. First Applications of DLCQ to QCD(3+1)

The application of the DLCQ method to QCD(3+1) will inevitably be difficult since meaningful numerical results will require Fock states containing two or more gluons. At the least, asymptotic freedom cannot appear in the coupling constant renormalization unless one allows for two or more gluons "in flight." A consistent renormalization program for the non-Abelian theory has not been completely worked out within the non-perturbative framework. However, as an initial exercise to test the power of the Lanczos method, Hollenberg *et al.* have diagonalized the unrenormalized light cone Hamiltonian for QCD(3+1) within the meson Fock classes $|q\bar{q}\rangle$ and $|q\bar{q}g\rangle$ only. Figure 29 shows the result for the lowest meson eigenvalue $\mathcal{M}^2$ as a function of QCD bare coupling g for several values of the quark mass. The Fock space was limited by taking the harmonic resolution $K = 6$; in addition, the ultraviolet cut-off used in this work limited the square of the Fock state invariant mass to 24 GeV^2. The maximum transverse momentum was taken at $Q_\perp = \frac{\pi N_\perp}{L_\perp} = 1$ GeV, with $N_\perp$ transverse points. The fact that the mass-squared spectrum turns negative at large coupling may possibly be cured by a consistent light-cone Hamiltonian renormalization procedure or the use of the Coulomb singularity trick used for positronium.[83] Another possibility is that negative eigenvalues of the P^- actually represents a cross-over with a negative P^+ spectrum.[91]

13. The Light-Cone Vacuum

In the introduction we discussed the remarkable feature that the perturbative vacuum in light-cone perturbation theory can also be an eigenstate of the full Hamiltonian. Let us review the arguments: By definition, the perturbative vacuum is annihilated by the free Hamiltonian: $H_0^{LC}\left|0\right\rangle = 0$. In gauge theory the interaction terms in H_{LC} are three-and four-point interactions; for example, in QED, the application on the vacuum of the interaction $H_I^{LC} = \int d^3\underline{x} e\bar{\psi}\gamma\cdot A\psi$ results in a sum of terms $b^\dagger(\underline{k}_1)a^\dagger(\underline{k}_2)d^\dagger(\underline{k}_3)\left|0\right\rangle$. Just as in the discussion of LCPTH, P^+ conservation requires $\sum_{i=1}^3 k_i^+ = 0$. However $k_i^+ = 0$, is incompatible with finite energy for the massive fermions. Thus the total light-cone Hamiltonian also annihilates the perturbative vacuum: $H^{LC}\left|0\right\rangle = 0$. In contrast, the state $H\left|0\right\rangle$ is a highly complex composite of pair fluctuations in equal-time quantization.

The apparent simplicity of the vacuum in light-cone quantization is in severe contradiction to normal expectations for the structure of the lowest mass eigenstate of QCD. In the instant form, the QCD vacuum is believed to be a highly structured quark-gluon condensate, which in turn is believed to be connected to color confinement, chiral symmetry breaking, the Goldstone pion, etc.[92] In the standard model, the $W^\pm$ and Z bosons acquire their mass through the spontaneous symmetry breaking of the scalar Higgs potential. Thus an immediate question is how one can obtain non-trivial vacuum properties in a light-cone formulation of gauge field theory.[93] This problem has recently been attacked from several directions. In the analyses of Hornbostel and Lenz *et al.*, one can trace the fate of the equal time vacuum as one approaches the $P_z \to \infty$ or equivalently rotate $\theta \to \pi/2$ as the evolution parameter $\tau = t\cos\theta + \frac{z}{c}\sin\theta$ approaches time on the light-cone. As shown in Refs. 23 and 93, one finds that for theories that allow spontaneously symmetry breaking, there is a degeneracy of light-cone vacua, and the true vacuum state can differ from the perturbative vacuum through the addition of zero mode quanta with $k^+ = k^- = k_\perp = 0$.

An illuminating analysis of the influence of zero modes in $QED(1+1)$ has been given by Werner, Heinzl and Krusche.[94] They show that although it is correct to impose the gauge condition $A^+ = 0$ on the particle sector of the Fock space, one must allow for $A^+ \neq 0$ if $k^+ = 0$. Allowing for this degree of freedom, one obtains a series of topological θ vacua on the light-cone which reproduce the known features of the massless Schwinger model including a non-zero chiral condensate. However, the effect of the infrared zero mode quanta decouples from the physics of zero charge bound states, so that the physical spectrum in one-space one-time gauge theories is independent of the choice of vacuum. The freedom in having a non-zero value for A^+ at $k^+ = 0$ can also be understood by using the gauge $\partial^+ A^+ \sim k^+ A^+ = 0$. [46]

It is thus anticipated that zero mode quanta are important for understanding the light-cone vacuum for QCD in physical space-time. In particular, the non-Abelian four-point interaction term

$$H_I^{LC} = -\frac{1}{2} g^2 \int d^3\underline{x} Tr\{[A^\mu, A^\nu][A^\mu, A^\nu]\}$$

plays a unique and an essential role, since $H_I^{LC}|0\rangle \neq 0$ as long as one allows for zero mode gluon fields in the Fock space. Thus the true light-cone vacuum $|\Omega\rangle$ is not necessarily identical to the perturbative vacuum $|0\rangle$. In fact the zero mode excitations of H_I^{LC} produce a color-singlet gluon condensate $\langle\Omega|G_{\mu\nu}G^{\mu\nu}|\Omega\rangle \neq 0$ of the type postulated in the QCD sum rule analyses. The effect of such condensates will be to introduce "soft" insertions into the quark and gluon propagators and their effective masses $m(p^2)$, and to modify the perturbative interactions at large distances. (See Fig. 30). Thus unlike the one-space one-time theory, the zero-mode gluon excitations do affect the color-singlet bound states. On the other hand, such zero mode corrections to vacuum cannot appear in Abelian QED(3+1) as long as a non-zero fermion mass appears in the free Hamiltonian.

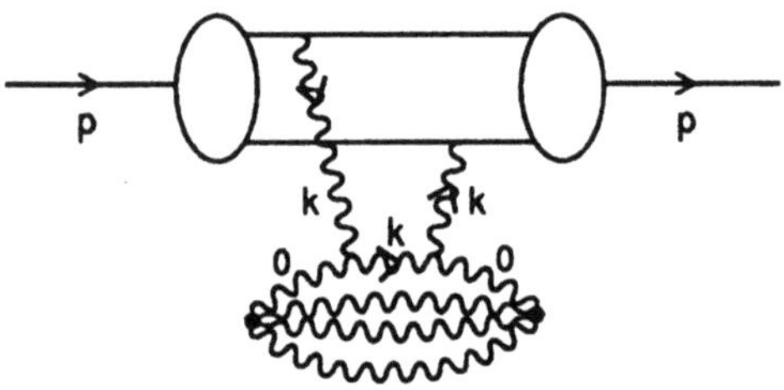

Fig. 30. Effect of a zero-mode gluon condensate on quark and gluon propagators.

14. Advantages of Discretized Light-Cone Quantization

As we have discussed in these lectures, the method of discretized light-cone quantization provides a relativistic, frame-independent discrete representation of quantum field theory amenable to computer simulation. In principle, the method reduces the light-cone Hamiltonian to diagonal form and has the remarkable feature of generating the complete spectrum of the theory: bound states and continuum states alike. DLCQ is also useful for studying relativistic many-body problems in relativistic nuclear and atomic physics. In the nonrelativistic limit the theory is equivalent to the many-body Schrödinger theory. As we have reviewed in these lectures, DLCQ has been successfully applied to a number of field theories in one-space and one-time dimension, providing not only the bound-state spectrum of these theories, but also the light-cone wavefunctions needed to compute structure functions, intrinsic sea-quark distributions, and the e^+e^- annihilation cross section.

Although our primary has been to apply light-cone methods to non-perturbative problems in QCD in physical space-time, it is important to first validate these techniques– particularly the renormalization program–in the much simpler Abelian theory of QED. In the proceeding sections we have quantized quantum electrodynamics on the light-cone in a discretized form which in principle allows practical numerical solutions for obtaining its spectrum and wavefunctions at arbitrary coupling strength α. We also have discuss a frame-independent and approximately gauge-invariant particle number truncation of the Fock basis which is useful both for computational purposes and physical approximations. In this method [28] ultraviolet and infrared regularizations are kept independent of the discretization procedure, and are identical to that of the continuum theory. One thus obtains a finite discrete representation of the gauge theory which is faithful to the continuum theory and is completely independent of the choice of Lorentz frame. Hopefully, these techniques will be applicable to non-Abelian gauge theories, including quantum chromodynamics in physical space-time.

The recent applications of DLCQ to the positronium spectrum are encouraging, but they also show formidable numerical difficulties as the number of Fock states and level of discreteness grows. Whether QCD can be solved using such methods — considering its large number of degrees of freedom is unclear.

Nevertheless, DLCQ has the potential for solving important non-perturbative problems in gauge theories. It has a number of intrinsic advantages:

- The formalism is independent of the Lorentz frame – only relative momentum coordinates appear. The computer does not know the Lorentz frame!
- Fermions and derivatives are treated exactly; there is no fermion doubling problem.
- The ultraviolet and infrared regulators are introduced in DLCQ as Lorentz invariant momentum space cut-offs of the continuum theory. They are thus independent of the discretization.
- The field theoretic and renormalization properties of the discretized theory are faithful to the continuum theory. No non-linear terms are introduced by the discretization.
- One can use the exact global symmetries of the continuum Lagrangian to pre-diagonalize the Fock sectors.
- The discretization is denumerable; there is no over-counting. The minimum number of physical degrees of freedom are used because of the light-cone gauge. No Gupta-Bleuler or Faddeev-Popov ghosts occur and unitarity is explicit.
- Gauge invariance is lost in a Hamiltonian theory. However, the truncation can be introduced in such a way as to minimize explicit breaking of the gauge symmetries.
- The output is the full color-singlet spectrum of the theory, both bound states and continuum, together with their respective light-cone wavefunctions.

There are, however, a number of difficulties that need to be resolved:

- The number of degrees of freedom in the representation of the light-cone Hamiltonian increases rapidly with the maximum number of particles in the Fock state. Although heavy quark bound states probably only involve a minimal number of gluons in flight, this is probably not true for light hadrons.

- Some problems of ultraviolet and infrared regulation remain. Although Pauli-Villars ghost states and finite photon mass can be used to regulate Abelian theories, it is not suitable method in non-Abelian theories.[72] This problem may possibly be avoided by working with finite but broken, supersymmetric theories. A cutoff in the invariant mass of the Fock state introduces extra renormalization terms compatible with the light-cone Lorentz symmetries.
- The renormalization procedure is not completely understood in the context of non-perturbative problems. However, a non-perturbative recursive representation for electron mass renormalization has been successfully tested in QED(3+1).
- The Coulomb singularity in the effective gluon-exchange potential is poorly approximated in the discrete form. An analytic trick must be used to speed convergence. Such a method has been tested successfully in the case of the positronium spectrum in QED(3+1).
- The light-cone gauge introduces extra divergences at $k^+ \to 0$ which in principle cancel between instantaneous gluon exchange and gluon propagation. However, this cancellation requires relating instantaneous potential terms to higher gluon number Fock states.
- The vacuum in QCD is not likely to be trivial since the four-point interaction term in $g^2 G^2_{\mu\nu}$ can introduce new zero-mode color-singlet states which mix with the free vacuum state. Thus a special treatment of the QCD vacuum is required. In the case of zero mass quarks, there may be additional mixing of the perturbative vacuum with fermion zero-modes. Since the zero-mode $k^\mu = 0$ states have no spatial structure, the light-cone vacuum is evidently much simpler than that of the equal-time theory. In the case of massless fermions, chiral symmetry could be spontaneous broken by fermion pair zero-modes which form a chiral condensate.

15. Acknowledgements

Much of the material presented in these lectures is based on collaborations with Kent Hornbostel, Andrew Tang, Peter Lepage, Matjaz Kaluža, Michael Krautgärtner, Thomas Eller, and Frank Wölz. We particularly thank Alex Langnau and Matthias Burkardt for many valuable discussions and suggestions. We also wish to thank Professors H. Mitter and his colleagues for their work in organizing an outstanding Winter School in Schladming. The work was supported by the Department of Energy under contract DE-AC03-76SF00515.

References

1 A.S. Kronfeld and D.M. Photiadis: Phys. Rev. **D31**, 2939 (1985).
2 G. Martinelli and C.T. Sachrajda: Phys. Lett. **B217**, 319 (1989).
3 V.L. Chernyak, A.R. Zhitnitskii: Phys. Rept. **112**, 173 (1984). See also M. Gari and N.G. Stephanis: Phys. Lett. **B175**, 462 (1986), and references therein.
4 G.P. Lepage and S.J. Brodsky: Phys. Rev. **D22**, 2157 (1980); Phys. Lett. **87B**, 359 (1979); Phys. Rev. Lett. **43**, 545, 1625(E) (1979).
5 See P. Kroll: Wuppertal University preprint WU-B-90-17 (1990), and references therein.
6 For a review, see S.J. Brodsky: SLAC-PUB 5529 (1991), to be published in the proceedings of the Lake Louise Winter Institute (1991).
7 I. Tamm: J. Phys (USSR) **9**, 449 (1945). S.M. Dancoff: Phys. Rev. **78**, 382 (1950).
8 We shall use the conventional term "light-cone quantization" as the equivalent to the front-form and null-plane quantization. For further discussion, see P.L. Chung, W.N. Polyzou, F. Coester, and B.D. Keister: Phys. Rev. **C37**, 2000 (1988), and references therein.
9 P.A.M. Dirac: Rev. Mod. Phys. **21**, 392 (1949).
10 Gribov has emphasized the interesting alternative possibility that hadrons in QCD are actually bound states of light quarks with *negative* kinetic energy. This removes the exclusion of negative k^+ in the Fock basis and thus would have profound consequences for light-cone quantization. See V.N. Gribov: Lund preprint, LU TP 91-7 (1991).
11 For a discussion of renormalization in light-cone perturbation theory, see S.J. Brodsky, R. Roskies and R. Suaya: Phys. Rev. **D8**, 4574 (1974), and also Ref. 4.
12 M. Burkardt, A. Langnau: SLAC-PUB-5394, (1990), and to be published.
13 J. R. Hiller: University of Minnesota preprints (1990), and Phys. Rev. **D43**, 2418 (1991).
14 S.D. Drell and T.M. Yan: Phys. Rev. Lett. **24**, 181 (1970).
15 S.J. Brodsky and S.D. Drell: Phys. Rev. **D22**, 2236 (1980).
16 G.P. Lepage and B.A. Thacker: CLNS-87/114, (1987). See also G.P. Lepage and W. Caswell: Phys. Lett. **167B**, 437 (1986).
17 S.J. Brodsky and I.A. Schmidt: Phys. Lett. **B234**, 144 (1990); Phys. Rev. **D43**, 179 (1991).
18 K. Hornbostel, S.J. Brodsky, H.C. Pauli: Phys. Rev. **D41**, 3814 (1990).
19 G.P. Lepage, S.J. Brodsky, T. Huang, P.B. Mackenzie: published in the *Proceedings of the Banff Summer Institute*, 1981.
20 S.J. Brodsky, G.P. Lepage: in *Perturbative Quantum Chromodynamics*, p. 93, edited by A.H. Mueller (World Scientific, Singapore, 1989).
21 H.C. Pauli and S.J. Brodsky: Phys. Rev. **D32**, 1993 (1985); Phys. Rev. **D32** , 2001 (1985).
22 T. Eller, H.C. Pauli, S.J. Brodsky: Phys. Rev. **D35**, 1493 (1987).
23 A. Harindranath and J.P. Vary: Phys. Rev. **D36**, 1141 (1987).
24 M. Burkardt: Nucl. Phys. **A504**, 762 (1989).
25 R.J. Perry, A. Harindranath, K.G. Wilson: Phys. Rev. Letters **65**, 2959 (1990), R.J. Perry, and A. Harindranath: Ohio State University preprint (1990).
26 D. Mustaki, S. Pinsky, J. Shigemitsu, K. Wilson: Ohio State University preprint (1990).
27 A.C. Tang: SLAC-351 (1990).
28 A.C. Tang, S.J. Brodsky, and H.C. Pauli: SLAC-PUB-5425 (1991), to be published in Phys. Rev. D.
29 L.C.L. Hollenberg, K. Higashijima, R.C. Warner, B.H. J. McKellar: KEK-TH-280, (1991).
30 M. Kaluža: University of Heidelberg thesis, and to be published (1990).
31 J.R. Klauder, H. Leutwyler, and L. Streit: Nuovo Cimento **59**, 315 (1969).
32 J.B. Kogut and D.E. Soper: Phys. Rev. **D1**, 2901 (1970).
33 F. Rohrlich: Acta Phys. Austriaca, Suppl. VIII, 2777 (1971).
34 H. Leutwyler: Nucl. Phys. **B76**, 413 (1974).
35 A. Casher: Phys. Rev. **D14**, 452 (1976).
36 S.J. Chang, R.G. Root, T.M. Yan: Phys. Rev. **D7**, 1133 (1973); S.J. Chang, T.M. Yan: Phys. Rev. **D7**, 1147 (1973).

37 S.J. Brodsky, C.R. Ji: SLAC-PUB-3747, (1985).
38 G. McCartor: Z. Phys. **C41**, 271 (1988); and SMU preprint SMUTH/91-02 (1991).
39 E.V. Prokhvatilov and V.A. Franke: Sovj. J. Nucl. Phys. **49**, 688 (1989).
40 V.A. Franke, Y.V. Novozhilov, and E.V. Prokhvatilov: Letters in Mathematical Physics **5**, 239 (1981).
41 A.M. Annenkova, E.V. Prokhvatilov, and V.A. Franke: Zielona Gora Pedag. Univ. preprint - WSP-IF 89-01 (1990).
42 V.A. Karmanov: Nucl. Phys. **B166**, 378 (1980).
43 V.A. Karmanov: Nuclear Physics, **A362**, 331 (1981).
44 V.N. Pervushin: Nuclear Physics **B15**, 197 (1990).
45 G. McCartor: Z. Phys. **C41**, 271, (1988).
46 F. Lenz: in the proceedings of the NATO Advanced Summer Institute on Nonperturbative Quantum Field Theory Cargese, France, Aug 8-18, 1989. Edited by D. Vautherin, F. Lenz, and J.W. Negele.: Plenum Press, N. Y. (1990).
47 S. Weinberg: Phys. Rev. **150**, 1313 (1966).
48 S.D. Drell, D. Levy, T.M. Yan: Phys. Rev. **187**, 2159 (1969); Phys. Rev. **D1**, 1035 (1970); Phys. Rev. **D1**, 1617 (1970).
49 L. Susskind: Phys. Rev. **165**, 1535 (1968); L. Susskind, G. Frye: Phys. Rev. **164**, 2003 (1967).
50 J.D. Bjorken, J.B. Kogut, D.E. Soper: Phys. Rev. **D3**, 1382 (1971); J.B. Kogut, D.E. Soper: Phys. Rev. **D1**, 2901 (1970).
51 General QCD analyses of exclusive processes are given in Ref. 4, S.J. Brodsky and G.P. Lepage, SLAC-PUB-2294, presented at the Workshop on Current Topics in High Energy Physics, Cal Tech (Feb. 1979), S.J. Brodsky, in the Proc. of the La Jolla Inst. Summer Workshop on QCD, La Jolla (1978), A.V. Efremov and A.V. Radyushkin: Phys. Lett. **B94**, 245 (1980), V.L. Chernyak, V.G. Serbo, and A.R. Zhitnitskii: Yad. Fiz. **31**, 1069 (1980), S.J. Brodsky, Y. Frishman, G.P. Lepage, and C. Sachrajda: Phys. Lett. **91B**, 239 (1980), and A. Duncan and A.H. Mueller: Phys. Rev. **D21**, 1636 (1980).
52 QCD predictions for the pion form factor at asymptotic Q^2 were first obtained by V.L. Chernyak, A.R. Zhitnitskii, and V.G. Serbo: JETP Lett. **26**, 594 (1977), D.R. Jackson: Ph.D. Thesis, Cal Tech (1977), and G. Farrar and D. Jackson: Phys. Rev. Lett. **43**, 246 (1979). See also A.M. Polyakov: Proc. of the Int. Symp. on Lepton and Photon Interactions at High Energies, Stanford (1975), and G. Parisi: Phys. Lett. **84B**, 225 (1979). See also S.J. Brodsky and G.P. Lepage: in High Energy Physics–1980, proceedings of the XXth International Conference, Madison, Wisconsin, edited by L. Durand and L.G. Pondrom: (AIP, New York, 1981); p. 568. A.V. Efremov and A.V. Radyushkin: Rev. Nuovo Cimento **3**, 1 (1980); Phys. Lett. **94B**, 245 (1980). V.L. Chernyak and A.R. Zhitnitsky: JETP Lett. **25**, 11 (1977); G. Parisi, Phys. Lett. **43**, 246 (1979); M.K. Chase, Nucl. Phys. **B167**, 125 (1980).
53 S.J. Brodsky, G.R. Farrar: Phys. Rev. **D11**, 1309 (1975).
54 A. Kronfeld and B. Nizic: Fermilab-Pub 91/64-T (1991). The Compton scattering data are from M. Shupe *et al.*: Phys. Rev. **D16**, 1921 (1979).
55 P. Stoler: Phys. Rev. Lett. **66**, 1003 (1991); and to be published in Phys. Rev. D.
56 C.E. Carlson and J.L. Poor: Phys. Rev. **D38**, 2758 (1988).
57 N. Isgur and C.H. Llewellyn Smith: Phys. Lett. **B217**, 535 (1989)
58 J. Hansper, R. Eckardt, and M.F. Gari: *et al.*, Ruhr-Universität Bochum preprint (1991).
59 A. Szczepaniak, L. Mankiewicz: Univ. of Florida Preprint (1991).
60 S.J. Brodsky: SLAC-PUB-5529 (1991).
61 J. Botts: Nucl. Phys. **B353**, 20 (1991).
62 P.V. Landshoff: Phys. Rev. **D10**, 1024 (1974).
63 S.J. Brodsky, G.P. Lepage, and P.B. Mackenzie: Phys. Rev. **D28**, 228 (1983). For a recent discussion of the scale-fixing problem in QCD, see J.C. Collins: ANL-HEP-CP-90-58 (1990).
64 W. Kwong, P.B. Mackenzie, R. Rosenfeld. and J.L. Rosner: Phys. Rev. **D37**, 3210,(1988).
65 A. Szczepaniak, E.M. Henley, S.J. Brodsky: Phys.Lett. **B243**, 287 (1990).
66 S.J. Brodsky, G.P. Lepage, and S.A.A. Zaidi: Phys. Rev. **D23**, 1152 (1981).

67 As noted by U. Dokshitser and A. Mueller (private communication), virtual loop corrections related to the Sudakov form factor due to the difference in the overall heavy quark mass scale M_b^2 and the hard scattering scale can also modify the hard scattering amplitude. However, such corrections are not expected to be large.
68 D. Bartoletta *et al.*: Phys. Rev. Lett. **62**, 2436 (1989).
69 S.G. Gorishny, A.L. Kataev, and S.A. Larin: Phys. Lett. **B212**, 238 (1988).
70 We follow here the notation and discussion of D. Soper: Phys. Rev. **D4**, 1620, (1971); and K. Hornbostel: CLNS-90-1038, (1990).
71 A. Langnau: to be published.
72 It also should be noted that in Gribov's approach to quark confinement, Pauli-Villars or dimensional regulation cannot even be used in principle for strong coupling problems in QCD because of the way it eliminates the negative energy sea. See Ref. 10.
73 S.J. Brodsky, Y. Frishman, G.P. Lepage and C. Sachrajda: Phys. Lett. **91B**, 239 (1980).
74 $SU(N)$ gauge theories, restricted to one spatial dimension and time have been studied extensively at large N; see e.g. G.'t Hooft: Nucl. Phys. **B75**, 461 (1974); C.G. Callan, N. Coote, and D. J. Gross: Phys. Rev. **D13**, 1649 (1976); M.B. Einhorn: Phys. Rev. **D14**, 3451 (1976); and I. Bars and M. B. Green: Phys. Rev. **D17**, 537 (1978).
75 C.J. Burden and C.J. Hamer: Phys. Rev. **D37**, 479 (1988), and references therein.
76 Y. Frishman and J. Sonnenschein: Nucl. Phys. **B294**, 801 (1987), and Nucl. Phys. **B301**, 346 (1988).
77 H. Bergknoff: Nucl. Phys. **B122**, 215 (1977).
78 For a recent discussion and further references, see C.S. Kim: Nucl. Phys. **B353**, 87 (1991).
79 J.J. Aubert, *et al.*: Nucl. Phys. **B213**, 31 (1983). See also E. Hoffmann and R. Moore: Z. Phys. **C22**, 71 (1983).
80 K. Hornbostel: private communication; S.J. Brodsky and K. Hornbostel: to be published.
81 S.J. Brodsky and P. Hoyer: SLAC-PUB-5422, (1991).
82 G.T. Bodwin, D.R. Yennie, and M.A. Gregorio: Rev. Mod. Phys. **56**, 723 (1985).
83 M. Krautgärtner, H.C. Pauli, and F. Wölz: Heidelberg preprint MPIH-V4-1991.
84 P.M. Morse and H. Feshbach: *Methods in Theoretical Physics*, 2 Vols, Mc Graw-Hill, New York, N.Y., 1953.
85 M. Sawicki: Phys. Rev. **D32**, 2666 (1985).
86 M. Sawicki: Phys. Rev. **D33**, 1103 (1986).
87 L. Mankiewicz: private communication.
88 H.A. Bethe and E.E. Salpeter: *Quantum Mechanics of One- and Two-Electron Atoms*, Springer, Heidelberg, 1957.
89 W. Dykshoorn, R. Koniuk, and R. Muñoz-Tapia: Phys. Rev. **A41**, 60 (1990).
90 W. Dykshoorn and R. Koniuk: Phys. Rev. **A41**, 64 (1990).
91 We thank V. Gribov for an illuminating discussion on this point.
92 See, for example, M.H. Thoma and H. J. Mang: Z. Phys. us **C44**, 349 (1989).
93 St. Glazek: Phys. Rev. **D38**, 3277 (1988).
94 Th. Heinzl, St. Krusche, and E. Werner: Regensburg preprint TPR 90-44.

Conceptual and Geometrical Problems in Quantum Gravity

C.J. Isham

Blackett Laboratory
Imperial College of Science and Technology
South Kensington, London SW7 2BZ, UK

ABSTRACT

A summary is given of some of the conceptual and geometrical problems that arise in various attempts to construct a quantum theory of gravity. Particular emphasis is placed on the canonical quantisation of general relativity and the problem of 'time'.

1. Introduction

1.1 Preamble

These notes are based on a course of lectures entitled "Conceptual and Geometrical Problems in Quantum Gravity" given at the 1991 Schladming Winter School in Theoretical Physics. The majority of students at the school had a background in elementary particle physics and this is reflected in the exposition in which aspects of the subject that derive from general relativity are more detailed than those that come from particle physics. Particular emphasis is placed on the canonical quantisation programme and the problems it engenders.

It must be emphasised that this is *not* meant to be a full-blown review of quantum gravity, not even of the conceptual and geometrical aspects of the subject. It is merely an account of some of the problems that arise in this area, and the treatment is necessarily cursory in a number of places. So far as references are concerned, I have concentrated mainly on more recent papers and, although some of the "classics" in the field are cited, the list—both modern and ancient—is very incomplete. Most of the papers mentioned are ones that I happened to find useful when preparing the notes. My genuine apologies to authors who feel unduly neglected by this process.

1.2 What is the Problem of Quantum Gravity?

Techniques of differential geometry are widely used in modern theoretical physics and there are many areas about whose geometrical aspects one might write. However, quantum gravity differs from most other branches of physics by the way in which it raises severe problems of a conceptual nature concerning the status of our conventional ideas on space, time and matter. Furthermore, because of the specific mathematical structure of general relativity, these conceptual problems are frequently linked to the more technical question of the applicability of standard geometrical ideas to the quantum theory. It is therefore quite appropriate to discuss geometrical and conceptual issues within the framework of a single course of lectures.

But what is the problem of quantum gravity? And how significant are the conceptual problems which arise? There is no easy answer to these questions, and certainly no consensus among the physicists who work in the field. The minimal requirement for a successful theory is that it should reproduce conventional classical general relativity and normal quantum theory in their appropriate domains. The key quantities in this respect are the *Planck length* $L_P = (\frac{G\hbar}{c^3})^{\frac{1}{2}} \approx 10^{-35}m$, (corresponding to a Planck energy of $10^{28}eV$) and the associated *Planck time* $t_P = L_p/c \approx 10^{-42}s$, where G is Newton's constant and c is the speed of light. Dimensional arguments suggest that quantum gravity effects will become important at these scales and, conversely, that standard general relativity and quantum theory should apply at distances and times significantly greater than these minute quantities.

However, even if there is agreement about the physics at large distances and times, there is still a wide variety of opinions about what happens at the Planck length itself and, in particular, the extent to which the conceptual and structural frameworks of general relativity and quantum theory are still applicable.

The conservative view is that nothing special happens at the Planck length. The opposite, iconoclastic, view is that fundamental changes are needed leading to a radical reassessment of the traditional ideas of space, time and quantum matter. For example, there is a recurrent idea that the use of ***continuum*** concepts (in particular, differential geometry) is inapplicable in this domain. Another example is Penrose's long-held view that quantum theory becomes ***non-linear*** at scales of the Planck length, and in such a way as to provide a natural mechanism for explaining the much-debated quantum mechanical problem of the reduction of the state vector (Penrose 1986).

One difficulty in making theoretical judgements on the nature of Planck length physics is that the agreed minimal requirement—that a theory of quantum gravity should reproduce classical general relativity at large scales—does not seem to place many restrictions on the overall theory. For example, within the context of the particle-physics approach to quantum gravity (see §3), there are a series of, more or less, rigorous theorems which show that ***any*** Lorentz-invariant theory of a spin-2 graviton coupled to a conserved energy-momentum tensor will necessarily yield the same low-energy scattering results as those obtained from the tree graphs of a weak-field perturbative expansion of the Einstein Lagrangian (*e.g.*, Gupta 1955, Feynman 1963, Weinberg 1964a,b 1965, Boulware and Deser 1975). [1] These theorems do not mean that it is easy to construct a quantum theory of gravity (it isn't!) but they do imply that not much can be learnt from the requirement that the low-energy limit must be general relativity. One result of this lack of ***a priori*** information about what is wanted of the theory is a notable absence of any comprehensive axiomatisation of the subject. In particular, there is nothing like the Wightman scheme, or the approaches based on C^*-algebras which have played such an important role in clarifying the content of normal quantum field theory. Indeed, it is by no means obvious that we would recognise a proper theory of quantum gravity even if it was presented to us on a revelatory plate!

1.3 Great Expectations

This uncertainty about the aims of quantum gravity has lead to a proliferation of approaches to the subject and of expectations of how the theory might be used. To a considerable extent these depend on the general techniques of the community to which a researcher belongs: in particular, there tend to be significant differences between the

[1] This result is often quoted, but there are loop holes. In particular, Wald (1986) demonstrated the existence of a consistent theory of a massless, interacting spin-2 field which is *not* generally covariant.

methods adopted by an elementary particle physicist (or quantum field theorist) and those employed by someone whose primary training has been in general relativity.

1.3.1 The Approach of a Particle Physicist

The emphasis here is mainly on techniques drawn from conventional, Poincaré-group based quantum field theory. Thus the key concept is *special* relativity and the idea of "gravitons" (quanta of the gravitational field) propagating in a fixed Minkowski spacetime. The minimal expectation of a quantum gravity theory (over and above the general requirements mentioned above) is that it should produce scattering amplitudes for gravitons and other particles that are free of irremovable divergences; thus the theory must be renormalisable or, better still, genuinely finite. The maximal expectation would be that such a theory is part of a general Grand Unified Theory with the gravitational sector playing an essential role.

The main technical approach in the early studies was an expansion of the spacetime metric tensor $g_{\mu\nu}(x)$ around the Minkowski metric $\eta_{\mu\nu}$ in the form $g_{\mu\nu}(x) = \eta_{\mu\nu} + \kappa h_{\mu\nu}(x)$ where $\kappa = (16\pi G/c^2)^{\frac{1}{2}}$ is the (dimensional) coupling constant. The Lorentz tensor field $h_{\mu\nu}(x)$ can be quantised using any of the standard techniques drawn from the relativistic quantum field theory of gauge systems. Such an expansion (to which we will return in §3) is anathema to most general relativists, but that has never upset the particle physics community to any noticeable degree!

More recently, the efforts of elementary particle physicists have been directed towards theories of superstrings and other extended objects. Most calculations in this area also employ the language of gravitons, although it should be emphasised that this is not an inevitable consequence of string theory: it is just that suitable non-perturbative methods have not yet been developed.

The main problem with the particle physics approach is how to do full justice to the concepts of space and time engendered by the classical theory of general relativity. For example, a perturbative expansion around any background metric [2] leaves open the question of the "real" causal structure of the theory; there is certainly no reason why it should be the same as that of the background space. Similarly, it is difficult to handle cosmological issues within such a framework. Simply using a standard classical

[2] In the modern background-field methods this need not just be Minkowski space.

cosmology (*e.g.*, the Robertson-Walker metric) as the background is not sufficient if one wishes to study, for example, the influence of quantum gravity on the initial spacetime singularity.

1.3.2 The Approach of a General Relativist

General relativists usually approach the problem of quantum gravity from a very different perspective, with the emphasis being placed on the geometrical features of the theory and the role of spacetime structure. In particular, there is a great reluctance to employ any quantisation scheme which is based on a special background like, for example, Minkowski spacetime. The assumption is that if such a space does have a special role, it should appear naturally as part of the structure of the theory itself and not just be put in by hand. The question of background structure is very important and is one to which we will return in §2.

A general relativist's minimal expectation of a quantum theory of gravity is likely to include the possibility of solving the problems posed by spacetime singularities, black holes, and similar features of the classical theory. In particular, one expects to be able to recover Hawking's famous results on the quantum production of particles by a classical black hole, and to extend them to answer questions concerning the back reaction of the created particles and the final state of the evaporating hole. A more ambitious goal is to apply the theory to cosmological issues and, in particular, to study the creation of the universe as a quantum theoretical process.

Various technical approaches have been employed over the years but there has always been a tendency to adopt a "back to basics" philosophy with the emphasis being placed on quantum theory in general rather than on the more specialised techniques of relativistic quantum field theory. The problems of particular interest are mainly those relating to the quantum status of the spacetime concepts of classical general relativity, and the extent to which conventional quantum theoretical ideas can be applied. In general, relativists tend to be more concerned with the conceptual issues which arise in quantum gravity than are their cousins who approach the subject from a background of elementary particle physics.

In addition to the sociological issues that generate such different approaches to the subject there is one question which, until it is answered satisfactorily, is guaranteed to maintain the uncertainty about what one is trying to do in quantum gravity:

> Is the central problem of quantum gravity one of (i) physics, (ii) mathematics, or (iii) philosophy?

In particular, how severe are the conceptual difficulties? Is it possible that one needs to get to grips with them *before* any serious technical development can be made?

This central question is considered in chapter 2 which lists some of the prima facie conceptual problems which arise in quantum gravity. Chapter 3 is a brief account of the particle physics approach, while in chapter 4 there is a discussion of the canonical quantisation of gravity which places more emphasis on the geometrical aspects of the subject. In chapter 5 we consider one of the central issues—the problem of time—and the final chapter is a short introduction to some of the special problems of quantum cosmology.

2. Prima Facie Problems in Quantum Gravity

2.1 General Remarks

In this chapter we will consider some of the prima facie problems that can arise in various general approaches to constructing a quantum theory of gravity. Some problems of this type stem from incompatibilities in the way the two theories are normally interpreted. Others are more technical in nature and arise when a structural feature of one of the theories is particularly difficult to maintain in the presence of the other.

One of the broadest problems of all is the extent to which a quantum theory of gravity can be expected to maintain:

1. the picture of spacetime afforded by general relativity;
2. the interpretative and structural frameworks of conventional quantum theory.

This question is central, but it is too broad to succumb to a frontal assault and is better served by a more circumspective approach. We will start therefore by considering two problems which, although very important, are more specific. They are

1. the role of the spacetime diffeomorphism group Diff($\mathcal{M}$);
2. the significance of background structure.

2.2 Spacetime Diffeomorphisms

The equations of general relativity are covariant with respect to the group Diff($\mathcal{M}$) of diffeomorphisms of the spacetime manifold $\mathcal{M}$. We will restrict our attention to diffeomorphisms with compact support, *i.e.*, those that are equal to the unit map outside some closed and bounded region of $\mathcal{M}$. Thus, for example, a Poincaré -group transformation of Minkowski spacetime is not deemed to belong to Diff($\mathcal{M}$). This restriction is imposed because the role of transformations with a non-trivial action in the asymptotic regions of $\mathcal{M}$ is quite different from those that act trivially.

It must be emphasised that Diff($\mathcal{M}$) means the group of active point transformations of $\mathcal{M}$. This should not be confused with the pseudo-group which describes the relations between overlapping pairs of coordinate charts (although of course there is a connection between them).

In some respects, the role of the diffeomorphism group in both classical and quantum general relativity is analogous to that of the gauge group in Yang-Mills theory. For example, in both cases the groups are associated with field variables that are non-dynamical.

However, in another, and very important, sense the two groups are quite different. Yang-Mills transformations occur at a fixed spacetime point whereas the diffeomorphism group moves points around. This is responsible for the famous Einstein "hole" problem (Stachel 1987, Earman and Norton 1986, Rovelli 1989) with its conclusion that invariance under Diff($\mathcal{M}$) means that the individual mathematical points in $\mathcal{M}$ have no intrinsic physical significance. This is closely related to the question of what constitutes an "observable" in general relativity—a surprisingly contentious issue that has generated much debate over the years and which is of particular interest in the context of quantum gravity.

For example, although the Riemann scalar curvature $R(x) := g^{\mu\nu}(x)R_{\mu\nu}(x)$ is a scalar function on $\mathcal{M}$, its value at any particular point $x \in \mathcal{M}$ can *not* be regarded as an observable. In the quantum case, this follows from the observation that if $U(f)$ is a unitary representation of $f \in \mathrm{Diff}(\mathcal{M})$, the action on the quantised metric field should lead to the transformation law

$$U(f)R(x)U(f)^{-1} = R(f^{-1}(x)) \tag{2.2.1}$$

provided that $R(x)$ can be defined as a proper operator function of $g_{\mu\nu}(x)$ and its derivatives. Thus if a physical observable is defined to be one that *commutes* with the action of the "gauge group" (as is the case in Yang-Mills theory), then $R(x)$ is evidently not such an object.

One way of constructing genuine invariants is to integrate scalar functions of $g_{\mu\nu}(x)$ over the entire spacetime such as

$$\int_{\mathcal{M}} R(x)\big(\det g(x)\big)^{\frac{1}{2}}\, d^4x; \quad \int_{\mathcal{M}} R^{\mu\nu}(x)R_{\mu\nu}(x)\big(\det g(x)\big)^{\frac{1}{2}}\, d^4x. \tag{2.2.2}$$

Note that these are highly *non-local* and therefore any quantisation scheme based on expressions of this type will be very different from a conventional quantum field theory. [3] Another, quite different, approach to the problem of constructing invariants has a long history in classical general relativity and has played an important role in much recent work on quantum gravity. The central idea is that although, for example, $R(x)$ is not an observable, the expression $R(X)$ *is* if X refers to the point on the manifold occupied by an actual physical particle. In other words, we must locate ourselves on the spacetime manifold with the aid of a "material reference system". Thus simple general relativity is incomplete as a physical theory because it ignores both the equations of motion of the reference system and the effects of its energy-momentum tensor on the gravitational field. This idea of using "physical" coordinates is very important and we will return to it in §4 and §5 in the contexts of the canonical quantisation of gravity and the problem of time.

---

[3] The nearest analogue would perhaps be a quantisation of Yang-Mills theory using the gauge-invariant, but non-local, Wilson loop variables as the basic quantum objects.

2.3 Background Structure

A key feature of any approach to quantum gravity is the background structure it assumes and the way it is used. This "background" can take various forms. Sometimes it involves a choice of a particular mathematical element in the theory. On other occasions it refers to the conceptual or interpretative framework that is assumed *a priori*.

An example of the former is any theory that employs a fixed manifold to represent space or spacetime, or in which a particular spacetime metric (*e.g.*, the Minkowski metric) is assigned a central role. [4] The conceptual background may include part, or all, of the interpretative structure of conventional quantum theory. This in turn may assume the existence of a background classical world within which measurements of observables are made. Indeed, it can be argued that the conventional Copenhagen interpretation of quantum theory presupposes as part of its background a *fixed* spacetime (in both the topological and the metrical sense) and that it is therefore intrinsically incompatible with the idea of quantum gravity! The existence of such a background is also frequently associated with a division of a part of the universe into a 'system' and an 'observer'—a split that is particularly difficult to maintain in the context of quantum cosmology.

There are several reasons why it is important to understand clearly the precise background structure that is assumed in any particular approach to quantum gravity. For example, in the case of a background manifold or metric it can be argued that an object of this type constitutes a special choice for what could, or should, be determined dynamically. This is closely related to what Smolin (1991) calls an "ideal" element. At best, such a situation is likely to yield only an approximation to the complete theory, and it could be extremely misleading.

A more insidious problem arises if background structure exists that is not consciously recognised as such. This may be part of a conceptual or interpretative framework, or it may be more technical: like the use of special reference frames to label spacetime points and thus (superficially) to avoid the Einstein hole problem. A property of all background structure is that once inserted into a theory it tends to stay

---

[4] In saying that the Planck length or Planck time have definite values, an implicit structure of this sort is needed to define the units in which length or time are expressed.

there. This does no harm if it is recognised for what it is, but if its initial presence is not made explicit the structure may be "rediscovered", accompanied perhaps by shouts of great triumph. For example, some of the claims that spacetime "emerges" from string theory look somewhat shaky once it is realised that much of the structure (*e.g.*, the manifold nature of spacetime) is imposed *a priori* in the very formulation of the theory.

2.4 How Should the Construction of a Quantum Theory of Gravity Be Approached?

Let us see how these issues arise in the context of the various general ways in which one might approach the problem of constructing a quantum theory of gravity. The discussion here is very brief and many of the topics will be taken up later in more detail.

2.4.1 Quantise General Relativity

Many of the most famous applications of quantum theory start with a classical theory which is then 'quantised'. Note that "quantise" is a verb: it is something that is done to a classical theory, presumably with the expectation of improving it in some way. Certainly this is how elementary quantum theory is taught: we 'quantise' the hydrogen atom, the simple harmonic oscillator *etc.* Even quantum chromodynamics [5] is based on the idea of quantising a given classical Lagrangian system.

In the case of general relativity we do not know if it is possible to quantise the simple Einstein action plus whatever additional fields we fancy, or whether a consistent theory requires a very special selection of matter. This latter idea has been a constant source for the hope that a quantum theory of gravity will lead to a unique grand unified theory of all the fundamental forces. The last major programme of this sort was N=8 supergravity.

One big advantage of quantising a given classical theory is that it is guaranteed that the theory can be recovered in an appropriate classical limit. In particular, the desire

---

[5] This is the first, and probably the only, connection between my course and the other lectures at the Schladming Winter School!

to reproduce general relativity at low energies should be easy to meet if one starts with the Einstein action $S = 1/\kappa^2 \int R(g)(\det g)^{\frac{1}{2}}\, d^4x$ and then quantises the metric using operator methods, functional integrals or whatever else can be drawn from the armoury of quantum field theory. This approach is also helpful in interpreting the quantum results in so far as the theory is understood to deal with small "quantum fluctuations" about some classical solution to the field equations.

Unfortunately, it is not a trivial matter to turn the metric tensor $g_{\mu\nu}(x)$ into an operator $\widehat{g}_{\mu\nu}(x)$, or to construct a functional integral over fields of this type. One major problem is that many of the techniques of conventional quantum field theory are simply not applicable.

The most obvious example is the entire edifice of Poincaré -group based quantum field theory which can only be invoked if a background Minkowski spacetime is employed from the outset. Without such a structure, even the notion of a 'graviton' becomes very problematic. Selecting any other background only makes matters worse since the theory of quantum fields in a general curved space has many ambiguous features.

Problems with the diffeomorphism group

Another major difficulty arises if one takes seriously the idea that the quantum theory should reflect the diffeomorphism group invariance of classical general relativity (*e.g.*, Rovelli and Smolin 1990). In a normal quantum field theory, the two-point function for a scalar field ϕ has the typical behaviour

$$W(x,y) := \langle 0|\,\widehat{\phi}(x)\widehat{\phi}(y)\,|0\rangle \sim \frac{\text{const.}}{(x-y)^2} \tag{2.4.1}$$

as the spacetime points x and y approach each other (note that the distance between x and y is measured using the background Minkowski metric). The fact that $W(x,y)$ depends only on the difference vector $x - y$ is a direct consequence of the invariance of the vacuum state $|0\rangle$ under the action of the translation subgroup of the Poincaré -group.

On the other hand, if $|0\rangle$ is $\text{Diff}(\mathcal{M})$ invariant, and if $\widehat{\phi}(x)$ transforms as in (2.2.1), then

$$W(x,y) := W\big(f(x), f(y)\big) \tag{2.4.2}$$

for all $f \in \text{Diff}(\mathcal{M})$. However, for any two pairs of points (x,y) and (x',y') that are sufficiently close to each other (for example, if all four points lie in a single coordinate chart) there is a diffeomorphism f such that $x' = f(x)$ and $y' = f(y)$. It follows that $W(x, y)$ is actually a *constant* for all points y in a sufficiently small neighbourhood of x.

In assessing the significance of this behaviour it should be noted that it is related to the fact that the value $\phi(x)$ of the scalar field ϕ at a point $x \in \mathcal{M}$ is not an observable in a Diff($\mathcal{M}$)-invariant theory. Nevertheless, it does show that the questions of short-distance behaviour and ultraviolet divergences are likely to be very different in quantum gravity than in a normal quantum field theory. Similarly, the concept of regularisation is difficult to implement without the explicit introduction of a background metric with which to give a numerical measure of the "nearness" of one spacetime point to another.

The question of Diff($\mathcal{M}$) invariance also produces problems in functional integral approaches to quantisation. One might try to construct a theory of quantum gravity by using functional integrals analogous to those employed in normal quantum field theory to produce the vacuum expectation values of the time-ordered product of a string of fields. For example, one might look at

$$G(x_1, x_2 \ldots x_n) = \int \mathcal{D}[g]\, R(x_1)R(x_2)\ldots R(x_n)e^{i\int_{\mathcal{M}} R(g)\sqrt{g}\, d^4x}. \qquad (2.4.3)$$

However, it is difficult to see what a "time-ordered" product could mean in the absence of any background metric to provide a preferred notion of spacelike and timelike. [6] But even if such a background *is* provided, the idea of time-ordering still seems inconsistent with Diff($\mathcal{M}$) invariance since whether a pair of points is timelike or spacelike can be changed by the action of a diffeomorphism. [7] A related problem appears in a spacetime operator approach to quantisation. In a conventional quantum field theory, a scalar field $\widehat{\phi}$ obeys the basic microcausality condition

$$[\widehat{\phi}(x), \widehat{\phi}(y)] = 0 \qquad (2.4.4)$$

[6] This particular problem can be superficially avoided by using a "Euclidean" formalism in which the functional integral is over metrics with a Riemannian, rather than Lorentzian, signature. The problem then is to relate these Riemannian n-point functions to physical amplitudes.

[7] Of course, if the background metric is transformed at the same time, then it is tautological that nothing changes. But this merely reflects the absence of any intrinsic physical meaning for the mathematical points in the spacetime manifold. The situation is very different if the points $x_1, x_2 \ldots x_n$ are labelled by the positions of *physical* particles, or the values of physical fields.

if x and y are spacelike separated. But what does this mean in a quantum theory of gravity? As emphasised by Fredenhagen and Haag (1987), for most [8] pairs of points $x, y \in \mathcal{M}$ there will exist at least one Lorentzian metric with respect to which they are *not* spacelike separated, and hence, in so far as all metrics are "virtually present" (or summed over in a functional integral) the right hand side of (2.4.4) will never vanish. Thus one of the bedrocks of conventional quantum field theory is removed at a stroke!

The same argument throws doubts on the use of canonical commutation relations (with respect to what metric is the hypersurface meant to be spacelike?) and on the whole C^*-algebra approach to quantum field theory.

Use of a Minkowski background metric

In the light of the above, it is not surprising that much of the work on quantum gravity has involved the introduction of a background Minkowski space with the intent of using conventional, Poincaré -group based quantum field theory. As we will see briefly in §3, this leads naturally to a perturbative approach based on writing the metric tensor as $g_{\mu\nu}(x) = \eta_{\mu\nu} + \kappa h_{\mu\nu}(x)$ and then quantising $h_{\mu\nu}(x)$ as a normal field with "gravitons" as quanta.

This is closely related to the possibility of using the background metric $\eta_{\mu\nu}$ to construct a Gaussian measure with the weight factor

$$\exp\left(-\int_{\mathcal{M}} \eta^{\mu\nu}\eta^{\alpha\beta}h_{\mu\alpha}(x)h_{\nu\beta}(x)\, d^4x\right). \tag{2.4.5}$$

On the other hand, the analogous construction using just the metric $g_{\mu\nu}(x)$ (*i.e.*, replacing both η and h with g) leads only to a constant under the spacetime integral!

Many advantages (and disadvantages) accrue from the introduction of such a background structure, but further discussion is deferred to §3 where a short survey is given of the approach.

A minimum length in quantum gravity?

Another problem with the use of local quantum fields is the possible existence of a natural "minimum" length/time related to the Planck units. An example of a recent suggestion of this sort is the work of Padmanabhan who argued that

1. geodesic distance is intrinsically bounded from below in quantum gravity (Padmanabhan 1985a,b);

---

[8] The qualification arises from the possible existence of global obstructions to the construction.

2. the uncertainty relations, plus the existence of the Schwarschild radius, impose lower bounds on measurements of both space and time (Padmanabhan 1987).

The latter is reminiscent of some of the recent studies on "quantum clocks" which play an important role in the analysis of time in the canonical quantisation of gravity. We will return to this topic in §5.

Another example is the suggestion by Greensite (1990) that a minimum length may arise in the context of a lattice approach to quantum gravity. And of course there are several indications that string theory may lead to a natural minimum length (§3.3).

Note that the idea of a minimum length can have several different meanings. One possibility is that a length can be *measured* only to an accuracy of the Planck value (presumably specified with respect to some background metric?) but that the underlying model of a continuum spacetime is nethertheless still valid. A more iconoclastic view is that the existence of a minimum length/time signals the *breakdown* of the continuum picture itself.

Ideas like this cast further shadows over the idea of trying to construct a theory of quantum gravity by quantising the point fields of classical general relativity. In particular, if time cannot be measured, or defined, to an accuracy greater than the Planck time, one needs to rethink the significance of "equal-time" commutators and the general quantum mechanical idea of a complete commuting algebra of "simultaneously" measurable observables.

The breakdown of the simple continuum picture would be even more serious. It suggests that the Diff($\mathcal{M}$) invariance of the theory is only a coarse-grained feature—an idea that is difficult to implement in practice, partly because of the absence of any finite-dimensional approximations to the diffeomorphism group. This also causes serious problems in the Regge calculus approach to quantum gravity, and contrasts strongly with the situation vis-a-vis the gauge group in Yang-Mills theory.

Quantum topology too?

But once the idea of dropping the classical continuum picture of spacetime has been entertained, a number of related possibilities come to mind. In particular, general relativity really involves a pair $(\mathcal{M}, g)$ where $\mathcal{M}$ is a manifold and g is a metric, and therefore, if a 'hat' is to be placed on the g, why should not the same be done to $\mathcal{M}$?

The intriguing idea that the manifold itself should be quantised goes back to the seminal ideas of Wheeler (1964). However, it is far from clear what the symbol $\widehat{\mathcal{M}}$ is supposed to mean. The mathematical framework of general relativity has a hierarchical form that can be written symbolically as

$$\text{set} \rightarrow \text{topology} \rightarrow \text{differential structure} \rightarrow \text{Lorentzian metric} \tag{2.4.6}$$

in which each step represents a structure superimposed on the previous one in the chain. *A priori*, a 'hat' might be inserted at any of these stages. The least adventurous approach is to keep $\mathcal{M}$ fixed and just quantise the metric. But there are other possibilities. For example:

1. One might remain within the category of differential geometry but let the manifolds become part of the quantum structure. This idea was developed at length by Hawking (1978,1979) and collaborators in the context of his "Euclidean" quantum gravity programme which is based on the use of path integrals over Riemannian, rather than Lorentzian, metrics. Thus a typical quantity would be something like

$$Z = \sum_{\mathcal{M}} \int_{\text{Riem}(\mathcal{M})} \mathcal{D}[g]\, e^{-\int_{\mathcal{M}} R(g)\sqrt{g}\, d^4x} \tag{2.4.7}$$

where the integral is over the set Riem($\mathcal{M}$) of all Riemannian metrics on $\mathcal{M}$ and the sum [9] is over all four-manifolds $\mathcal{M}$. A recent application of this idea has been the theory of wormholes and their possible implications for determining some of the basic coupling constants (*e.g.*, the cosmological constant) of nature (Coleman 1988, Hawking 1988, Hawking and LaFlamme 1988).

   The use of topologically complicated manifolds $\mathcal{M}$ raises intriguing possibilities, one of which is that information may be lost if, for example, particles fall into the event horizon of a (virtual) black hole. An observer external to the black hole would interpret this as a transition from a pure quantum state to one that is mixed, leading to what Hawking called the "\$-matrix": the pure→mixed analogue of a normal S-matrix (Hawking 1982, 1984).

2. A less ambitious project might be to keep the topology of $\mathcal{M}$ fixed but to allow quantum fluctuations in the different differential structures that are compatible with this topology. It is known that four-dimensional manifolds (including

[9] In practice it is very difficult to know what this means, not least because it is not possible to classify four-manifolds in any algorithmic way.

Minkowski space) do admit such exotic structures although, to the best of my knowledge, no one has yet proposed a scheme for quantising them.

3. A much more adventurous approach is to quantise at the topology stage. This means constructing a theory in which the basic point set $\mathcal{M}$ is fixed but where the topology (in the technical sense of a collection of open sets) is quantised. This involves quantum fluctuations over a much wider class of topological space than those that can carry the structure of a differentiable manifold (*e.g.*, Isham 1989, 1990, 1991). A more modest version of this scheme is to restrict attention to those topological spaces whose topology can be obtained from a distance function (Alvarez 1988, Alvarez *et al* 1990, Isham *et al* 1990).

4. Perhaps the most radical idea of all is to place the 'hat' on the first entry in the chain in (2.4.6) and to quantise the set of spacetime points. This might be interpreted in various ways.
   a) It could mean that the *number* of points in the set $\mathcal{M}$ is subject to quantum fluctuations. This is technically feasible in the topological quantisation scheme mentioned above and is particularly relevant in models of spacetime that employ a finite number of spacetime points in any bounded region.
   b) A more radical proposal is to interpret the "quantisation of points" as saying that the very concept of a spacetime point loses its meaning.

   This might perhaps be suggested by the ideas of a minimum length mentioned earlier. Another motivation stems from the long-standing suggestion of Penrose (1975) who, in the context of his twistor programme, advocates a projective picture in which the null-lines in spacetime are the primary entities and with a spacetime 'point' defined to be the intersection of a family of such lines. If the quantisation of gravity leads to quantum fluctuations in the null-cones, then the null-lines will be spread out and the "intersection" will no longer correspond to a single point.

Ideas on quantising sets, topologies and manifolds can be (and have been) developed in a number of directions and there is a substantial gnostic underworld of people (including myself) who work in this area. However, these ideas tend to be very speculative and rather difficult to relate to conventional approaches to quantum gravity. For this reason I have elected not to develop the theme further in these notes and I will concentrate instead on the more "respectable" conceptual and geometrical problems in quantum gravity.

However, whether an approach be speculative or mainstream, it is clear that the idea of "quantising" classical general relativity opens up a number of important issues. In particular, a major question in any approach of this type will always be:

How are the classical concepts of space and time *modified* in the transition *to* the quantum theory?

2.4.2 Quantise Some Other Theory That Can Give General Relativity as the Low-Energy Limit

I have spent some time discussing the issues that arise in "quantising" the classical theory of general relativity since this is the approach to quantum gravity that has received most attention over the years. Another possibility is to remain with the idea of quantising a classical system but start with something other than general relativity. The key step here is to find a system whose quantum theory is well-defined and which yields classical general relativity as a low-energy limit, even though that is not the starting point.

A good example of a scheme of this type is the "induced gravity" programme which was popular in the early 1980s. In this approach, the action principle includes the usual coupling of matter fields to the metric tensor, but *not* the Einstein-Hilbert action of pure gravity. The crucial idea is that the latter will arise as a counterterm from the quantised matter fields although the gravitational field itself is *not* quantised.

A much more sophisticated example is the theory of closed superstrings in which the graviton occurs as just one of an infinite number of particles associated with the quantised string (§3.2).

Note that theories of this type still involve an act of "quantisation", but the classical theory that is quantised is not general relativity. This is particularly clear in the Polyakov approach to string theory in which, genus by genus in the target space, one is faced with the quantisation of a conventional, two-dimensional non-linear σ-model.

The central philosophy of such approaches is that general relativity is only a low-energy "phenomenological" theory; rather as the standard four-dimensional non-linear σ-model can be regarded as an effective Lagrangian for calculating the low-energy results of QCD. Consequently, it is *wrong* to assign any fundamental operator status to the metric tensor. If correct, this position means that much past (and current) work in quantum gravity was (and is) very misguided!

The possibility also arises that if the metric tensor is only a "phenomenological" concept, then perhaps the same is true of the spacetime manifold? This raises the intriguing question of where such an "effective" manifold could come from. In practice, almost all theories of this type introduce the spacetime manifold as part of the background structure, even though the metric tensor is a derived concept.

In any event, it is clear that a key question in this general approach to the problem of quantum gravity will always be

> How are the classical concepts of space and time *recovered* in the transition *from* the quantum theory?

2.4.3 "General-Relativise" Quantum Theory

The verb "general-relativise" will not be found in any of the standard dictionaries, but the broad idea should be clear enough. Rather than starting with classical general relativity which is then quantised, one begins instead with normal quantum theory and studies the extent to which it can be made compatible with the ideas of general relativity.

This approach is intriguing but it has not been developed much other than the substantial effort that has been devoted to quantum field theory in a fixed spacetime.[10] An exception is the interesting paper by Fredenhagen and Haag (1987) who tackle the problem of making a quantum theory invariant under spacetime diffeomorphisms (see also Bannier 1988).

[10] This has been a very active area of research and one which raises its own conceptual and geometrical problems. However, for reasons of space, in these notes I will only discuss quantum gravity proper.

2.4.4 General Relativity and Quantum Gravity Both Emerge From Something Quite Different

In this perception of the problem of quantum gravity, the relation between the sought-for theory and general relativity (and/or quantum theory) is deemed to be analogous to the relation between quantum theory and classical mechanics. Thus what is called for is a radically new theoretical and conceptual structure which—although reducing to normal quantum theory and general relativity in their appropriate domains—is nevertheless quite different from both.

It is this exciting possibility that entices many physicists to work in quantum gravity in the first place. But it is not easy to effect a fundamental transformation in the foundations of physics, especially in the absence of any experimental or observational evidence to point in the right direction!

2.4.5 The Semi-Classical Option

At the opposite extreme is the idea, dating back to Moller (1962), that perhaps it is not necessary to quantise the gravitational field, but only the matter to which it couples. This "semi-classical" approach can, and has, been interpreted in several different ways, one of which is the induced gravity programme alluded to earlier.

However, the original idea was to study the system of coupled equations

$$G_{\mu\nu}(g) = {}_t\langle\psi|\, T_{\mu\nu}(\widehat{matter}, g)\, |\psi\rangle_t \tag{2.4.8}$$

and

$$i\hbar\frac{d}{dt}\, |\psi\rangle_t = H(\widehat{matter}, g)\, |\psi\rangle_t \tag{2.4.9}$$

in which the source of the gravitational field is the expectation value in some special state $|\psi\rangle_t$ of the energy-momentum of the quantised matter.

There was renewal of interest in this approach in the early 1980s, although the results were not conclusive (for a survey see Kibble 1981). One problem is that if the matter is chosen to be a quantised field, the right hand side of (2.4.8) can only be defined after the quantum energy-momentum tensor has been regularised and renormalised (Randjbar-Daemi *et al* 1980)—a procedure that is ambiguous in the presence of the non-trivial metric field. This process introduces counter-terms that involve higher powers of the Riemann curvature (as does the induced gravity scheme).

The two equations (2.4.8-9) are strongly cross-coupled and the effective equations for the metric tensor $g_{\mu\nu}$ are far more non-linear than those of normal general relativity. This has lead to claims that the system is intrinsically unstable (Horowitz and Wald 1978, Horowitz 1980, Suen 1989a,b). However, these calculations were mainly perturbative in nature and have been challenged recently by Simon (1990); see also Fulling (1990). The question is an important one and should be studied further.

From the perspective of conventional quantum theory, the coupled equations (2.4.8-9) are rather peculiar. In particular, the effective equation for $|\psi\rangle$ is *non-linear*, and hence the superposition principle is lost. Whether or not this is a problem depends on one's attitude towards conventional quantum theory. For example:

1. The theory is difficult to interpret in the absence of the superposition principle.
2. What is the status of different states $|\psi\rangle$ that satisfy the cross-coupled equations (2.4.8-9)? Are we supposed to select one particular state via a theory of quantum cosmology?
3. On the other hand, it has frequently been suggested that linear quantum theory is only an approximation to a deeper non-linear theory and that a non-linear term could provide a physical mechanism for state-vector reduction. Perhaps the semi-classical approach to quantum gravity is the source of this non-linearity?

One might wonder if the whole idea of quantising everything but the gravitational field is simply inconsistent—perhaps along the lines of the famous Bohr-Rosenfeld (1933) argument which showed that the electromagnetic field *has* to be quantised if it is to couple consistently to the current generated by quantised matter. However, Rosenfeld (1963) himself pointed out that there is no direct analogue in the gravitational case since the proof for electromagnetism involves taking to infinity the ratio e/m of the charge e to the inertial mass m of a test particle—a procedure that is impossible in the gravitational case since the analogue of e is the gravitational mass whose ratio to the inertial mass is fixed by the equivalence principle. There have been several attempts since then (*e.g.*, Eppley and Hannah 1977, Page and Geilker 1981) to clarify the situation but it is still somewhat unclear.

The semi-classical approach has reappeared in recent years in the guise of a Born-Oppenheimer/WKB approximation to quantum gravity. Once again, the quantum matter effectively couples to a classical gravitational field but with the difference that, unlike (2.4.8-9), this is only intended to be an *approximation* to the, as yet unknown, full theory of quantum gravity. We will return later to the WKB approximation in the context of the problem of 'time' in canonical quantum gravity (§5).

2.5 Other Prima Facie Questions in Quantum Gravity

There are many other questions that arise when one considers carefully what is involved in trying to construct a quantum theory of gravity. For example, what role is played by *complex* metrics, or those that are *degenerate*? Both possibilities have arisen several times in recent work on quantum gravity, and they pose non-trivial problems for the geometrical interpretation of the theory.

But perhaps the most substantial of the issues that we have not yet addressed is the whole question of quantum cosmology and, in particular, the problems that arise when attempts are made to apply the interpretative framework of quantum theory to the universe in its entirety.

1. The conventional Copenhagen interpretation of quantum theory places great emphasis on the role of "measurement" and "probability", often interpreted in a relative frequency sense. But who can stand outside the entire universe to measure it? And what is an ensemble of universes?
2. Theories of the quantum creation of the universe (QCU) aspire to construct a *unique* quantum state in terms of some simple quantum boundary conditions near the big-bang. Is such an idea compatible with the standard uses of quantum theory?
3. As long as we do not probe too deeply, the world around us is distinctly classical. How did such a world emerge from a totally quantum mechanical origination region?
4. QCU theories involve the idea of the "beginning" of time. Is such a concept compatible with either general relativity or conventional quantum theory?
5. Is there a potential Gödel-type paradox in proposing a theory that, if correct, must presumably predict itself in some way?

These questions are very non-trivial and we will not attempt to address them until the final section of the notes.

3. The Particle Physics Approach

3.1 Quantisation of General Relativity

In this section we will consider some of the approaches to quantum gravity that derive from the community of elementary particle physicists. The discussion is brief and is focused on the conceptual and geometrical issues that arise in schemes of this type. More detailed accounts of the technical material can be found in the recent review by Alvarez (1989), two of my own reviews (Isham 1985, 1987), and in many of the older accounts of quantum gravity (*e.g.*, the various articles in Isham *et al* 1975, 1981).

3.1.1 The Graviton

In the particle physics approach to quantum gravity, the basic entity is the *graviton*—the quanta of the gravitational field. This particle is perceived as propagating on a background Minkowski spacetime, and—like all elementary particles—is associated with a specific representation of the Poincaré group which is labelled by its mass m and spin s. The possible values of m and s are sharply limited by the functions which the graviton is to serve:

1. The t-channel exchange of a particle of mass m can give rise to a static force of the form e^{-mr}/r^2 where r is the distance between the two particles. Thus the graviton must be massless to secure the usual gravitational inverse-square law.
2. The spin of a graviton cannot be half-integral because the Pauli exclusion principle makes it impossible to construct a classical-sized field from a coherent superposition of fermions.
3. Weinberg (1964, 1965) showed that a particle whose spin is greater than two will not produce a static force, and hence the only options are $s = 0, 1$ or 2. However, $s = 1$ gives a repulsive force between like particles (*e.g.*, the spin-1 photon serves this function in electromagnetism), and therefore only $s = 0$ or $s = 2$ is possible.

Zero spin is associated with a scalar field $\phi(x)$ whereas $s = 2$ comes from a (symmetric) Lorentz tensor field $h_{\mu\nu}(x)$. The obvious implication is that $s = 0$ and $s = 2$ correspond respectively to Newtonian gravity and general relativity.

The general theory of relativistic (*i.e.*, special relativity) wave equations shows that a free, massless spin-2 field $h_{\mu\nu}$ must satisfy the field equation

$$h_{\mu\nu,\alpha}{}^{\alpha} - h_{\mu}{}^{\alpha}{}_{,\alpha\nu} - h_{\nu}{}^{\alpha}{}_{,g\mu} + h^{\rho}{}_{\rho,\mu\nu} + \eta_{\mu\nu}(h^{gg\beta}{}_{,\alpha\beta} - h^{\alpha}{}_{\alpha,}{}^{\beta}{}_{\beta}) = 0 \tag{3.1.1}$$

up to an arbitrary redefinition of the form

$$h_{\mu\nu} \rightarrow h_{\mu\nu} + \lambda\eta_{\mu\nu}h^{\alpha}{}_{\alpha} \tag{3.1.2}$$

with $\lambda \neq 1gver4$ (to avoid the new field being traceless). Note that all indices in these expressions are raised and lowered with the Minkowski metric $\eta_{\mu\nu}$.

An important property of the field equations (3.1.1) is their invariance under the gauge transformations

$$h_{\mu\nu} \rightarrow h_{\mu\nu} + \xi_{\nu,\mu} + \xi_{\mu,\nu} \tag{3.1.3}$$

where $\xi_{\mu}(x)$ is an arbitrary Lorentz vector field. This invariance is a consequence of the massless nature of the graviton and is necessary to project out the lower spin ghosts which are otherwise associated with the tensor field $h_{\mu\nu}$.

3.1.2 The Derivation From General Relativity

The connection of the above with general relativity is as follows. If we take the Einstein-Hilbert action

$$S[g] = \frac{1}{\kappa^2}\int R(g(x))(-\det g(x))^{\frac{1}{2}}\, d^4x \tag{3.1.4}$$

and write the metric tensor as

$$g_{\mu\nu}(x) = \eta_{\mu\nu} + \kappa h_{\mu\nu}(x), \tag{3.1.5}$$

then the lowest order (in κ) contribution to the field equations for $h_{\mu\nu}(x)$ is precisely (3.1.1). Furthermore, the redefinition (3.1.2) simply corresponds to using $g_{\mu\nu}(-\det g)^{\lambda}$ as the basic field rather than $g_{\mu\nu}$— a change that should have no effect on on-shell S-matrix elements.

The gauge transformation (3.1.3) also has a simple geometrical interpretation: it is just the effect induced on $h_{\mu\nu}$ (via (3.1.5) by an infinitesimal diffeomorphism of Minkowski space generated by the vector field ξ.

As was remarked in §2, several significant advantages are gained by introducing a background Minkowski metric and hence reducing the problem of quantum gravity to one of conventional quantum field theory. For example:

1. The background metric provides a reference framework in which to discuss "short-distance" behaviour, operator-product expansions, regularisation, and related topics.
2. The background metric gives a fixed causal structure which can be used to define
   (a) microcausal spacetime commutation relations for spacetime fields;
   (b) equal-time commutation relations for canonical fields;
   (c) a good notion of "time-ordering" for use in a functional integral or general LSZ formalism.
3. Some of the prima facie problems with Diff($\mathcal{M}$) invariance are converted into standard problems for gauge-invariant quantum field systems. These can be treated in ways that are familiar from Yang-Mills theory (gauge-fixing, ghost particles, BRS techniques *etc*).
4. The underlying Minkowski structure can be used to give a sensible definition of physical observables that is related to the *asymptotic* behaviour of the fields. The crucial point is that the gauge group generators ξ have compact support and therefore do not affect the fields asymptotically. The use of asymptotic observables played an important role in Bryce DeWitt's seminal investigations into the structure of spacetime-covariant quantum gravity (DeWitt 1965, 1967a,b).

3.1.3 Geometric and Conceptual Objections

Unfortunately there are several objections to the introduction of this background structure which throw the entire scheme into some doubt:

1. There is no reason to suppose that the causal structure of the Minkowski metric is the physically correct one. Indeed, it has often been suggested that a non-perturbative treatment could lead to light cones that are displaced from those of

the background Minkowskian space. The status of the initial microcausal structure is then uncertain.

2. The Minkowskian background also fixes the *topology* of spacetime to be that of a trivial vector space. It is thus particularly difficult to discuss cosmological problems, spacetime singularities, black holes and event horizons, or any other features of classical general relativity that involve non-trivial topological structure. For example, it would not be easy to construct a theory of the quantum creation of the universe in terms of a few gravitons propagating in Minkowski space!
3. The expansion (3.1.5) is a poor one when viewed from the geometrical perspective of classical general relativity. In particular, the object on the left hand side will only be a genuine metric tensor (*i.e.*, an invertible matrix with signature (-1,1,1,1)) for small values of the field $h_{\mu\nu}(x)$. However, naive quantisation methods involve, for example, integrating over *all* values of $h_{\mu\nu}$. In effect, rather than quantising on the space PRiem($\mathcal{M}$) of pseudo-Riemannian metrics, we are only quantising in the tangent space to the particular metric $\eta_{\mu\nu}$.

This latter problem is reminiscent of the use of normal coordinates in a non-linear σ-model and suggests that the quantum theory cannot be globally correct. Even if a parametrisation is chosen for which the values of $h_{\mu\nu}$ are unrestricted (*e.g.*, $g_{\mu\nu}(x) = \big(\exp h(x)\big)_{\mu\nu}$), it is very unclear if, or how, one should "patch" together the quantisations on the different tangent spaces of PRiem($\mathcal{M}$) (assuming it is possible to do the quantisation around any background metric other than $\eta_{\mu\nu}$). [11] To see how non-trivial this problem can be, consider the quantisation of a system whose phase space $\mathcal{P}$ is the (compact) two-dimensional sphere S^2. The Lie group $SU(2)$ acts transitively and symplectically on this space, and there are good arguments for saying that a correct quantisation of the system involves finding an irreducible representation of this group. In other words, $SU(2)$ plays the same role for $\mathcal{P} = S^2$ as does the Weyl-Heisenberg group for $\mathcal{P} = \mathbb{R}^2$. However, note that each such representation of $SU(2)$ has a *finite* dimension whereas the non-trivial representations of the Weyl-Heisenberg group are all infinite dimensional.

[11] Of course, background field methods are apparently designed to perform a quantisation around any background. However, this "quantisation" is rather formal and cannot be made rigorous in terms of a unique Hilbert space and identification of, say, particle observables. This is the unsolved (and probably insoluble) general problem of quantum field theory in a curved background.

Now imagine performing a "tangent-space" quantisation on S^2. Each tangent space of S^2 is a copy of $\mathbb{R}^2$ and is therefore naturally quantised using the Weyl-Heisenberg group and its infinite dimensional representation. Thus one is faced with the problem of patching together a collection of infinite dimensional quantum theories (one for each tangent plane on S^2) in such a way as to reproduce the finite-dimensional state space of the globally-correct quantisation. This is not a trivial task!

3.1.4 Non-Renormalisability

In spite of these reservations there is no doubt that, had this approach to quantising gravity been successful, the resulting theory would have been of intense interest. However, as is well-known, the theory is not renormalisable and hence fails, even in its own terms. This result is not very surprising. If the expansion (3.1.5) is inserted into the Einstein-Hilbert action (3.1.4), the ensuing Lagrangian for $h_{\mu\nu}(x)$ has interactions that are non-polynomial, derivatively-coupled, and with a dimensional coupling constant κ—each of which is a recipe for non-renormalisability for a quantum field theory in four spacetime dimensions.

Early power-counting estimates indicated that the theory should indeed be non-renormalisable, and this was rendered more plausible by a number of explicit calculations made in the mid 1970s of the one-loop graphs for gravity plus various matter fields (for purely kinematical reasons, the pure gravity one-loop graphs are finite on-shell). But it was not until quite recently that lengthy calculations by Goroff and Sagnotti (1985, 1986) of the two-loop contributions finally disposed of the faint remaining hope that gravity might have been well-behaved, notwithstanding all the evidence to the contrary.

Tweak the theory

Various schemes have been suggested whereby this sad state of affairs might be averted. A conservative view is that the infinite quantities can be controlled by "tweaking" the theory slightly using some, not too drastic, modification of conventional general relativity.

One suggestion which has received persistent attention over the years is to use an "$R + R^2$" Lagrangian, *i.e.*, augment the original Einstein-Hilbert action with terms involving the square of the Riemann curvature. The linearised theory becomes fourth-order in derivatives and, consequently, the free propagator changes from a $1/p^2$ form to something like

$$\frac{1}{p^2(\alpha + p^2)}. \tag{3.1.6}$$

This faster fall-off in momentum reduces the severity of the ultraviolet divergences and leads to a theory that is renormalisable (Stelle 1977) and, in some versions (Fradkin and Tseytlin 1981, 1982), even asymptotically free (for further details see the review articles cited earlier). Unfortunately, the momentum behaviour in (3.1.6) also produces ghosts and non-unitarity, and the community at large has not been persuaded to accept the suggestion (Tomboulis 1984) that these are merely an artifact of an inappropriate form of perturbation theory.

The theory of supergravity was another major programme aimed at removing the ultraviolet divergences with the hope that the infinities produced by the additional fermi loops would cancel those of the bosonic graviton loops. Some infinities can indeed be removed in this way, but the discovery of supersymmetric counterterms (at 3 loops for $N = 1$, and at 8 loops for $N = 8$) blighted this particular approach to constructing a fundamental theory of quantum gravity. [12] One of the attractive feature of this approach is that, had it worked, it would have yielded a definite prediction for the fundamental matter Lagrangian which must be attached to general relativity. This could have provided a basis for a true grand unified field theory.

Non-perturbative approaches

Another frequent response to the non-renormalisability of quantum gravity has been the claim that the problem stems from the use of weak-field perturbation theory. This statement has several interpretations, one of which is that it is still feasible to use a Minkowski background and associated quantum field theory but this must be augmented with some sort of non-perturbative [13] summation of the terms that arise. The various attempts/suggestions that have been made along these lines include:

---

[12] To the best of my knowledge, there are no explicit calculations showing that these counterterms do actually occur. So a miracle could still happen, although the odds must be strongly against it in the absence of any good theoretical reason why the coefficients of the counterterms should vanish.

[13] In every branch of physics, almost any calculation that is feasible involves *some* sort of approximation method, and the same is doubtless true of quantum gravity. Thus "non-perturbative" does not mean literally "without the use of any sort of perturbative technique" but only that weak-field expansions like (3.1.5) are to be avoided.

1. Trying to sum infinite sets of Feynman graphs to give a finite result. This has often been seen as an implementation of the old idea (Pauli 1967) that ultraviolet divergences can be removed by including quantum gravity effects which lead to a smearing of the lightcone and an effective cut-off at the Planck length.
2. Various sorts of $1/N$ expansion.
3. Large coupling constant expansions in the inverse of κ.
4. Weinberg's idea of *asymptotic* safety in which the renormalised coupling constants are forced to lie in a finite-dimensional subspace of the infinite-dimensional space of all possible values (Weinberg 1979).
5. Klauder's work on techniques for handling non-renormalisable quantum field theories in general (*e.g.*, Klauder 1979).

For our purposes, the most relevant property of approaches of this type is that they still use a Minkowskian background structure. They are therefore subject to the objections raised earlier, with the caveat that some of these may be addressed (albeit indirectly) by the non-perturbative nature of the methods.

However, another interpretation of the failure of weak-field perturbation theory is that the entire programme is wrong, including the use of a Minkowski background. Some of the more interesting suggestions in this direction (and still involving the idea of "quantising" the classical theory of general relativity) are as follows.

1. There have been a number of attempts to set up lattice versions of quantum gravity. Some of these use the type of lattice employed in Yang-Mills theory; others employ simplicial approximations and Regge Calculus. A number of technical difficulties arise in the process (especially in regard to invariance under diffeomorphisms) but perhaps the most serious worry is the absence of any *a priori* reason for supposing the continuum limit to be well behaved.

   However, serious attempts of this sort are still made, one of the most striking of which is DeWitt's ongoing programme for evaluating a Lorentzian functional integral version of quantum gravity with the aid of large scale computing techniques.
2. The canonical theory of quantum gravity has also been studied at length using techniques which avoid weak-field perturbation theory. These include minisuperspace approximations and various schemes involving the Ashtekar variables.

The work on canonical quantum gravity is very important but, since it cannot be said to fall under the category of a "particle physics approach", discussion is deferred to later (§4). On the other hand, string theory certainly does come under this heading, and undoubtedly constitutes one of the more important current approaches to the problem of quantum gravity.

3.2 String Theory

The idea that a quantum theory of gravity can be constructed using closed strings emerged in the early 1970s from studies of the zero-slope limit of the dual resonance model (Scherk and Schwarz 1974, Yoneya 1974) and is frequently cited as one of the main reasons for studying string theory. It is therefore surprising that only a small fraction of the huge literature on superstring theory has been devoted directly to the types of issue that are of interest in quantum gravity. However, even this residue is still quite large and I will restrict myself here to some of the main ideas and, in particular, to those that relate directly to the issue of geometrical and/or conceptual problems in quantum gravity. A very useful review of some of these matters is Horowitz (1989).

3.2.1 The Polyakov Action

In its current formulation, string theory is an example of the second class of approach discussed in §2, *i.e.*, it still involves a "quantisation" of a given classical system but the system that is quantised is not general relativity. Instead we have the Polyakov action

$$S[q,X] := \frac{1}{4\pi\alpha'}\int_W q^{ij}(\sigma)\partial_i X^\mu(\sigma)\partial_j X^\nu(\sigma) g_{\mu\nu}(X(\sigma))\big(\det q(\sigma)\big)^{\frac{1}{2}}\, d^2\sigma \tag{3.2.1}$$

where q_{ij} is a metric on the two-dimensional manifold W (known as the "world-sheet"), $X : W \rightarrow \mathcal{M}$ are the string fields which map W into the spacetime manifold $\mathcal{M}$, and $g_{\mu\nu}$ is a background metric on $\mathcal{M}$. The constant α is related to the string tension and is assumed to be of the order of the Planck length.

The classical system is invariant under the conformal transformations $q_{ij}(\sigma) \mapsto F(\sigma)q_{ij}(\sigma)$, $F(\sigma) > 0$, but this will be lost when the two-dimensional fields q_{ij} and X^μ are quantised unless the metric $g_{\mu\nu}$ satisfies an equation which is effectively the vanishing of the trace of the energy-momentum tensor of the two-dimensional quantum field theory. These equations have the form

$$0 = R_{\mu\nu} + \frac{1}{2}\alpha' R_{\mu\alpha\beta\gamma} R_{\nu}{}^{\alpha\beta\gamma} + \ldots \tag{3.2.2}$$

where ... refers to higher-order powers and derivatives of the Riemann curvature. Other background spacetime fields may also be introduced (in particular, massless dilaton and 2-form fields) and generate modified forms of (3.2.2).

Conformal invariance also requires the dimension of $\mathcal{M}$ to be equal to some critical value which depends on the background fields that are present as well any two-dimensional spinor fields that may be added to the system (*e.g.*, in a superconformal theory).

The field equations (3.2.2) are the string theory replacement for the classical equations of general relativity. Note that, in so far as $g_{\mu\nu}$ is determined by a solution to a set of dynamical equations, it is not strictly correct to think of it as part of the background structure. On the other hand, it does function in this way whenever calculations are made of quantum fluctuations around this metric.

3.2.2 Exact Solutions and Spacetime Singularities

A number of attempts have been made to simplify the field equations (3.2.2) and/or to construct exact solutions. For example:

1. The number of extra terms (*i.e.*, over and above conventional general relativity) can be drastically reduced by imposing invariance under world-sheet supersymmetry transformations. The presence of one supersymmetry generator removes the R^2 and R^3 terms, and two such generators means there are no higher-order terms at all. The latter condition can only be satisfied if the metric $g_{\mu\nu}$ is complex and Kähler. An example is the well known Calabi-Yau space solutions of Candelas *et al* (1985).
2. Exact solutions can be found if the target space $\mathcal{M}$ is a group manifold (Witten 1984).
3. Plane wave solutions have been found Güven (1987) and Amati and Klimcik (1989).
4. Solutions of a cosmological type have been found by Antoniadis *et al* (1988, 1989).

From the viewpoint of traditional general relativity, some of the most interesting work is that of Horowitz and Steif who have discussed a number of exact solutions that contain genuine spacetime singularities (Horowitz and Steif 1990a, Horowitz 1990a,b).

An important question in this context is precisely what is meant by a "singularity". The usual definition in general relativity involves the properties of geodesics, which can be thought of as the paths of hypothetical test particles. However, this seems inappropriate in a theory in which the basic entities are *loops*, and Horowitz and Steif have developed a definition of a singularity which refers to the properties of a first-quantised string propagating in the spacetime, rather than a point particle.

3.3 A Minimum Length in String Theory?

From one perspective, the existence of singular solutions to the string-induced equations for $g_{\mu\nu}$ is perhaps disappointing since one of the hopes for quantum gravity is that it will remove the spacetime singularities that are such an endemic feature of classical general relativity. The string result may seem all the more surprising because of the persistent claim that there is a minimum length in the theory; indeed, that is often said to be the basic reason why the quantum amplitudes should be free of the ultraviolet divergences that plague conventional quantum field theory. The idea of a minimum length comes from a variety of sources:

1. The string is basically an extended object whose internal quantum fluctuations stop it degenerating to a point.
2. In some of the compactification schemes, the scattering amplitudes are invariant under the transformation $R \rightarrow \frac{1}{R}$ where R is the radius of the internal space in units of the Planck length. This suggests the existence of a minimum value for this radius.

   However, it is not clear if this is property is generic or if is only a feature of some special solutions. Horowitz and Steif (1990c) have recently shown that a compactified form of the plane-wave solutions mentioned above is *not* invariant under this transformation, although of course this does not rule out the existence of a more generalised form of duality operation.
3. There have been a series of calculations of high energy scattering amplitudes in string theory. Some of these are at large s and fixed angle (Gross and Mende 1987,

1988) while others are at small t (*i.e.*, large distance) and large s (Amati *et al* 1987, 1988, 1989). The results suggest the existence of a minimum length that can be reached by the string, and which has been associated with some sort of generalised uncertainty relation in the form (Veneziano 1989a)

$$\Delta x > \frac{\hbar}{\Delta p} + \alpha' \Delta p \tag{3.3.3}$$

See also Veneziano (1989b), Fabbrichesi and Veneziano (1989) and Konish *et al* (1990).

It should be noted that, as with previous discussions about a minimum length, it is not clear whether the implication is that strings can only *probe* to a minimum length (but leaving intact the notion of a spacetime continuum) or whether the phenomenon signals the breakdown of the entire continuum picture. This issue cannot be decided from within the Polyakov approach since the presence of a continuum manifold $\mathcal{M}$ is part of the background structure.

The central difficulty is the same as that which arises when trying to assess the significance of the existence of singular solutions to the effective field equations. Namely, the next step in the string calculations involves the computation of quantum fluctuations around the classical background solution, and in that sense we are back once more to the weak-field scenario of the old approaches to quantum gravity. It is true that many of the high-energy calculations cited above involve non-perturbative methods for summing the various contributions but, nevertheless, what is lacking at the moment is any complete, non-perturbative alternative to the Polyakov approach which could be used to tackle the types of problem that are of real interest in quantum gravity (spacetime singularities, quantum topology, quantum cosmology *etc*).

For example, there have been several suggestions that there may be a phase change in the theory at the Planck length, with the graviton becoming a meaningless concept. But this is a very difficult thing to study using the current formulation of string theory.

The perturbative results are important and encouraging, but in some respects the situation is similar to what would have been the case had the original particle physics approach been successful. In particular, many of the challenges faced by modern string theory are reflected in the very non-trivial question:

> What would have been the next stage of the theoretical development if ordinary quantum gravity had been perturbatively ultraviolet finite?

One can only hope that the on-going attempts to face some of these challenges will meet with success in the years to come.

4. The Canonical Quantisation of Gravity

4.1 Introductory Remarks

4.1.1 Advantages of the Canonical Approach

The development of the canonical quantisation of gravity has been mainly at the hands of physicists with backgrounds in general relativity. As a result, the techniques employed differ considerably from those discussed in the previous chapter and are likely to be relatively unfamiliar to a particle physics audience. My account of this material will therefore be more detailed than was the case for the "graviton" approaches.

There are several reasons why the canonical theory may be preferred to approaches based on more conventional quantum field theory. For example,

1. Many of these "conventional" approaches use formal functional integral techniques in which a number of serious difficulties are swept under the carpet as problems to do with the integration measure, the contour of integration *etc.* On the other hand, canonical quantisation is usually discussed in an operator-based framework, with the advantage that problems appear in a more explicit (and honest!) way.
2. It is easier to develop techniques in canonical quantisation that do not depend on a background metric. This offers new possibilities for non-perturbative approaches to quantum gravity.
3. For related reasons, canonical quantisation is better for discussing quantum cosmology, spacetime singularities and similar topics.
4. Canonical methods place more emphasis on the geometrical structure of general relativity. In particular, it is easier to address the issue of the extent to which such structure is, or should be, maintained in the quantum theory.

5. Many of the deep conceptual problems in quantum gravity become more transparent in a canonical approach. This applies in particular to the problems of 'time' and the applicability of the interpretative framework of conventional quantum theory.

It should be emphasised that some of these advantages arise only in comparison with weak-field perturbation theory, and this does *not* mean that a canonical scheme is intrinsically superior to one in which spacetime fields are used from the outset (*i.e.*, the so-called "covariant" approaches). For example, there exist *bona fide* approaches to quantum gravity that involve the calculation of a functional integral like

$$Z = \int \mathcal{D}[g]\, e^{iS(g)} \tag{4.1.1}$$

using methods other than weak-field perturbation theory. Indeed, many of the non-perturbative techniques that have been developed in canonical quantum gravity have a fairly obvious translation into a framework of this type.

On the other hand, there *are* fundamental differences between the canonical and the covariant approaches to general relativity—especially in regard to the ultimate status of the concepts of space and time—and it is necessary to commit oneself, at least partially, to one position or the other. However, there is no conclusive technical or conceptual argument to decide the point and it is largely a question of personal taste.

4.1.2 Prima Facie Problems with a Canonical Theory of Gravity

A canonical approach to a standard field theory has three important ingredients:

1. Canonical variables are defined on spacelike hypersurfaces. All the points on such a surface are at "equal-time" and the dynamical equations specify how the canonical variables evolve from one equal-time hypersurface to another.
2. The quantised canonical variables obey canonical commutation relations on these surfaces. In particular, if ϕ is the canonical field variable then

$$[\,\hat{\phi}(x), \hat{\phi}(x')\,] = 0 \tag{4.1.2}$$

for all points x, x' in such a surface.

3. If the theory has a gauge group then some of the potential canonical variables are manifestly non-dynamical and can be removed before quantisation takes place.

The gauge group of general relativity is Diff($\mathcal{M}$) and one of the important steps in the canonical construction (see below) is to show that, roughly speaking, the four components $g_{0\mu}$, $\mu = 0\ldots3$ of the metric tensor are gauge variables and have no dynamical evolution. Thus the canonical configuration variable in general relativity is the metric g_{ab} where $a, b = 1, 2, 3$. However, there are several obvious difficulties in trying to apply the ideas of canonical quantisation to g_{ab} and its conjugate:

1. A time parameter in general relativity is one of the coordinates on the spacetime $\mathcal{M}$ and, like all coordinates, it may only be defined in a local region. This could mean that
   a) the time variable t is only defined for some finite range of values for t; or
   b) the subset of points t=const may not be a complete three-dimensional submanifold of $\mathcal{M}$.
2. In a normal field theory, the idea of an "equal-time" hypersurface is synonymous with one that is "spacelike" But the situation in quantum gravity is different. Whether or not a particular hypersurface is spacelike depends on the Lorentzian metric, but this is precisely what is being quantised! This harks back to the discussion of microcausality in §2.4 (cf (2.4.4)) and implies that the meaning, and validity, of the gravitational analogue of (4.1.2) needs some thought.

4.2 Quantum Field Theory in a Curved Background

4.2.1 The Canonical Formalism

Let us start by considering briefly the quantisation of a scalar field ϕ propagating in a background Lorentzian metric $g_{\mu\nu}$ on a spacetime manifold $\mathcal{M}$. The action for the system is

$$S[\phi] = -\int_{\mathcal{M}} \left(\frac{1}{2}g^{\mu\nu}(x)\partial_\mu\phi(x)\partial_\nu\phi(x) + V(\phi(x))\right)\left(-\det g(x)\right)^{\frac{1}{2}} d^4x \tag{4.2.1}$$

where $V(\phi)$ is an interaction potential for ϕ (which could include a mass term).

This system has a well-posed classical Cauchy problem only if $(\mathcal{M}, g)$ is a *globally-hyperbolic* pair (Hawking and Ellis 1973). In particular, this means that $\mathcal{M}$ is topologically equivalent to the Cartesian product $\Sigma \times \mathbb{R}$ where Σ is some three-manifold

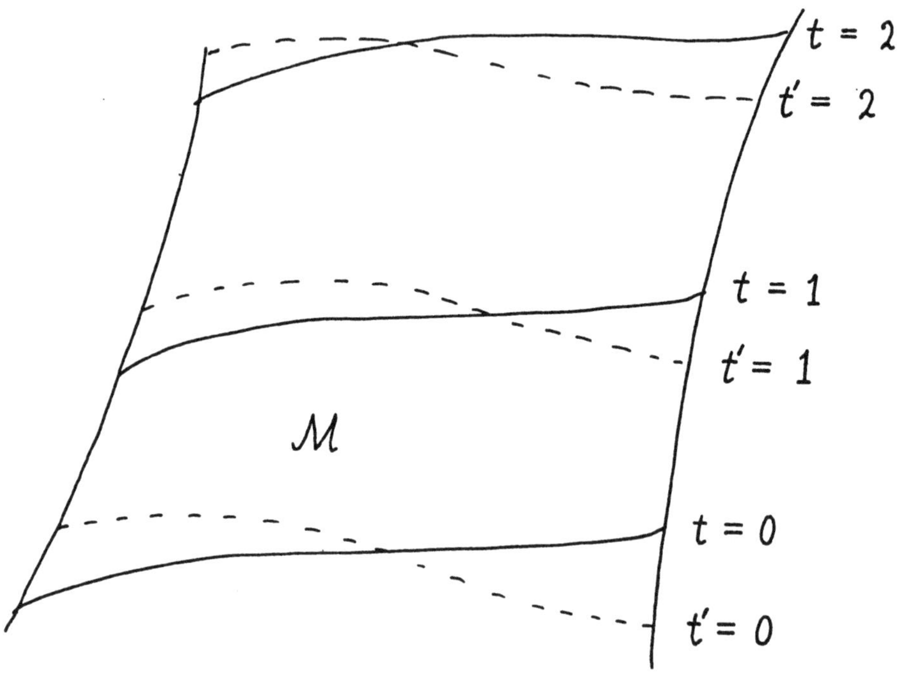

Fig. 4.1. Different global times correspond to different foliations

and $\mathbb{R}$ is a global time direction. We will assume that $(\mathcal{M}, g)$ is such that a global time structure of this sort exists. Technically, we can say that $\mathcal{M}$ has been *foliated* into a one-parameter family of hypersurfaces that are spacelike with respect to the background Lorentzian metric $g_{\mu\nu}$. Note that different choices of such a global time correspond to different foliations (Figure 1.1).

The canonical variables are defined on Σ and consist of the scalar field ϕ and its conjugate variable π (a scalar density) which is essentially the time-derivative of ϕ. Classically, these constitute a well-defined set of Cauchy data and satisfy the Poisson bracket relations

$$\{\phi(x), \phi(x')\} = 0 \tag{4.2.2}$$

$$\{\pi(x), \pi(x')\} = 0 \tag{4.2.3}$$

$$\{\phi(x), \pi(x')\} = \delta(x, x'). \tag{4.2.4}$$

The Dirac δ-function is defined such that the smeared fields satisfy

$$\{\phi(f), \pi(h)\} = \int_{\Sigma} f h \, d^3x \tag{4.2.5}$$

where the test-functions f and h are respectively a scalar density, and a scalar, on the three-manifold Σ.

The formal canonical quantisation of this system follows the usual rules of replacing Poisson brackets with operator commutators so that

$$[\widehat{\phi}(x), \widehat{\phi}(x')] = 0 \tag{4.2.6}$$

$$[\widehat{\pi}(x), \widehat{\pi}(x')] = 0 \tag{4.2.7}$$

$$[\widehat{\phi}(x), \widehat{\pi}(x')] = i\hbar\delta(x, x') \tag{4.2.8}$$

which can be made rigorous by smearing and exponentiating in the usual way.

4.2.2 Representations of the Canonical Algebra

Most treatments of conventional quantum field theory use Fock space representations of the canonical field algebra. These representations are based on the idea of "quanta" of the quantum field and are particularly useful in situations where weak-field perturbative methods are appropriate. However, we are striving to escape from this strait-jacket and it is more productive to develop an approach to the representations of (4.2.6-8) that is not based directly on Fock space.

The starting point is the familiar canonical commutation relations of the standard quantum mechanics of a particle moving in three dimensions. These are

$$[\widehat{x}_i, \widehat{x}_j] = [\widehat{\pi}_i, \widehat{\pi}_j] = 0 \tag{4.2.9}$$

$$[\widehat{x}_i, \widehat{\pi}_j] = i\hbar\delta_{ij} \tag{4.2.10}$$

where $i, j = 1, 2, 3$. Modulo certain mathematical niceties, all irreducible, self-adjoint representations of this algebra are unitarily equivalent to the familiar wave-function representation

$$(\widehat{x}_i\psi)(\underline{x}) := x_i\psi(\underline{x}) \tag{4.2.11}$$

$$(\widehat{\pi}_j\psi)(\underline{x}) := -i\hbar\partial\psi ver\partial x_j(\underline{x}) \tag{4.2.12}$$

on the Hilbert space $L^2(\mathbb{R}^3, d^3x)$ of all functions that are square-integrable with respect to the volume element d^3x. Thus the inner product is

$$\langle\psi_1|\psi_2\rangle = \int_{\mathbb{R}^3} \psi_1^*(x)\psi_2(x)\, d^3x \tag{4.2.13}$$

and the interpretation of a normalised state ψ is that, if a measurement is made of the position of the particle, the probability of finding it in some integrable region $B \subset \mathbb{R}^3$ is

$$\mathrm{Prob}(x \in B; \psi) = \int_B |\psi(x)|^2 \, d^3x. \tag{4.2.14}$$

This suggests that it might be possible to take the state space of the quantum field theory to be a set of functionals Ψ on the space $\mathcal{Q}$ of all classical fields, with the canonical operators defined as

$$(\widehat{\phi}(x)\Psi)[\phi] := \phi(x)\Psi[\phi] \tag{4.2.15}$$

$$(\widehat{\pi}(x)\Psi)[\phi] := -i\hbar\frac{\delta\Psi}{\delta\phi(x)}[\phi]. \tag{4.2.16}$$

Thus the inner product would be

$$\langle\Psi_1|\Psi_2\rangle = \int_{\mathcal{Q}} \Psi_1^*[\phi]\Psi_2[\phi]\, d\mu[\phi] \tag{4.2.17}$$

and

$$\mathrm{Prob}(\phi \in B; \Psi) = \int_{\mathcal{Q}} |\Psi[\phi]|^2 \, d\mu[\phi] \tag{4.2.18}$$

is the probability that a measurement of the field on Σ will find it in the subset B of $\mathcal{Q}$.

This analysis can be made rigorous after smearing the fields with functions from an appropriate test function space, and gives the following results:

1. Equation (4.2.16) only yields a hermitian operator if the measure $d\mu$ on $\mathcal{Q}$ is an infinite-dimensional version of the Lebesgue measure d^3x. However, no such measure exists. Rather, one needs to employ the infinite-dimensional analogue of a weighted measure, with (4.2.16) being replaced by

$$(\widehat{\pi}(x)\Psi)[\phi] := -i\hbar\frac{\delta\Psi}{\delta\phi(x)}[\phi] + i\rho(x)\Psi[\phi] \tag{4.2.19}$$

where $\rho(x)$ is a function that compensates for the weight factor in the measure $d\mu$.

2. The support of a state functional Ψ is typically on *distributions* rather than smooth functions. Thus the Hilbert space is actually $\mathcal{L}^2(\mathcal{Q}', d\mu)$ where $\mathcal{Q}'$ is the topological dual of the infinite dimensional topological vector space $\mathcal{Q}$.

3. There are an infinite number of unitarily inequivalent representations of the canonical commutation relations (4.2.6-8). Each can be written on a Hilbert space of the

form $\mathcal{L}^2(\mathcal{Q}', d\mu)$ with the measure $d\mu$ determining which representation is which. For example, the familiar Fock space representations are associated with Gaussian measures on $\mathcal{Q}'$.

One of the major problems in quantum field theory on a curved background is knowing what representations to choose and what their physical interpretations are.

One consequence of the appearance of the space of distributions $\mathcal{Q}$ is that the value $\widehat{\phi}(x)$ of the field at any particular point $x \in \Sigma$ is likely to be ill-defined. For example, the operator $\widehat{\phi}^2(x)$ acts on $\Psi[\phi]$ as

$$(\widehat{\phi}^2(x)\Psi)[\phi] = (\phi(x))^2\Psi[\phi] \tag{4.2.20}$$

and the right hand side may not exist if Ψ is non-vanishing on fields ϕ that are distributions.

Another source of divergences is the appearance of expressions like $(\pi(x))^2$ in, for example, the hamiltonian of the system. This becomes the operator $-\hbar^2\frac{\delta^2}{\delta\phi(x)\delta\phi(x)}$ which is singular on many functionals Ψ. As we will see, both types of divergence can appear in the canonical quantisation of gravity.

4.3 The Canonical Structure of General Relativity

4.3.1 Introduction of the Foliation

Let us now consider how these ideas extend to the canonical formalism of general relativity itself. The early history of this subject is well-known and culminated in the investigations of Arnowitt, Deser and Misner (*e.g.,* Arnowitt *et al* 1962). However, these original studies involved the selection of a specific coordinate system on the spacetime manifold, and some of the global issues were thereby obscured. A more geometrical, global, approach was developed by Kuchar in a series of papers (Kuchar 1972, 1976a,b,c, 1977) and this will be adopted here. A good general review of canonical quantisation from this perspective is Kuchar (1981).

The starting point is a four-dimensional manifold $\mathcal{M}$ (a model for spacetime) and a three-dimensional manifold Σ which we assume to be compact [14] and which will

[14] This is the usual situation in cosmology. If the space is asymptotically flat then some of the expressions to be used will need to be augmented by surface terms.

play the role of physical three-space. Furthermore, the topology of $\mathcal{M}$ is assumed to be such that it can be foliated by a one-parameter family of copies of Σ. Thus there exists a family of embeddings $X_t : \Sigma \to \mathcal{M}$, with $t \in \mathbb{R}$ and such that the induced map

$$X : \Sigma \times \mathbb{R} \to \mathcal{M}; \quad X(x,t) := X_t(x) \tag{4.3.1}$$

is a diffeomorphism of $\Sigma \times \mathbb{R}$ with $\mathcal{M}$. The space of all embeddings of Σ in $\mathcal{M}$ is denoted $\mathrm{Emb}(\Sigma, \mathcal{M})$ and can be regarded as an infinite-dimensional manifold. [15]

Since X is a diffeomorphism from $\Sigma \times \mathbb{R}$ to $\mathcal{M}$, its inverse $X^{-1} : \mathcal{M} \to \Sigma \times \mathbb{R}$ is also a diffeomorphism and can be written in the form

$$X^{-1}(y) = \big(S(y), T(y)\big) \in \Sigma \times \mathbb{R} \tag{4.3.2}$$

where $S : \mathcal{M} \to \Sigma$ and $T : \mathcal{M} \to \mathbb{R}$. The map T is a global *time function* [16] and gives the natural time coordinate associated with the foliation since $T(X_t(x)) = t$ for all $x \in \Sigma$.

Note that the existence of such a foliation imposes a substantial *a priori* limitation on the possible topologies for the background manifold $\mathcal{M}$. One consequence is that the topology of Σ is also a fixed part of the background structure. In particular, it is not possible to discuss the idea of quantum topology fluctuations of Σ. On the other hand, nothing has been said about any background metric on $\mathcal{M}$, and none is intended.

As we have seen, the foliation provides a natural choice of time variable and we will use this in what follows. However, the foliation is not unique and there exist many other diffeomorphisms of $\Sigma \times \mathbb{R}$ with $\mathcal{M}$. Each corresponds to a possible choice of time function, as illustrated by the two time slicings in Figure 1.1.

If we wanted to be more mathematically precise we should probably restrict our attention to those foliations whose leaves (*i.e.*, equal-time slices) can be deformed smoothly into the leaves of the given foliation X. Similarly, the set of Lorentzian metrics g that might be placed on $\mathcal{M}$ should be limited to those for which there exists at least one foliation whose leaves are all spacelike with respect to g.

---

[15] $\mathrm{Emb}(\Sigma, \mathcal{M})$ is a subset of the space $C^\infty(\Sigma, \mathcal{M})$ of *all* smooth maps from Σ to $\mathcal{M}$. The latter can be equipped with the structure of an infinite-dimensional manifold in several well-known ways and $\mathrm{Emb}(\Sigma, \mathcal{M})$ becomes an open subset. As such, it inherits a differential structure, and this is the one we will use.

[16] Strictly speaking, one should not talk about a 'time' function until $\mathcal{M}$ has been equipped with a Lorentzian metric such that each subspace $X_t(\Sigma)$ is spacelike. Therefore, for the moment, our use of the word is topological rather than metrical.

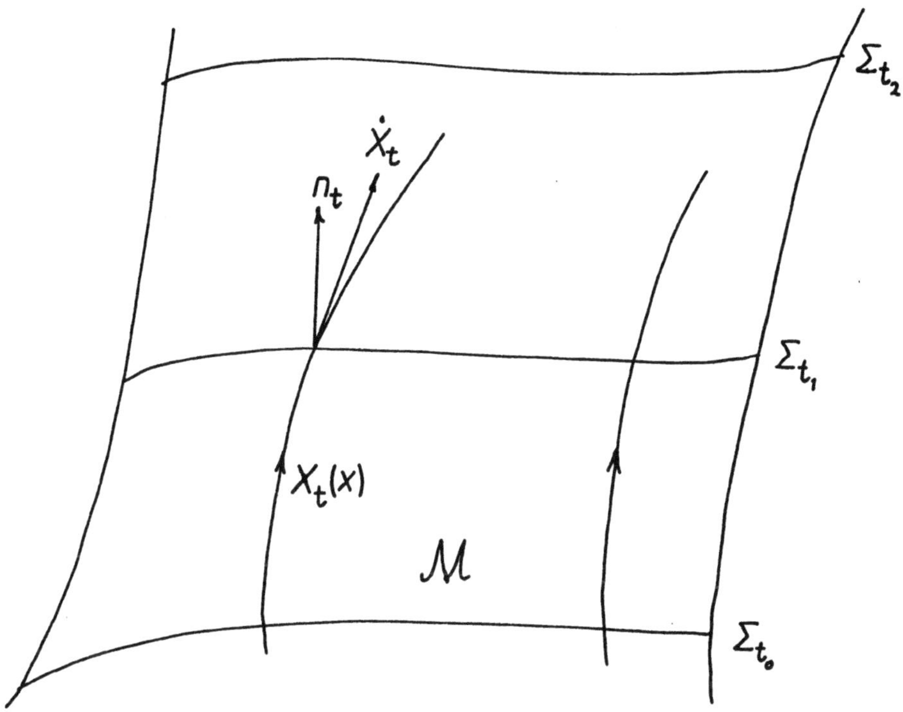

Fig. 4.2. The flow lines of the foliation

4.3.2 The Deformation Vector

The canonical analysis of general relativity proceeds as follows. We fix a foliation $X : \Sigma \times \mathbb{R} \approx \mathcal{M}$ and consider the situation in which $\mathcal{M}$ carries a Lorentzian metric g that satisfies the vacuum Einstein field equations $G_{\mu\nu} = 0$ and is such that each leaf $\Sigma_t := X_t(\Sigma)$ is a submanifold of $\mathcal{M}$ that is spacelike with respect to g. The problem is to find a set of canonical variables for this system and an associated set of first-order differential equations that determine how the variables evolve from one leaf to another.

The map $t \mapsto X_t$ is a curve in $\mathrm{Emb}(\Sigma, \mathcal{M})$ and therefore, for each $t \in \mathbb{R}$, it has a tangent vector, denoted $\dot{X}_t$, which is called the *deformation* vector of the foliation. This vector belongs to the tangent space $T_{X_t}\mathrm{Emb}(\Sigma, \mathcal{M})$ which is the set of all tangent vectors to the infinite-dimensional manifold $\mathrm{Emb}(\Sigma, \mathcal{M})$ at the particular embedding X_t.

From another perspective, for each $x \in \Sigma$ the map $t \mapsto X_t(x)$ is a curve in $\mathcal{M}$. As such, it also has a tangent vector, this time belonging to the tangent spaces of $\mathcal{M}$ (Figure 4.2). Note that, for any particular t, the tangent vector associated with the point $x \in \Sigma$ lies in the tangent space $T_{X_t(x)}\mathcal{M}$ at the point $X_t(x) \in \mathcal{M}$.

This is characteristic of elements that are tangent to function spaces, and in general we have

$$T_f C^\infty(\Sigma, \mathcal{M}) = \{v : \Sigma \to T\mathcal{M} \mid v(x) \in T_{f(x)}\mathcal{M}\} \tag{4.3.3}$$

where $f : \Sigma \to \mathcal{M}$. Note that an element v in $T_f C^\infty(\Sigma, \mathcal{M})$ is a "hybrid" object in the sense that to each point $x \in \Sigma$ it associates a vector $v(x)$ in the tangent space to $\mathcal{M}$ at the point $f(x) \in \mathcal{M}$. In component form this vector would be written as $v^\mu(x)$ where $\mu = 0 \dots 3$ refers to a coordinate system on $\mathcal{M}$.

A function space has cotangent vectors as well as tangent vectors. Specifically, the space of cotangent vectors at the map $f : \Sigma \to \mathcal{M}$ is defined to be

$$T_f^* C^\infty(\Sigma, \mathcal{M}) = \{\ell : \Sigma \to T^*\mathcal{M} \mid \ell(x) \in T^*_{f(x)}\mathcal{M}\}. \tag{4.3.4}$$

In coordinate form, an element $\ell \in T_f^* C^\infty(\Sigma, \mathcal{M})$ associates to each point $x \in \Sigma$ a covariant vector $\ell_\mu(x)$ at the point $f(x) \in \mathcal{M}$.

An important example of an object of this type is the *normal vector* to the foliation at the embedding $X_t \in \mathrm{Emb}(\Sigma, \mathcal{M})$. This is defined to be the unique vector $n_t \in T^*_{X_t}\mathrm{Emb}(\Sigma, \mathcal{M})$ which satisfies the equations

$$n_\mu(x,t) X^\mu{}_{,a}(x,t) = 0 \tag{4.3.5}$$

$$g^{\mu\nu}(X(x,t)) n_\mu(x,t) n_\nu(x,t) = -1 \tag{4.3.6}$$

where $a = 1, 2, 3$ refers to some coordinate system on Σ. Note that the cumbersome expression $(n_t)_\mu(x)$ has been written as $n_\mu(x,t)$. Equation (4.3.5) defines what it means to say that $n_t(x)$ is normal to the hypersurface Σ_t at the point $X_t(x)$ in $\mathcal{M}$. It does not depend on the choice of coordinate system. [17] Equation (4.3.6) is a normalization condition on n and emphasises that this vector is *timelike* with respect to the Lorentzian metric on $\mathcal{M}$.

The normal vector is a functional of both the Lorentzian metric $g_{\mu\nu}$ and the foliation X. This can been seen clearly in the explicit expression

$$n_\mu(x,t) = \left. \frac{T_{,\mu}}{g^{\kappa\lambda} T_{,\kappa} T_{,\lambda}} \right|_{y = X(x,t)} \tag{4.3.7}$$

that relates n to $g_{\mu\nu}$ and the global time function T.

---

[17] In a more abstract notation, (4.3.5) is equivalent to the equation $X_t^*(n_t(x)) = 0$ where $X_t^* : T^*_{X_t(x)}\mathcal{M} \to T^*_x\Sigma$ is the usual pull-back on cotangent spaces.

4.3.3 The Lapse and Shift functions

Any vector in $T_{X_t}\mathrm{Emb}(\Sigma, \mathcal{M})$ can be decomposed into two components, one of which lies along the surface Σ_t, and one that is parallel to n_t. In particular, we can write

$$\dot{X}^\mu(x,t) = N(x,t)g^{\mu\nu}(X(x,t))n_\nu(x,t) + N^a(x,t)X^\mu{}_{,a}(x,t) \tag{4.3.8}$$

where N and N^a are known respectively as the *lapse* function and the *shift* vector. We see that $N_t \in C^\infty(\Sigma, \mathcal{M})$ where $N_t(x) := N(t,x)$, and $N_t \in T_{X_t}\mathrm{Emb}(\Sigma, \mathcal{M})$ where $N_t^a(x) := N^a(x,t)$, $a = 1, 2, 3$.

Note that, like the normal vector, the lapse and shift are functionals of both the spacetime metric $g_{\mu\nu}$ and the foliation. It is important to realise that this relationship can be partially inverted with, for a fixed foliation, the lapse and shift functions being identified as parts of the metric tensor g on $\mathcal{M}$. This can be seen most clearly by studying the "pull-back" of g by the foliation $X : \Sigma \times \mathbb{R} \to \mathcal{M}$. This gives a Lorentzian metric X^*g on $\Sigma \times \mathbb{R}$ whose components are

$$(X^*g)_{\alpha\beta}(x,t) := g_{\mu\nu}(X(x,t))X^\mu{}_{,\alpha}(x,t)X^\nu{}_{,\beta}(x,t) \tag{4.3.9}$$

where g, β=0–3 denotes a coordinate system on $\Sigma \times \mathbb{R}$.

In particular, coordinates can be chosen that are "adapted" to the foliation in the sense that $x^{\alpha=0}(x,t) := t$, and $\alpha = 1, 2, 3$ refers to some coordinate system on Σ. It can then be shown that (4.3.9) is equivalent to the system of equations

$$(X^*g)_{00}(x,t) = N^a(x,t)N^b(x,t)\,{}_t\gamma_{ab}(x) - \left(N_t(x)\right)^2 \tag{4.3.10}$$

$$(X^*g)_{0a}(x,t) = N^b(x)\,{}_t\gamma_{ab}(x) \tag{4.3.11}$$

where

$${}_t\gamma_{ab}(x) := g_{\mu\nu}(X(x,t))X^\mu{}_{,a}(x,t)X^\nu{}_{,b}(x,t) \tag{4.3.12}$$

is the pull-back by $X_t : \Sigma \to \mathcal{M}$ of the Lorentzian metric g on $\mathcal{M}$ to give a Riemannian metric ${}_t\gamma$ on Σ.

It is clear from (4.3.10-11) that the shift vector and lapse function are essentially the g_{0a} and g_{00} parts of the spacetime metric. It was in terms of this coordinate approach that the original ADM approach was formulated. As we will see shortly, the Einstein field equations do not lead to any dynamical development for these variables that appear as Lagrange multipliers.

4.3.4 The Canonical Form of General Relativity

The variable conjugate to γ.

The Riemannian metric on Σ defined in (4.3.12) plays the role of the canonical configuration variable. Geometrically, we can say that the metric ${}_t\gamma$ measures the "intrinsic" curvature of the spacelike surface Σ_t.

The variable conjugate to ${}_t\gamma$ is closely related to the *extrinsic curvature* of Σ_t which is a measure of how Σ_t looks when seen from the perspective of the Lorentzian metric on $\mathcal{M}$. In coordinate terms, this object is defined as

$$\begin{aligned} K_{ab}(x,t) &:= -n_{\nu;\mu}(x,t)X^{\nu}{}_{,a}(x,t)X^{\mu}{}_{,b} & (4.3.13)\\ &= \frac{1}{2N}\left(-\frac{\partial\gamma_{ab}}{\partial t}+N_{a|b}+N_{b|a}\right) \\ &= \frac{1}{2N}\left(-\frac{\partial\gamma_{ab}}{\partial t}+(L_N\,{}_t\gamma)_{ab}\right) & (4.3.14)\end{aligned}$$

where $n_{\nu;\mu}$ and $N_{a|b}$ denote covariant differentiation with respect to $g_{\mu\nu}$ on $\mathcal{M}$ and ${}_t\gamma$ on Σ. The symbol $(L_N\,{}_t\gamma)_{ab}$ means the Lie derivative of the metric ${}_t\gamma$ along the vector field N^a.

The key step in deriving the canonical form of the action principle is to express the pull-back of the Einstein lagrangian density by the foliation $X:\Sigma\times\mathbb{R}\to\mathcal{M}$ as a function of the extrinsic curvature, the metric γ and the lapse and shift functions. The result is

$$\begin{aligned} X^*(R(g)(-\det g)^{\frac{1}{2}}) &= N(\det\gamma)^{\frac{1}{2}}(K_{ab}K^{ab}-(K^a_a)^2+R^{(3)}(\gamma)) \\ &\quad -2\frac{d}{dt}((\det\gamma)^{\frac{1}{2}}K_a{}^a)-((\det\gamma)^{\frac{1}{2}}(K_a{}^aN^b-\gamma^{ab}N_{,a}))_{,b} \qquad (4.3.15)\end{aligned}$$

where $R^{(3)}(\gamma)$ is the curvature scalar of the metric γ on Σ and the t label on γ is implicitly understood. Note that the extrinsic curvature K is to be thought of as the explicit functional (4.3.14) of γ_{ab}, $\dot\gamma_{ab}$ and N.

The spatial divergence term vanishes since Σ is assumed to be compact. However, the term with the total time derivative must be removed to produce a genuine action principle. The resulting Einstein lagrangian for matter-free gravity can be written as

$$L(t)=\frac{1}{\kappa^2}\int_\Sigma N(\det\gamma)^{\frac{1}{2}}(K_{ab}K^{ab}-(K_a{}^a)^2+R^{(3)}(\gamma))\,d^3x \qquad (4.3.16)$$

where $\kappa^2:=16\pi G/c^2$.

The canonical analysis now proceeds in the standard way. The variable conjugate to γ_{ab} (with respect to the natural time parameter t associated with the foliation) is

$$\pi^{ab} := \frac{\delta L}{\delta \dot{\gamma}_{ab}} = -\frac{(\det \gamma)^{\frac{1}{2}}}{\kappa^2}(K^{ab} - \gamma^{ab} K_c{}^c) \tag{4.3.17}$$

which can be inverted in the form

$$K^{ab} = -\kappa^2 (\det \gamma)^{-\frac{1}{2}} (\pi^{ab} - \frac{1}{2}\gamma^{ab}\pi_c{}^c). \tag{4.3.18}$$

However, the dependence of (4.3.16) on the lapse function and shift vector is such that

$$\pi_N := \frac{\delta L}{\delta \dot{N}} = 0; \quad \pi_{N^a} := \frac{\delta L}{\delta \dot{N}^a} = 0 \tag{4.3.19}$$

which are called the ***primary constraints*** of the theory. They justify the claim made earlier that the time-behaviour of N and $\boldsymbol{N}$ is not determined by the Einstein equations.

The constraint equations

The Einstein field equations can be usefully separated into the equations $G_{0\mu}(g) = 0$ and $G_{ab} = 0$. Expressed in terms of the canonical variables defined above, the $G_{ab} = 0$ equations are genuine first-order dynamical equations that equate $\dot{\gamma}_{ab}$ and $\dot{\pi}^{ab}$ with specific functions of γ, π, N and $\boldsymbol{N}$. However, the equations $G_{0\mu} = 0$ are very different. They are equivalent to the equations

$$C_a(\gamma, \pi) = 0 \tag{4.3.20}$$

$$C_\perp(\gamma, \pi) = 0 \tag{4.3.21}$$

where

$$C_a(x) = -\pi_a{}^b{}_{|b}(x) \tag{4.3.22}$$

$$C_\perp(x) = \frac{\kappa^2}{(\det \gamma(x))^{\frac{1}{2}}}(\gamma_{ab}(x)\gamma_{cd}(x) - \frac{1}{2}\gamma_{ac}(x)\gamma_{bd}(x))\pi^{ac}(x)\pi^{bd}(x) - \frac{(\det\gamma(x))^{\frac{1}{2}}}{\kappa^2}R(\gamma(x)) \tag{4.3.23}$$

and are therefore ***constraints*** on the canonical variables $\gamma_{ab}(x)$ and $\pi^{ab}(x)$.

It is useful to note that the constraints and the true equations of motion can all be derived from the first-order action

$$S[\gamma,\pi,N,N] := \int dt \int_{\Sigma} d^3x \left(\pi^{ab}\dot{\gamma}_{ab} - NC_{\perp}(\gamma,\pi) - N^a C_a(\gamma,\pi)\right) \tag{4.3.24}$$

in which γ_{ab}, π^{cd}, N and N are to be varied as independent variables. This shows clearly that N and N are Lagrange multipliers with no genuine dynamical status.

The importance of the constraints

The constraints (4.3.20-21) are of major importance in both the classical and the quantum theories of gravity. In particular, the initial value problem for general relativity is distinctly more complicated than for a normal field theory since the Cauchy data on an initial hypersurface must satisfy the constraints. Of the various results that have been obtained, the following are the most significant for our purposes:

1. If a Lorentzian metric $g_{\mu\nu}$ on a manifold $\mathcal{M} \simeq \Sigma \times \mathbb{R}$ satisfies the vacuum Einstein field equations then, as discussed above, the constraint equations (4.3.20-21) are satisfied on any spacelike hypersurface in $\mathcal{M}$.
2. Conversely, if a Lorentzian metric $g_{\mu\nu}$ on a manifold $\mathcal{M} \simeq \Sigma \times \mathbb{R}$ satisfies the constraint equations (4.3.20-21) on all spacelike hypersurfaces then $g_{\mu\nu}$ necessarily satisfies all ten Einstein field equations $G_{\mu\nu}(g) = 0$.

This last result is very significant since it implies that there is a sense in which the dynamical aspects of the Einstein equations are already contained in the constraints. This plays a crucial role in the quantum theory.

4.3.5 The Dirac Algebra of the Constraints

The classical canonical algebra of the system is expressed by the basic Poisson brackets

$$\{\gamma_{ab}(x), \gamma_{cd}(x')\} = 0 \tag{4.3.25}$$

$$\{\pi^{ab}(x), \pi^{cd}(x')\} = 0 \tag{4.3.26}$$

$$\{\gamma_{ab}(x), \pi^{cd}(x')\} = \delta^c_{(a}\delta^d_{b)}\delta(x,x') \tag{4.3.27}$$

which leads to the fundamental relations (Dirac, 1965)

$$\{C_a(x), C_b(x')\} = -C_b(x)\partial^{x'}_a \delta(x,x') + C_a(x')\partial^x_b \delta(x,x') \tag{4.3.28}$$

$$\{C_a(x), C_{\perp}(x')\} = C_{\perp}(x)\partial^x_a \delta(x,x') \tag{4.3.29}$$

$$\{C_{\perp}(x), C_{\perp}(x')\} = \gamma^{ab}(x)C_a(x)\partial^{x'}_b \delta(x,x') - \gamma^{ab}(x')C_a(x')\partial^x_b \delta(x,x'). \tag{4.3.30}$$

Using the smeared variables

$$C(\eta) := \int_\Sigma C_\perp(x)\eta(x)\, d^3x, \quad C(\xi) := \int_\Sigma C_a(x)\xi^a(x)\, d^3x \tag{4.3.31}$$

where η and ξ are a scalar function and vector field on Σ, these equations can be written as

$$\{C(\xi_1), C(\xi_2)\} = C([\xi_1, \xi_2]) \tag{4.3.32}$$

$$\{C(\xi), C(\eta)\} = C(L_\xi \eta) \tag{4.3.33}$$

$$\{C(\eta_1), C(\eta_2)\} = C(\xi) \tag{4.3.34}$$

where, in (4.3.34), $\xi^a := \gamma^{ab}(\eta_1\eta_{2,b} - \eta_2\eta_{1,b})$.

The geometrical interpretation of these expressions is as follows.

1. The Lie algebra of the diffeomorphism group Diff(Σ) is generated by the vector fields on Σ with minus the commutator of a pair of vector fields

$$[\xi_1, \xi_2]^a := \xi_1^b \xi_{2,b}^a - \xi_2^b \xi_{1,b}^a \tag{4.3.35}$$

playing the role of the Lie bracket. Thus (4.3.32) shows that the map $\xi \to -C(\xi)$ is a homomorphism of the Lie algebra of Diff(Σ) into the Poisson bracket algebra of the theory. Calculation of the Poisson brackets of $C(\xi)$ with $\gamma_{ab}(x)$ and $\pi^{cd}(x)$ confirms the interpretation of $-C(\xi)$ as a generator of spatial diffeomorphisms. For this reason, $C_a(x)$ are sometimes known as the "transverse" constraints.
2. Similarly, $C_\perp$ can be interpreted as generating deformations of a hypersurface *normal* to itself as it is embedded in $\Sigma \times \mathbb{R} \simeq \mathcal{M}$. Thus $C_\perp(x)$ are known as the "longitudinal" constraints.

Two important things should be noted about the "gauge" algebra (4.3.32-34).

1. It is *not* the Lie algebra of Diff($\mathcal{M}$) even though this was the invariance group of the original theory.
2. The presence of the γ^{ab} factor in the right hand side of (4.3.34) means that is not a genuine Lie algebra at all.

These two features are closely related since the Dirac algebra (4.3.32-34) is essentially the Lie algebra of Diff($\mathcal{M}$) *projected* along, and normal to, a spacelike hypersurface. The significance of this in the quantum theory will emerge later.

4.4 Canonical Quantisation

4.4.1 Reduction to True Canonical Form

We will now consider the quantisation of the system with the first-order action principle (4.3.24) in which the canonical variables $\gamma_{ab}(x)$ and $\pi^{cd}(x)$ are subject to the constraints (4.3.20-21) where C_a and $C_\perp$ are given by (4.3.22-23).

The quantisation of any constrained system is a complicated matter and several different approaches can be taken. The two key questions are

1. How are the constraints to be handled?
2. What is the quantum status of Lagrange multipliers like N and $\mathbf{N}$?

One approach is to reduce the theory to a true canonical form (*i.e.*, with no constraints and associated Lagrange multipliers) *before* the theory is quantised. The main steps are as follows:

1. Solve the constraints classically [18] for $4\infty^3$ of the $12\infty^3$ variables $\left(\gamma_{ab}(x), \pi^{cd}(x)\right)$.
2. Impose a suitable gauge. This can be done in several ways but it will always lead to
   a) the elimination of $4\infty^3$ of the $8\infty^3$ variables $\left(\gamma_{ab}(x), \pi^{cd}(x)\right)$ that remain after the first step;
   b) a set of equations that enable N and $\mathbf{N}$ to be determined as explicit functions of the remaining canonical variables.
3. Find a canonical action that reproduces the dynamical equations for the remaining $4\infty^3$ "true" canonical variables.
4. Impose canonical commutation relations on these variables and then proceed the usual way.

This programme is not easy to implement and it is unattractive for several reasons. For example,

1. The constraints (4.3.20–21) cannot be solved in a closed form. [19] It is possible to use weak-field perturbative methods but this merely takes us back to the framework we are trying to avoid.

[18] The notation "$n\infty^3$" means n variables at each spatial point x in the three-manifold Σ.

[19] The use of the Ashtekar variables (§4.5) might change this situation. They lead to powerful new techniques for tackling the constraints at a *quantum* level and perhaps they could also be used effectively in the classical theory.

2. Choosing a gauge breaks the gauge invariance of the theory and, as usual, the task is then to show that the final quantum results are independent of the choice. However, the Diff($\mathcal{M}$) invariance of general relativity is so important that many people prefer a formalism in which a choice of gauge is deferred as long as possible and, in particular, to *after* the quantisation has been performed. Arguably, this makes it easier to discuss the overall significance of the diffeomorphism group in the quantum theory.
3. The programme violates the geometrical structure of general relativity by, for example, removing parts of the metric tensor. This makes it very difficult to explore the significance of spacetime geometry in the quantum theory.

We will return to some of these issues in §5.1 in our discussions of the problem of time in quantum gravity.

4.4.2 Quantisation Before Constraining

The canonical commutation relations

For these reasons, much interest has been devoted to the alternative approach in which all $12\infty^3$ variables $(g_{ab}(x), \pi^{cd}(x))$ are assigned quantum status, and questions of gauge fixing and/or solving the constraints are tackled at the quantum level. The starting point is the operator version of the Poisson algebra (4.3.25-27) in the form of the canonical commutation relations

$$[\,\widehat{\gamma}_{ab}(x), \widehat{\gamma}_{cd}(x')\,] = 0 \tag{4.4.1}$$

$$[\,\widehat{\pi}^{ab}(x), \widehat{\pi}^{cd}(x')\,] = 0 \tag{4.4.2}$$

$$[\,\widehat{\gamma}_{ab}(x), \widehat{\pi}^{cd}(x')\,] = i\hbar\delta^{c}_{(a}\delta^{d}_{b)}\delta(x, x'). \tag{4.4.3}$$

Several things need to be said about this algebra:

1. We are working in the "Schrödinger" representation in which no time dependence is carried by the canonical variables. As we will see later, there is more to this than meets the eye.
2. Like the Poisson brackets (4.3.25-27), these operator relations appear to employ the components of γ and its conjugate π in some coordinate system. However, this dependence is superficial and can be removed by rewriting the expressions in the smeared form

$$[\widehat{\gamma}(h), \widehat{\gamma}(h')] = 0 \tag{4.4.4}$$

$$[\widehat{\pi}(k), \widehat{\pi}(k')] = 0 \tag{4.4.5}$$

$$[\widehat{\gamma}(h), \widehat{\pi}(k)] = i\hbar \int_{\Sigma} h^{ab}(x) k_{ab}(x)\, d^3x \tag{4.4.6}$$

where h and k are a tensor density and a tensor respectively. The right hand side of (4.4.6) is coordinate independent even though I have chosen to express it in coordinate form.

3. Equation (4.4.1) is a form of microcausality. However, since we have made no *specific* choice of a time parameter (this would involve a partial gauge fixing), it is not clear what this means in terms of the usual ideas of an "equal-time" hypersurface, or indeed, how the notion of spacetime structure (as opposed to spatial structure) appears at all. The most that can be said at this stage is that, whatever the final spacetime interpretation will be, (4.4.1) guarantees that the points of Σ can be regarded as being spacelike separated. We will return later to this issue in the context of the problem of 'time' in quantum gravity.

The affine commutation relations

Another problem with (4.4.1-3) is of a more geometrical nature and arises from the fact that the classical object $\gamma_{ab}(x)$ is not merely a symmetric covariant tensor: it is also a *metric* tensor *i.e.*, , at each $x \in \Sigma$ it is an invertible matrix with signature (1,1,1). It is very relevant to ask:

1. Is there a well-defined *operator* analogue of this requirement?
2. If so, do we expect the canonical operator $\widehat{\gamma}_{ab}(x)$ to be such an object?
3. Is such a requirement compatible with the commutation relations (4.4.1-3)?

The first question is easy to answer. A natural analogue of the positive-definiteness of the classical metric is to require that the smeared quantum operator satisfies the condition

$$\widehat{\gamma}(v \otimes v) > 0 \tag{4.4.7}$$

for all non-vanishing vector-density fields v^a (of an appropriate weight) on Σ.

However, whether or not we expect this to be true is a different matter. There have been several suggestions in recent years that some degree of degeneracy should be allowed, *i.e.*, there may exist vector fields v^a that are non-vanishing but for which the smeared operator in (4.4.7) is nevertheless 0. For example:

1. Semi-definite operators of this type arise naturally in certain representations (Isham and Kakas 1984b) of the affine commutation relations (see below).
2. Ashtekar's formulation of general relativity admits degenerate metrics, even at the classical level (Ashtekar 1987).
3. Witten (1988) suggested that there might exist a sector of quantum gravity in which $g_{\mu\nu} = 0$ whilst maintaining Diff($\mathcal{M}$) invariance. This is an extreme case of degeneracy!
4. Horowitz (1990) has looked at a first-order form of general relativity in which the field equations admit a Lorentzian signature metric but with changes in the spatial topology. These solutions involve a degenerate metric.

The answer to the third question is unequivocal: the conventional commutation relations (4.4.4-6) are manifestly incompatible with (4.4.7) or, indeed, with the weaker requirement that the smeared metric operator be only *semi*-definite. This is analogous to the problem of constructing the quantum theory of a particle that is constrained to move on the positive real line $\mathbb{R}_+$. It is easy to show that the conventional commutation relations $[\widehat{x}, \widehat{\pi}] = i\hbar$ imply that the spectrum of $\widehat{x}$ is the *whole* real line, which is inconsistent with the requirement $x > 0$.

In this case, it is well known (Klauder and Aslaksen, 1970) that the answer is to replace the conventional relations with the *affine* algebra defined by

$$[\widehat{x}, \widehat{p}] = i\hbar\widehat{x}. \tag{4.4.8}$$

The appropriate affine algebra for general relativity has the form (Klauder 1970, Pilati 1982, 1983, Isham 1984, Isham and Kakas 1984a,b)

$$[\widehat{\gamma}_{ab}(x), \widehat{\gamma}_{cd}(x')] = 0 \tag{4.4.9}$$

$$[\widehat{p}_a{}^b(x), \widehat{p}_c{}^d(x')] = i\hbar\big(\delta_a^d \widehat{p}_c{}^b(x) - \delta_c^b \widehat{p}_a{}^d(x)\big)\delta(x, x') \tag{4.4.10}$$

$$[\widehat{\gamma}_{ab}(x), \widehat{p}_c{}^d(x')] = i\hbar\delta^d_{(a}\widehat{\gamma}_{b)c}(x)\,\delta(x, x'). \tag{4.4.11}$$

Representations of this algebra can be constructed in which the spectrum of the smeared metric operator is guaranteed to satisfy (4.4.7), at least in a semi-definite form.

4.4.3 The Imposition of the Constraints

Dirac constraint quantisation

The key question now is how the constraint equations (4.3.20-21) are to be handled. The essence of the Dirac approach is to impose them as constraints on the physically allowed states in the form

$$\mathcal{C}_a(\widehat{\gamma}, \widehat{\pi})\Psi = 0 \tag{4.4.12}$$

$$\mathcal{C}_\perp(\widehat{\gamma}, \widehat{\pi})\Psi = 0. \tag{4.4.13}$$

We recall that, in the classical theory, the constraints are *equivalent* to the dynamical equations in the sense that if they are satisfied on all spatial hypersurfaces of a Lorentzian metric $g_{\mu\nu}$ then $g_{\mu\nu}$ necessarily satisfies the Einstein field equations. This is reflected in the quantum theory by the assumption that the operator constraints (4.4.12-13) are the *sole* technical content of the theory of quantum gravity.

This is closely related to the following fundamental observation. According to the first-order action (4.3.24), the canonical hamiltonian associated with general relativity is

$$H[N, \mathcal{N}] := \int_\Sigma (N(x)\mathcal{C}_\perp(x) + N^a(x)\mathcal{C}_a(x))\, d^3x \tag{4.4.14}$$

where N and $\mathcal{N}$ are regarded as external c-number functions. However, (4.4.14) has a rather remarkable implication for the putative "Schrödinger equation"

$$i\frac{d}{dt}\Psi_t = \widehat{H}[N, \mathcal{N}]\Psi_t \tag{4.4.15}$$

since if the state Ψ_t satisfies the constraint equations (4.4.12-13), we see that it has no time dependence at all! Similarly it is not meaningful to speak of a "Schrödinger" or a "Heisenberg" picture since the matrix elements between physical states of a Heisenberg-picture field will be the same as for the field in the Schrödinger picture.

This so-called "frozen formalism" caused much confusion when it was first discovered since it seems to imply that nothing happens in a quantum theory of gravity. However, this is now understood to reflect the absence of any intrinsic definition of 'time' in general relativity; indeed, since we have not gauge-fixed the theory, no such coordinate has been selected. It *is* possible to speak of time evolution in quantum gravity, but the subject is subtle (not to say contentious) and is discussed at length in the next chapter.

4.4.4 The Dirac Operator Algebra

Many problems arise when attempting to implement the Dirac scheme, one of the most serious of which is knowing to what extent the classical Poisson bracket algebra (4.3.28-30) can, or should be, maintained in the quantum theory. The principle problems are

1. The constraints (4.3.22) and (4.3.23) are highly non-linear functions of the canonical variables and involve non-polynomial products of field operators evaluated at the same point. Such formal expressions are ill-defined and require significant regularisation and renormalisation. This is the canonical analogue of the ultraviolet divergences which mar the particle physics approach and, as in that case, the problem is very non-trivial.
2. Severe operator-ordering problems arise since the constraints contain products of $\widehat{\gamma}$ and $\widehat{\pi}$. There have been many discussions of whether a consistent ordering can be found, but the problem is complicated by a number of factors. For example:
   a) The problem is not merely one of defining the constraint operators. It is also necessary to decide where to place the $\widehat{\gamma}^{ab}(x)$ operator that appears on the right hand side of the operator version of (4.3.30). One important condition is that we do not wish the commutators of the given first-class constraints to generate *further* constraints on the physical state vectors.
   b) It is natural to assume that the constraint operators $\widehat{C}_\perp(x)$ and $\widehat{C}_a(x)$ should be hermitian. However, this has been challenged several times (Komar 1979, Kuchar 1986, 1987, Hajicek and Kuchar 1990a,b) and the issue clearly has important implications for the operator ordering problem. The possibility of using a non-hermitian operator can be partly justified by noting that the Hilbert space structure on the space that carries the representation of the canonical (or affine) algebra may be only distantly related to the Hilbert space structure which ought to be imposed on the *physical* states (*i.e.*, those that satisfy the constraints).
   c) The operator products in the constraints are so singular that any suggested ordering is anyway likely to be very ambiguous (Friedmann and Jack 1988).
   d) It is possible that an anomaly *should* be present in the theory as a genuine Planck length effect. But very little is known about, for example, central extensions of the Dirac algebra.

3. As mentioned above, it is very unclear how the inner product on the physical states is to be constructed from the Hilbert space structure on the original space $\mathcal{H}$ that carries the representation of the canonical, or affine, commutation relations. There are various aspects to this question.
   a) The set of physical states $\mathcal{H}_{\rm phys}$ is unlikely to be simply a subspace of $\mathcal{H}$. This is because, if the constraint operators *are* hermitian, the operator constraints (4.4.12-13) can be read as saying that the physical states are simultaneous eigenstates of $\widehat{C}_\perp(x)$ and $\widehat{C}_a(x)$ with a zero eigenvalue. However, if (as seems likely) the spectra of these operators are *continuous* then the resulting generalised eigenvectors will have an infinite norm and hence will not belong to the Hilbert space $\mathcal{H}$ itself.
   b) If the constraint operators are *not* hermitian the problem becomes even worse since very little can be said about the eigenvectors of an arbitrary non-hermitian operator.
   c) One way of trying to solve this, and related, problems, is to insist that, in some appropriate sense, the results obtained by quantising and then constraining are equal to those in which the constraints are solved before quantising (Kuchar 1986a,b). However, it is arguable whether it is either necessary or desirable that the acts of quantising and constraining *should* commute in this way.
4. From one perspective it is easy to see why the original Diff($\mathcal{M}$) invariance translates into the Dirac algebra in the canonical decomposition. Indeed, it is difficult to see *a priori* how the diffeomorphism group of the spacetime $\mathcal{M}$ could ever be represented by fields that are defined purely on a three-manifold Σ. However, it is natural to wonder if the original group *can* be recovered in some way, and this has been the subject of several different studies.
   a) In the approach of Bergmann and Komar (1972), the group is extended enormously by allowing the generating vector fields to be functionals of the canonical data.
   b) More recently, Isham and Kuchar (1985a,b) suggested a method which involves allowing the embedding functions $X_t : \Sigma \rightarrow \mathcal{M}$ to become canonical variables. This leads to a genuine representation of Diff($\mathcal{M}$). We will return to this idea in our discussion of 'time' in quantum gravity.

4.4.5 Representations on Functionals $\Psi[\gamma]$.

The general idea

In attempting to find concrete representations of the canonical algebra (4.4.1-3) it is natural to try an analogue of the scalar field representation (4.2.15-16). Thus the state vectors are functionals $\Psi[\gamma]$ of Riemannian metrics γ on Σ and the canonical operators are defined as

$$(\widehat{\gamma}_{ab}(x)\Psi)[\gamma] := \gamma_{ab}(x)\Psi[\gamma] \tag{4.4.16}$$

$$(\widehat{\pi}^{cd}(x)\Psi)[\gamma] := -i\hbar\frac{\delta\Psi}{\delta\gamma_{cd}(x)}[\gamma]. \tag{4.4.17}$$

These equations have been widely used in the canonical approach to quantum gravity. However, they are deceptive in several respects and it is important to note the following points:

1. As we have remarked already, the canonical algebra (4.4.1–3) is incompatible with the positive (or semi-positive) definiteness of a classical Riemannian metric. In particular, it *not* correct to say that the domain space of the state functionals is Riem(Σ): that is only possible if the affine commutation relations are used.
2. There are precise analogues of the general problems mentioned in connection with the scalar field operators in (4.2.15-16). In particular:
   a) There are no Lebesgue measures on, or associated with, Riem(Σ). Thus a weighted measure is needed with a compensating term added to (4.4.17) as in (4.2.19).
   b) If any measures do exist one would expect them to be concentrated on *distributional* versions of metrics. However, this concept has no obvious meaning since the space Riem(Σ) is *not* a vector space and therefore there is no natural way of defining its dual. One of the great virtues of using the affine commutation relations is that the representation theory of this algebra leads to an appropriate definition of what might be meant by a distributional metric (Isham 1984, Isham and Kakas 1984b).
3. Some of the representations of the affine relations (4.4.9-11) involve state vectors that are functionals of γ or their appropriate distributional analogues. However, several additional possibilities arise that are not present in the representations (4.4.16-17) of the naive commutation relations. For example,

a) There are whole classes of representations in which the state vectors are concentrated on distributional analogues of *degenerate* metrics (Isham and Kakas 1984b).

b) There are representations in which the state vector $\Psi[\gamma]$ has an *index* which acts rather like the internal spin for a relativistic particle.

The $\widehat{C}(\xi)\Psi = 0$ constraint

Let us now consider the constraints (4.4.12-13) in the context of this representation on functionals $\Psi[\gamma]$. From a physical perspective it is easy to see the need for these constraints. Modulo the remarks above, the domain space of the state functionals is Riem(Σ), and to specify a metric $\gamma_{ab}(x)$ at a point $x \in \Sigma$ requires six numbers (the components of the metric in some coordinate system). However, the true gravitational system should have only *two* degrees of freedom per spatial point (the two directions of circular polarization of a gravitational wave, the two helicities of a massless graviton, *etc*) and therefore four of the six degrees of freedom need to be lost. This is precisely what is achieved by the imposition of the constraints (4.4.12–13). (Of course, the same counting argument applies if the system is reduced to true canonical form before quantising).

The easiest constraints to handle are those in the first set

$$(\widehat{C}(\xi)\Psi)[\gamma] = 0. \tag{4.4.18}$$

We saw earlier that the classical functions $C(\xi)$ are the infinitesimal generators of the diffeomorphism group of Σ, and the same might be expected to apply here. The key question in this context is whether the operator ordering problems can be solved so that the algebra (4.3.32) is preserved at the quantum level. Experience suggests that this is usually fairly straightforward and, indeed, one powerful way of *solving* the operator ordering problem for the $\widehat{C}(\xi)$ generators is to insist that they form a hermitian representation of the Lie algebra of Diff(Σ).

It is therefore reasonable to expect the quantum theory to carry a unitary representation of Diff(Σ). In this language, the constraint equation (4.4.18) becomes

$$(D(f)\Psi)[\gamma] = \Psi[\gamma] \tag{4.4.19}$$

where $D(f)$ is the unitary operator representing $f \in$ Diff(Σ).

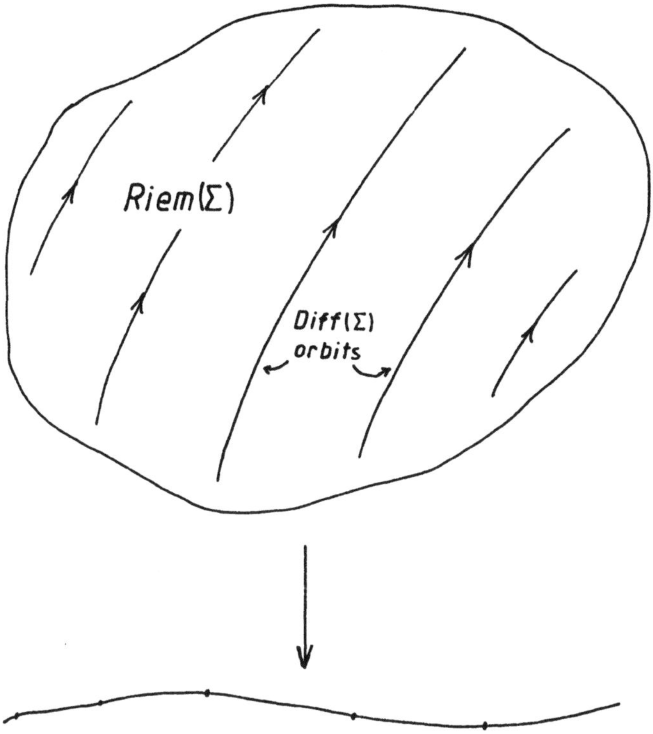

Fig. 4.3. Superspace as the space of orbits of Diff(Σ) on Riem(Σ)

The naive representation (4.4.16-17) suggests that the unitary operator $D(f)$ acts on Ψ as

$$(D(f)\Psi)[\gamma] := \Psi[f^*\gamma] \tag{4.4.20}$$

where $f^*\gamma$ is the usual pull-back of γ by the diffeomorphism $f : \Sigma \to \Sigma$. Then the constraint equation (4.4.19) is equivalent to

$$\Psi[f^*\gamma] = \Psi[\gamma] \tag{4.4.21}$$

for all $f \in$ Diff(Σ) and $\gamma \in$ Riem(Σ).

The implications of (4.4.21) are analogous to those in conventional Yang-Mills gauge theory. The group Diff(Σ) acts as a group of transformations on the space Riem(Σ) of Riemannian metrics on Σ with $f \in$ Diff(Σ) sending $\gamma \in$ Riem(Σ) to $f^*\gamma$. This leads to a bundle [20] picture in which Diff(Σ) is fibered by the orbits of the Diff(Σ) action (Figure 4.2).

[20] This is not quite true since the existence of metrics with isometry groups means that the action of Diff(Σ) is not strictly free. This problem can be solved by, for example, restricting attention to those diffeomorphisms that leave fixed some particular frame at a base point in Σ.

John Wheeler (1964, 1968) called the base space Riem(Σ)/Diff(Σ) of this bundle *superspace* and it has played an important role in canonical quantum gravity. This is because the constraint equation (4.4.21) can be interpreted as saying that the state functional Ψ is constant on the orbits of Diff(Σ) and therefore passes to a function on superspace. Thus, according to Wheeler, it is superspace that is the true domain space of the quantum gravity state vectors. However, as usual, several cautionary remarks are in order.

1. Equation (4.4.18) only really implies that Ψ is constant under *infinitesimal* transformations and therefore it could transform non-trivially under "large" diffeomorphisms that cannot be connected continuously to the identity map. This leads to a gravitational analogue (Friedman and Sorkin 1980, Isham 1981) of the θ states that have been much studied in Yang-Mills theory.
2. The action in (4.4.20) is only unitary if the Hilbert space measure $d\mu[\gamma]$ on the domain space is itself invariant under Diff(Σ). It is most unlikely that any such measures exist. The best that could be hoped for is that the measure is *quasi-invariant* under the action of Diff(Σ) (*i.e.*, the transformed and untransformed measures have the same sets of measure zero), in which case a Jacobian factor is needed on the right-hand side of (4.4.20). This will alter the form of the constraint (4.4.21) and the simple idea of the state functional being a constant on the Diff(Σ) orbits is no longer correct.
3. If the domain space of Ψ is really a space of distributional metrics then the picture changes rather dramatically. The action Diff(Σ) on a space of this type is likely to be very non-free and a bundle picture is not appropriate.

4.4.6 The Wheeler-DeWitt Equation.

The basic equation

We must now consider the final constraint (4.4.13). Unlike the $C(\xi)$ constraint, this has no simple group-theoretic interpretation since, as we have remarked already, the presence of the explicit γ factor on the right hand side of (4.3.30) means that (4.3.28-30) is not a genuine Lie algebra. Thus the operator ordering problem becomes much harder. However, if we choose as a simple example the ordering in which all the π^{cd} variables are placed to the right of the γ_{ab} variables, the constraint (4.4.13) becomes

$$-\hbar^2\kappa^2 \mathcal{G}_{abcd}(\gamma)\frac{\delta^2\Psi}{\delta\gamma_{ac}\delta\gamma_{bd}}[\gamma] - \frac{(det\gamma)^{\frac{1}{2}}}{\kappa^2}R(\gamma)\Psi[\gamma] = 0 \tag{4.4.22}$$

where

$$\mathcal{G}_{abcd}(\gamma) := (\gamma_{ab}\gamma_{cd} - \frac{1}{2}\gamma_{ac}\gamma_{bd})(\det\gamma)^{-\frac{1}{2}} \tag{4.4.23}$$

is the DeWitt metric on Riem(Σ) (DeWitt 1967a).

Questions and problems

Equation (4.4.22) is the famous Wheeler-DeWitt equation. It is the heart of the canonical theory of quantum gravity from which everything must be extracted. Needless to say, a number of problems and questions arise that need to be considered early on. For example:

1. The ordering chosen in (4.4.22) is a simple one but there is no particular reason why it should be correct. One popular alternative is to write the "kinetic energy" term as a covariant functional Laplacian using the DeWitt metric (4.4.23). This has the advantage (?) that this part of the operator is then invariant under redefinitions of coordinates on Riem(Σ). Of course, a key issue in discussing the ordering of $\widehat{C}_\perp(x)$ is whether one expects these operators to be *hermitian*.
2. The Wheeler-DeWitt equation contains products of functional differential operators evaluated at the same spatial point and is therefore likely to produce $\delta^3(0)$ singularities when acting on a wide variety of possible state functionals. (*cf* the remarks made in §4.2 concerning a scalar field theory). Thus regularisation will almost certainly be needed.
3. A major question is how to approach the problem of solving the Wheeler-DeWitt equation. One obvious tactic is to deal with it as a functional differential equation *per se*. However, whether or not this is really valid depends on the general interpretation of the constraint equations $\widehat{C}_\perp(x)\Psi = 0$.
   a) If this equation means that Ψ is an eigenvector of the hermitian operators $\widehat{C}_\perp(x)$ with eigenvalue 0 then, as with standard eigenfunction problems for differential operators, some sort of boundary value conditions need to be imposed on Ψ, and the theory itself is not too informative about what these might be.
   b) In fact, there is a tendency to solve the constraint equation as a functional differential equation without checking that 0 is a genuine eigenvalue. This can

lead to highly misleading results. For example, consider the equation in one-dimensional quantum mechanics

$$(\hat{p}^2 + \hat{x}^2 + \lambda \hat{x}^4)\psi = 0. \tag{4.4.24}$$

Under the substitution $\hat{p} = -i\hbar d/dx$ this becomes an ordinary differential equation in x that is second order and which therefore has a two-parameter family of solutions. Now suppose the momentum representation is employed with $\hat{x} = i\hbar d/dp$. Then (4.4.24) becomes a fourth-order differential equation in p and, as such, it has a *four*-parameter family of solutions. But the x and p representations of quantum theory are supposed to be equivalent! So what is going on?

The answer is that 0 is *not* a genuine eigenvalue of the hermitian operator $\hat{p}^2 + \hat{x}^2 + \lambda\hat{x}^4$, and therefore the solutions to the differential equation (4.4.24) have no intrinsic significance. Unfortunately, this simple fact is often overlooked in the quantum gravity literature, especially in the context of minisuperspace (see below) approximations to the theory. Thus to get a true eigenvalue problem it might first be necessary to *renormalise* the Wheeler-DeWitt operator.

4. One of the hardest problems is to decide what the Wheeler-DeWitt equation means in physical terms. In particular, the notions of 'time' and 'time-evolution' must be introduced in some way. The central idea is one we have mentioned several times already: time must be defined as an *internal* property of the gravitational system (plus matter) rather than being identified with some external parameter in the universe. Indeed, the "frozen formalism" interpretation of (4.4.15) is one way of re-emphasing this idea. We will return to this important topic in the next chapter.

Perturbative approaches to the Wheeler-DeWitt equation

Practical considerations dictate the need to develop effective perturbation methods for solving the Wheeler-DeWitt equation, and in a way which preserves as much of the geometrical structure as possible. For example:

1. One could introduce a background metric on Σ and perform a weak-field perturbative expansion. However, this is a very non-geometrical approach and takes us back to the type of framework we are trying to avoid.

2. The Wheeler-DeWitt equation (4.4.22) looks like the sum of a kinetic energy plus a potential energy. This analogy is deficient in several respects but it does suggest a form of perturbation theory in which the coefficient in front of the $R(\gamma)$ term is regarded as the coupling constant. This corresponds to a strong-coupling expansion in powers of $1/G$ and has been the subject of some study.
3. WKB approximations have been used quite widely, especially in the context of the problem of time in quantum gravity. We will discuss this approach in the next chapter.
4. One of the most popular forms of perturbation theory is the so-called "minisuperspace" technique. This involves freezing all but a finite number of the infinite degrees of freedom in Riem(Σ) and quantising the small number that remain. The Wheeler-DeWitt equation becomes a second-order partial differential equation that can be tackled using conventional means. This approach is particularly popular in studies of quantum cosmology since the finite degrees of freedom can be chosen in a way that is adapted to the classical models of cosmology. A simple example will be given shortly.
   However, it should be emphasised that this cannot be regarded as a genuine approximation scheme since there is no way of estimating the effect of the infinite number of degrees of freedom that are dropped. On the other hand, minisuperspace models are very useful for discussing conceptual problems, including the interpretation of the state vector, the problem of time, and related issues.

Other approaches to solving the Wheeler-DeWitt equation

Several techniques have been suggested for finding solutions to the Wheeler-DeWitt equation other than tackling it as a functional differential equation. Some of these involve projecting out a solution Ψ_{phys} from a "seed" state Ψ. For example, one can try a formal trick (Barvinsky and Ponomariov 1986, Barvinsky 1986) and define

$$\Psi_{\text{phys}} := \int \mathcal{D}[N]\,\mathcal{D}[N^a]\,\exp\left(i\int_{\mathcal{M}} (N(x)\widehat{\mathcal{C}}_\perp(x) + N^a(x)\widehat{\mathcal{C}}_a(x))\,d^4x\right)\Psi. \qquad (4.4.25)$$

Note that this procedure does not specifically require the use of state vectors that are functionals of γ.

Another example is the suggestion of Woodward (1989) that a gauge-fixed formalism should be used with

$$\Psi_{\text{phys}} := U(N|\mathcal{C})\Psi \tag{4.4.26}$$

where

$$U(N|\mathcal{C}) := \sum_{n=0}^{\infty} \frac{i^n}{n!} \int d^4x_1\, N^{\mu_1}(x_1) \ldots \int d^4x_n\, N^{\mu_n} \mathcal{C}_{\mu_1}(x_1) \ldots \mathcal{C}_{\mu_n}(x_n) \tag{4.4.27}$$

in which $N^0 \equiv N$ and N^a are q-number functions of the canonical variables that depend on the gauge chosen.

The use of functional integrals

A technique that has been much explored in recent years is the use of functional integrals to generate solutions to the Wheeler-DeWitt equation. This is motivated by the well-known fact that, in ordinary quantum mechanics, a solution to the time-dependent Schrödinger equation can be generated in the form

$$\psi(x,t) := \int \mathcal{D}x[s] \exp \frac{i}{\hbar} \int_{t_0}^{t} L(s)\, ds \tag{4.4.28}$$

where the path-integral is over all paths that end at the point x at time t. The solution thus obtained to $i\hbar\partial\psi/\partial t = \widehat{H}\psi$ depends on the value $x(t_0)$ of the path at the initial time t_0.

The analogue of this for the Wheeler-DeWitt equation was first studied in some detail by Teitelboim (1982, 1983a,b). More recently, it has played an important role in the Hartle-Hawking approach to quantum cosmology in which the starting point is a four-manifold $\mathcal{M}$ whose boundary is the disjoint union of a pair of three-manifolds, one of which is the space Σ. A state function is then constructed in the form

$$\Psi[\gamma] := \int \mathcal{D}[g]\, e^{-S_E[g]} \tag{4.4.29}$$

where $S_E[g]$ is the "Euclideanised" form of the Einstein-Hilbert action and the functional integral is over all Riemannian four-metrics g on $\mathcal{M}$ which (i) induce the metric γ on Σ, and (ii) satisfy some other boundary conditions on the second three-manifold. It can be shown (Halliwell 1988, 1990) that $\Psi[\gamma]$ satisfies the Wheeler-DeWitt equation (at least, formally).

The solution obtained depends on the boundary conditions on the second three-manifold. The Hartle-Hawking (1983) ansatz for quantum cosmology is that the four-manifold $\mathcal{M}$ should have just a *single* boundary Σ with the integral being over all metrics $g_{\mu\nu}$ which induce the metric γ_{ab} on Σ.

4.4.7 A Simple Minisuperspace Model

Further discussion of the Wheeler-DeWitt equation is assisted by having access to a simple minisuperspace model of canonical gravity. We will use a standard cosmological model in which the spacetime geometries are restricted to be one of the usual homogeneous and isotropic Robertson-Walker metrics

$$ds^2 := -N(t)^2 dt^2 + a(t)^2 \omega_{ab} dx^a dx^b \tag{4.4.30}$$

where ω_{ab} is the metric for a three-space of constant curvature k. The case $k = 1$ is associated with Σ being a three-sphere, while $k = 0$ and $k = -1$ correspond to the flat and hyperbolic cases respectively. Note that N^a does not appear in this expression: in effect, we are employing a $\mathrm{Diff}(\Sigma)$ gauge in which $N^a = 0$. This is a common feature of minisuperspace models and does not impose any significant limitations on what we can do.

Spacetimes of this type produce a non-vanishing Einstein tensor $G_{\mu\nu}$ and therefore require a source of matter. Much work in recent times has employed a scalar field ϕ with the Lagrangian (4.2.1), and we will use this here. After some calculation, it can be shown that the coupled equations for the variables a, ϕ and N can all be derived from the first-order action [21] (cf (4.3.24))

$$S[a, \pi_a, \phi, \pi_\phi, N] := \int \left(\pi_a \dot{a} + \pi_\phi \dot{\phi} - NC\right) dt \tag{4.4.31}$$

with

$$C := \frac{-\pi_a^2}{24a} - 6ka + \frac{p_\phi^2}{2a^3} + a^3 V(\phi) \tag{4.4.32}$$

and where π_a, a, π_ϕ, ϕ and N are all to be varied independently.

To derive the Wheeler-DeWitt equation for this system we need first to confront some of the problems we have mentioned already. In particular:

1. the classical variable a satisfies the inequality $a \geq 0$;
2. the classical expression $\pi_a^2/24a$ will lead to operator ordering problems in the quantum theory.

The first problem can be tackled in several different ways:

1. Ignore the problem and impose standard commutation relations

---

[21] For simplicity, I have chosen units in which $16\pi G/c^4 = 1$.

$$[\widehat{a}, \widehat{\pi}_a] = i\hbar \tag{4.4.33}$$

even though we know that this leads to a spectrum for $\widehat{a}$ which is the entire real line. The Hilbert space will be $L^2(\mathbb{R}, da)$ with the operators defined in the usual way as

$$(\widehat{a}\psi)(a) := a\psi(a) \tag{4.4.34}$$

$$(\widehat{\pi}_a\psi)(a) := -i\hbar\frac{d\psi}{da}(a). \tag{4.4.35}$$

The problem in this approach is to give some physical meaning to the negative values of a.

2. Insist on using the Hilbert space $L^2(\mathbb{R}_+, da)$ of functions which are concentrated on $\mathbb{R}_+$ but keep the definitions (4.4.34-35). The conjugate momentum $\widehat{\pi}_a$ is no longer self-adjoint but nevertheless it is possible to arrange for the operator $\widehat{C}_\perp$ to be a self-adjoint function of $\widehat{a}, \widehat{\pi}_a, \widehat{\pi}_a^\dagger, \widehat{\phi}$ and $\widehat{\pi}_\phi$.
3. Do a canonical transformation at the classical level to a new variable Ω defined by $a = e^\Omega$. This new variable ranges freely over the entire real line and can therefore be quantised as part of a conventional set of commutation relations using the Hilbert space $L^2(\mathbb{R}, d\Omega)$. The conjugate variable is $\pi_\Omega := e^{-\Omega}\pi_a$ and the constraint becomes

$$C := e^{-3\Omega}\left(\frac{-\pi_\Omega^2}{24} + \frac{\pi_\phi^2}{2}\right) - 6ke^\Omega + e^{3\Omega}V(\phi). \tag{4.4.36}$$

4. Use the affine relations (4.4.8). The affine momentum p_a is related classically to the canonical momentum π_a by $p_a := a\pi_a$ and the constraint becomes

$$C := \frac{1}{a^3}\left(\frac{-p_a^2}{24} + \frac{\pi_\phi^2}{2}\right) - 6ka + a^3V(\phi). \tag{4.4.37}$$

A self-adjoint representation of the affine commutation relations $[\widehat{a}, \widehat{p}_a] = i\hbar\widehat{a}$ can be defined on the Hilbert space $L^2(\mathbb{R}_+, da/a)$ with

$$(\widehat{a}\psi)(a) := a\psi(a) \tag{4.4.38}$$

$$(\widehat{p}_a\psi)(a) := -i\hbar a\frac{d\psi}{da}(a). \tag{4.4.39}$$

Note that the transformation $a := e^\Omega$ sets up an equivalence between this approach and the previous one.

To get a sample Wheeler-DeWitt equation, let us choose method 3 with the Ω variable satisfying standard commutation relations. Then, ignoring the operator ordering problem, we get

$$\left(\hbar^2 e^{-3\Omega}\left(\frac{1}{24}\frac{\partial^2}{\partial\Omega^2} - \frac{1}{2}\frac{\partial^2}{\partial\phi^2}\right) - 6ke^{\Omega} + e^{3\Omega}V(\phi)\right)\psi(\Omega,\phi) = 0. \tag{4.4.40}$$

This simple model illustrates many of the features of the full canonical theory of general relativity. We will consider the model further in the chapter dealing with the problem of time in quantum gravity. For a recent treatment using path integral techniques see Linden and Perry (1991).

4.5 The Ashtekar Variables

4.5.1 The Basic Ideas

One of the potentially most important ways of trying to solve the Wheeler-DeWitt equation is via the use of the Ashtekar canonical variables. The constraints (4.3.22–23) are much simpler when expressed in terms of these new variables, and a whole range of possibilities is opened up for constructing non-perturbative quantisations of gravity. The Ashtekar programme is complex and many-faceted, and is fully worthy of a course of lectures in its own right. I can only touch on a few of the more relevant features here.

The introduction of triads

The basic canonical variable in the Ashtekar programme is not a metric tensor γ_{ab} on Σ but a *triad*. A triad $e^{(i)}(x)$, $i = 1,2,3$ is a set of three linearly-independent cotangent vectors at the point $x \in \Sigma$ with components $e_a^{(i)}(x)$, $a = 1,2,3$. The metric tensor γ_{ab} can be written in terms of these triad fields as

$$\gamma_{ab}(x) = \sum_{i,j=1}^{3} e_a^{(i)}(x)e_b^{(j)}(x)\delta_{ij} \tag{4.5.1}$$

$$= \mathrm{Tr}\big(e_a(x)e_b(x)\big). \tag{4.5.2}$$

In (4.5.2) the index i on $e_a^{(i)}(x)$ is viewed as belonging to the Lie algebra $\mathcal{L}(SO3)$ of the Lie group $SO3$. Thus $e_a(x) := e_a^{(i)}(x)\lambda_i$ where λ_i, $i = 1,2,3$ are generators of $SO3$ normalised so that $\mathrm{Tr}(\lambda_i\lambda_j) = \delta_{ij}$.

Triads are used extensively whenever fermions are introduced into general relativity and result in an extra gauge group $C^\infty(\Sigma, SO3)$. The action of $\Lambda \in C^\infty(\Sigma, SO3)$ on a triad is

$$e_a^{(i)}(x) \mapsto e_a^{(j)}(x)\Lambda_j{}^i(x) \tag{4.5.3}$$

which leaves invariant the metric tensor γ_{ab}.

In the canonical formalism, each triad has a conjugate variable $\pi^a_{(i)}(x)$ and the analogue of the canonical commutation relations (4.4.1–3) is

$$[\widehat{e}_a^{(i)}(x), \widehat{e}_b^{(j)}(x')] = 0 \tag{4.5.4}$$

$$[\widehat{\pi}^a_{(i)}(x), \widehat{\pi}^b_{(j)}(x')] = 0 \tag{4.5.5}$$

$$[\widehat{e}_a^{(i)}(x), \widehat{\pi}^b_{(j)}(x')] = i\hbar\delta^i_j\delta^b_a\delta(x,x'). \tag{4.5.6}$$

with appropriate affine equivalents. In Dirac constraint quantisation, the presence of the extra gauge invariance is reflected in the existence of the constraints

$$\mathcal{G}_i(\widehat{e}, \widehat{\pi})\Psi = 0 \tag{4.5.7}$$

which supplements the constraints (4.4.12-13).

The Ashtekar-Sen connection

The heart of the Ashtekar programme is a connection which is the triadic version of one first introduced by Sen (1982). This object has the two possible forms (Ashtekar 1986,1987) [22]

$$^{\pm}A_a(x) := \Gamma_a(x) \pm iK_a(x) \tag{4.5.8}$$

where the $\mathcal{L}(SO3)$-valued connection $\Gamma_a(x)$ is the usual Christoffel symbol written in a triadic form, and K_a is the triadic form of the extrinsic curvature (4.3.18) with $K_a := K_a{}^{(i)}\lambda_i$.

One of the first key steps in Ashtekar's programme is his proof that the variables $\tilde{e}^a_{(i)} := e^a_{(i)}(\det\gamma)^{\frac{1}{2}}$ and $^{\pm}A_a$ form a classical canonical pair with the Poisson brackets

$$\{\tilde{e}^a_{(i)}(x), \tilde{e}^b_{(j)}(x')\} = 0 \tag{4.5.9}$$

$$\{^{\pm}A_a^{(i)}(x), {}^{\pm}A_b^{(j)}(x')\} = 0 \tag{4.5.10}$$

$$\{\tilde{e}^a_{(i)}(x), {}^{\pm}A_b^{(j)}(x')\} = \pm i\delta^j_i\delta^a_b\delta(x,x'). \tag{4.5.11}$$

---

[22] I am being somewhat cavalier here by dropping various dimensional coupling constants and assorted normalisation factors.

The constraints

The constraints of general relativity become dramatically simpler when expressed in terms of these new variables. Specifically,

$$\mathcal{G}_i = {}^{\pm}\mathcal{D}_a \tilde{e}^a_{(i)} \tag{4.5.12}$$

$$\mathcal{C}_a \approx \mathrm{Tr}(\tilde{e}^{b\,\pm}F_{ba}) \tag{4.5.13}$$

$$\mathcal{C}_\perp \approx \mathrm{Tr}(\tilde{e}^a \tilde{e}^{b\,\pm}F_{ab}) \tag{4.5.14}$$

where $\approx$ means equal modulo the constraints, ${}^{\pm}\mathcal{D}$ denotes the covariant derivative using ${}^{\pm}A$, and ${}^{\pm}F_{ab}$ are the usual field strengths for the connections ${}^{\pm}A$

$${}^{\pm}F_{ab} := \partial_a {}^{\pm}A_b - \partial_b {}^{\pm}A_a - [\,{}^{\pm}A_a, {}^{\pm}A_b\,]. \tag{4.5.15}$$

Various things can be said about (4.5.12-14).

1. The constraints are now polynomial functions of the canonical variables. This is the central feature that sustains the hope that the quantum version of the constraints may be more tractable than in conventional canonical quantum gravity.
2. From a spacetime perspective, ${}^{+}A$ is the restriction to the three-surface Σ of the *self-dual* part of the usual spacetime connection. Similarly, ${}^{-}A$ comes from the anti self-dual part. This has lead to new ways of constructing "half-flat" solutions to the classical Einstein equations (Ashtekar *et al* 1988).
3. I have presented the ideas in a strictly canonical form but there is also a covariant lagrangian formalism whose canonical reduction leads directly to the Ashtekar variables (Jacobson and Smolin 1988b). This has recently been extended in several ways to throw new light on the structure of generally covariant gauge theories (*e.g.*, Capovilla *et al* 1989).
4. The formalism introduces *complex* general relativity in an essential way. There are two options:
   a) e and A are both complex, corresponding to a fully complexified form of general relativity;
   b) e is real and A is complex, corresponding to a hybrid theory in which e and A are not true conjugates. In either case it is necessary to restore the reality of the theory at some stage. This is non-trivial step in the quantum theory.
5. The Ashtekar formalism allows *degenerate* triads/metrics and is hence a genuine extension of conventional general relativity.

4.5.2 Quantisation in the A-Representation

In setting up the quantisation of the theory it is necessary to make a choice of either ${}^{+}A$ or ${}^{-}A$ as the canonical variable. It is not helpful to try to use *both* since the Poisson bracket of ${}^{+}A$ with ${}^{-}A$ does not have any simple form. We will select the ${}^{+}A$ variable and drop the ${}^{+}$label in what follows.

The canonical commutation relations for the pair $(\tilde{e}, A)$ are

$$[\widehat{\tilde{e}}^{a}_{(i)}(x), \widehat{\tilde{e}}^{b}_{(j)}(x')] = 0 \tag{4.5.16}$$

$$[\widehat{A}^{(i)}_{a}(x), \widehat{A}^{(j)}_{b}(x')] = 0 \tag{4.5.17}$$

$$[\widehat{\tilde{e}}^{a}_{(i)}(x), \widehat{A}^{(j)}_{b}(x')] = -\hbar\delta^{j}_{i}\delta^{b}_{a}\delta(x, x'). \tag{4.5.18}$$

Note that, even if we use the hybrid form (so that $\widehat{\tilde{e}}$ is hermitian), the operator $\widehat{A}_a$ will not be hermitian, not even in a formal sense. This means the representation theory of (4.5.16-18) is rather ill-defined.

Nevertheless, proceeding by analogy with ordinary quantum field theory and conventional canonical quantum gravity, one can use the commuting algebra (4.5.17) of the A-variables to study representations on functionals Ψ of A in the form (cf. (4.4.16-17))

$$(\widehat{A}^{(i)}_{a}(x)\Psi)[A] = A^{(i)}_{a}(x)\Psi[A] \tag{4.5.19}$$

$$(\widehat{\tilde{e}}^{a}_{(i)}(x)\Psi)[A] = -\hbar\frac{\delta\Psi}{\delta A^{(i)}_{a}(x)}[A]. \tag{4.5.20}$$

Furthermore, to compensate for the complex nature of A, Ψ is required to be a *holomorphic* functional of A, thus restoring the correct count of degrees of freedom.

Note that the variable which is diagonalised is $\widehat{A}_a(x)$—which is like the canonical momentum in the normal theory—whereas the triad variable acts with a functional derivative. Thus the A-representation is analogous to the functional Fourier transform of the normal representation (4.4.16–17), and hence the geometrical interpretation of $\Psi[A]$ is very different from that of the familiar functional $\Psi[\gamma]$.

Imposition of the constraints

In order to study the Dirac constraint quantisation of this system it is first necessary to select an ordering for the operator versions of the constraints (4.5.12-14). Jacobson and Smolin (1988) chose the expressions

$$\widehat{\mathcal{G}}_i(x) = \mathcal{D}_a \frac{\delta}{\delta A_a^{(i)}(x)} \tag{4.5.21}$$

$$\widehat{\mathcal{C}}_a(x) = F_{ab}^{(i)}(x) \frac{\delta}{\delta A_b^{(i)}(x)} \tag{4.5.22}$$

$$\widehat{\mathcal{C}}_\perp(x) = \epsilon^{ij}{}_k F_{ab}^{(k)}(x) \frac{\delta^2}{\delta A_a^{(i)}(x) \delta A_b^{(j)}(x)} \tag{4.5.23}$$

which have the virtue that the operators $\widehat{\mathcal{G}}_i$ and $\widehat{\mathcal{C}}_a$ correctly generate the Lie algebras of the gauge group $C^\infty(\Sigma, SO3)$ and $\mathrm{Diff}(\Sigma)$ respectively.

The remarkable thing about this form for the constraints is that the Wheeler-DeWitt equation $\widehat{\mathcal{C}}_\perp(x)\Psi = 0$ has a number of *exact* solutions (Jacobson and Smolin 1988a):

1. The functional $\Psi[A] = 1$ satisfies all the constraints. $\Psi = 1$ may not seem very exciting, but it is the first exact solution to the equations of quantum gravity! Of course, any understanding of the physical significance of this result requires a knowledge of the measure on the space of connections A, and no serious study of this has yet been made.
2. A large class of holomorphic gauge-invariant functionals (*i.e.*, invariant under the $C^\infty(\Sigma, SO3)$ gauge group) is given by

$$H_\eta[A] := \mathrm{Tr}(\mathcal{P} \exp \oint_\eta A) \tag{4.5.24}$$

   where the path-ordered line integral is over a smooth loop η in Σ.
   Then a formal solution to the Wheeler-DeWitt equation is provided by the product

$$\Psi_{\{\eta_a\}}[A] := \prod_{a \in I} H_{\eta_a}[A] \tag{4.5.25}$$

   over the finite index set I where $\{\eta_a \mid a \in I\}$ is a set of smooth, non-intersecting loops.
3. Solutions can also be obtained using intersecting curves. They involve particular linear combinations of the states that correspond to the different ways in which such curves can be spliced.

These results are impressive, but there are various problems.

1. The proof that the functionals (4.5.25) satisfy the Wheeler-DeWitt equation exploits only the algebraic properties of F_{ab}, specifically the fact that it is skew-

symmetric in the indices a and b. This suggests that these particular solutions cannot capture the full content of the Wheeler-DeWitt equation.

2. The operator products in (4.5.21-23) are ill-defined and contain factors of $\delta^3(0)$. The problem of how to regularise these operators is a difficult one and is currently the subject of some debate. For example, to define a point-split operator it is necessary to introduce a background metric and curve, and the ultimate effect of this unwanted background structure is not clear.
3. The explicit dependence of the solutions on loops in Σ means they are not Diff(Σ) invariant. It is ironic that, in this formalism, the Wheeler-DeWitt equation can be solved fairly easily but the Diff(Σ) constraints—which are regarded as being relatively innocuous in the conventional approach—should now appear so intractable!

4.5.3 The Rovelli-Smolin Variables

One of the most important recent advances in the use of the Ashtekar variables is the development by Rovelli and Smolin (1988) of a new canonical algebra for general relativity. The basic objects are gauge-invariant functions defined on the phase space of the system by

$$T_\eta[\tilde{e}, A] := \mathrm{Tr}(\mathcal{P}\exp\oint_\eta A) \tag{4.5.26}$$

$$T^a_{\eta,s}[\tilde{e}, A] := \mathrm{Tr}(\mathcal{P}\tilde{e}^a(\eta(s))\exp\oint_\eta A) \tag{4.5.27}$$

where $\eta : S^1 \rightarrow \Sigma$ is a smooth loop in Σ and $s \in S^1$ is a particular parameter value. There are analogues for any number of points on the loop. For example, the function with two insertions is

$$T^{a_1 a_1}_{\eta;s_1,s_2}[\tilde{e}, A] := \mathrm{Tr}(\tilde{e}^{a_1}(\eta(s_1))\mathcal{P}_\eta(s_1, s_2)\tilde{e}^{a_2}(\eta(s_2))\mathcal{P}_\eta(s_2, s_1)) \tag{4.5.28}$$

where $\mathcal{P}_\eta(s_1, s_2)$ is the parallel transport operator from the point $\eta(s_1)$ to the point $\eta(s_2)$ along the loop η.

The first important property of these "T-variables" is that they generate *all* functions on the phase space. The second, and crucial, property is that they have a closed Poisson bracket algebra which can be written somewhat symbolically as

$$\{T^n, T^m\} \sim T^{n+m-1} \tag{4.5.29}$$

where the superscript represents the number of insertions (thus (4.5.26), (4.5.27) and (4.5.28) are T^0, T^1 and T^2 functions respectively). Note that the T^0 and T^1 variables generate a closed subalgebra of the complete T-algebra.

The central idea of the Rovelli-Smolin programme is to use this algebra as the basic algebra whose representations determine the quantum theory. In particular, they construct a type of Fock space quantisation in which the analogue of a "n-particle" state is a function $\psi(\eta_1, \eta_2 \ldots \eta_n)$ of n loops. The T^0 variable acts like a creation operator, for example

$$(T_\eta \psi_2)(\xi) := \psi_2(\eta, \xi), \tag{4.5.30}$$

whereas the T^1 operators map each n-loop sector into itself. By this means Rovelli and Smolin succeed in constructing a deformation (with powers of $\hbar$ as coefficients) of the classical T-algebra.

The next major step is to construct the constraints (4.5.13-14) as a limit of sequences of these T-variables. Rovelli and Smolin then show that the longitudinal constraints $\widehat{C}_\perp(x)\Psi = 0$ (*i.e.,* the Wheeler-DeWitt equation) can be satisfied provided the Fock-space functions $\psi(\eta_1, \eta_2 \ldots \eta_n)$ are concentrated on smooth, non-intersecting loops.

Of course, these states are not Diff(Σ) invariant because the diffeomorphism group moves the loops around. However, a state that *is* Diff(Σ)-invariant can be found by requiring the n-loop functions $\psi(\eta_1, \eta_2 \ldots \eta_n)$ to be constant on the Diff(Σ) orbits, which are the *link* classes of the manifold Σ. Thus, by these means, Rovelli and Smolin claim to have constructed exact solutions to *all* the constraints of quantum gravity.

This is a most impressive result but, as always, there are various difficulties tucked away in the formalism.

1. The Fock space construction is rather formal. In particular, it is not known what inner product should be placed on the states.
2. The classical T-variables are not real and, correspondingly, there is no reason to expect the quantum operators to be hermitian. This has several drawbacks:
   a) The problem of deciding what inner product to choose is made much harder.
   b) It is difficult to construct any general representation theory of the T-algebra. This means it is also hard to assess the significance of the particular Fock-like representation that has been found.

3. The constraints are singular limits of the T variables and generate the usual unpleasant problems of operator-ordering and regularisation.
4. It is difficult to relate the T-variables to anything in conventional general relativity. This makes it difficult to interpret the quantum results.

Much more could be said about this approach and the best ways of tackling the problems which arise. The latter are certainly non-trivial but, nevertheless, the Rovelli-Smolin algebra is one of the most intriguing ways yet suggested of solving the constraints of canonical quantum gravity.

5. The Problem of Time in Quantum Gravity

5.1 General Remarks

5.1.1 The Major Issues

The question of 'time' is one of the deepest issues that must be faced in a quantum theory of gravity. It is intimately related to a whole variety of interpretational problems, including

1. the status of the notion of *probability* and the extent to which it can be said to be *conserved*;
2. the associated concepts of *causality* and *unitarity*;
3. the extent to which *spacetime* is a meaningful concept;
4. the way in which our "classical" world *emerged* from some primordial quantum event (in a theory of quantum cosmology).

These difficulties stem from the variety of roles which time plays in the technical and conceptual frameworks of conventional physics:

1. Time is "Heraclitean" in the sense that "time is that which allows contradictory things to occur" (Unruh and Wald 1989). For example, a particle cannot be at different positions at the same time, but it can be at the same position at different times.
2. Time is the parameter with respect to which change is measured.
3. In the case of quantum theory we have in addition that

a) the scalar product of states is defined at a single time;

b) complete sets of commuting observables are defined at a single time;

c) measurements are carried out at a single time, and states specify the probabilities of the results of such measurements.

Several important questions arise concerning these properties of time in the case of a quantum theory of gravity.

1. To what extent can time continue to play these roles?
2. In particular, is the concept of time *exact*, or can it only be implemented in some *approximate*, perhaps semi-classical, sense? If so, the notions of probability, unitarity, causality and spacetime will also only be approximate.

   But if 'probability' is not an exact concept does this not mean that quantum theory itself is only approximate? If so, all existing approaches to quantum gravity seem doomed from the outset!
3. In classical general relativity, time is only a local coordinate, and a manifold may need to be covered by more than one coordinate chart. How, if at all, is this reflected in the quantum theory of gravity?

5.1.2 The Role of Background Structure

In conventional quantum physics, 'time' is not an observable in the normal sense since it is not represented by any operator but is instead treated as a background parameter which, as in classical physics, is used to mark the evolution of the system. This is true not only of non-relativistic quantum theory but also of relativistic particle dynamics and quantum field theory. Thus 'time' is arguably part of the classical background which plays such a crucial role in the "Copenhagen" interpretation of the theory.

It is not surprising therefore that, within any particular approach to quantum gravity, the nature of the problem of time depends strongly on the type of background structure that is present. For example, there is no particular difficulty in the particle physics schemes since time can be measured with respect to the background Minkowski metric, thus taking us back to the familiar realm of special relativity and quantum field theory.

Of course, whether or not such a measure of time is *physically* correct is another matter. This is part of the general question of the extent to which the spacetime concepts of classical general relativity can be properly captured by a weak-field perturbation around a Minkowskian background. For example, the peculiar behaviour of the lightcones (and therefore time) at the event horizon of a black hole is not the easiest thing to reproduce in a graviton-based picture.

Similar remarks apply to extensions of the particle physics approach in which the background is a more general spacetime manifold $\mathcal{M}$ with the gravitational fields—real or virtual—being required to be asymptotically flat. In such a situation it may be possible to define an asymptotic time which can be used to discuss a graviton S-matrix but, once again, it is not clear that this limited notion of time is sufficient to address all the questions we might wish to raise in a quantum theory of gravity.

In the canonical approach to quantum gravity the question of 'time' is very different. The background is now the *three*-dimensional manifold Σ. If Σ is non-compact then, as above, the asymptotic structure might be used to define an absolute time that is linked to the surface terms ignored in the discussions in §4.

However, in the "cosmological" case in which Σ is compact, the technical heart of the theory is the constraints on the $12\infty^3$ canonical variables $\big(\gamma_{ab}(x), \pi^{cd}(x)\big)$ defined on Σ, and the notion of time (and therefore spacetime) has to be *extracted* in some way from these variables, or from any matter fields or particles that may be added to the system. This is the question with which we will be concerned in the rest of this chapter.

The way in which time is related to the canonical variables depends rather on the approach to canonical quantisation that is being adopted.

1. In the scheme in which the system is reduced to true canonical form before it is quantised (§4.4.1), the definition of time is linked to the imposition of a gauge. This can take the form of
   a) an explicit definition of time as some functional of the canonical variables;
   b) a more implicit relation in which a gauge is chosen by requiring some function of the canonical variables to vanish, or
   c) by specifying the lapse function (or its time derivative) to be some specific functional of the canonical variables.
2. In the alternative approach, all $12\infty^3$ variables $\big(\gamma_{ab}(x), \pi^{cd}(x)\big)$ are quantised and the constraints are imposed as restrictions on the state vectors (§4.4.2). This leads

to the Wheeler-DeWitt equation whose interpretation (which we have studiously ignored so far) is deeply related to the meaning of time. Two conflicting views arise at this point:

a) It is possible to extract a concept of time from the Wheeler-DeWitt equation that is exact and which applies at all levels of the theory.

b) The notion of time is only valid in some semi-classical sense which can be extracted from WKB solutions to the Wheeler-DeWitt equation.

5.2 Time as a Classical Variable

5.2.1 The Main Idea

The idea is to employ the approach to canonical quantum gravity discussed in §4.4.1 in which the theory is reduced to a true canonical form before it is quantised. This involves gauge-fixing which, in the present context, means defining a time variable in terms of the gravitational or matter content of the theory.

For example, in a cosmological model, one might like to define time as the functional

$$T[\gamma] := \int_{\Sigma} \left(\det \gamma(x)\right)^{\frac{1}{2}} d^3x \tag{5.2.1}$$

which says that

1. the manifold Σ is always a surface of equal time;
2. the value of this time is equal to the volume of the universe.

Thus, rather than referring to, for example, the simultaneous value of a scalar field ϕ and the volume of the universe at some background time t, one speaks instead of the value of ϕ when the volume of the universe takes on various values.

Such a choice works in a minisuperspace model in which the Diff(Σ) action has already been gauge-fixed (*e.g.*, by choosing $N^a = 0$), but in general we need to remove $1\infty^3$ variables, which involves dropping the requirement that Σ is automatically a hypersurface of equal time. For example, one could define [23]

$$T(x;\gamma] := \left(\det \gamma(x)\right)^{\frac{1}{2}} \tag{5.2.2}$$

[23] The notation $T(x;\gamma]$ means that T is a function of x and a functional of γ.

which says that the value of time at a point $x \in \Sigma$ is equal to $\left(\det\gamma(x)\right)^{\frac{1}{2}}$. This "multi-fingered" time (Kuchar, 1972) is analogous to the Tomonaga multi-time formulation of conventional quantum field theory.

The most general definition of a time function is a functional $T(x;\gamma,\pi]$ on the *phase space* of the system with the interpretation that the value of 'time' at the point $x \in \Sigma$ is equal to $T(x;\gamma,\pi]$. The canonical variables of matter fields can also appear as arguments of T.

5.2.25 Reduction to Canonical Form

The key step is to reduce the constrained system to canonical form. In the Hamiltonian framework we are adopting this involves the following steps.

1. Find the momentum π_T which is conjugate to T so that
$$\{T(x), \pi_T(x')\} = \delta(x, x') \tag{5.2.3}$$
for each point $x \in \Sigma$.
2. At each $x \in \Sigma$, solve the classical constraint $C_\perp(x) = 0$ for π_T as a functional of T and the canonical variables. This functional becomes the Hamiltonian density $H_T(x;\gamma,\pi,T]$ for a conventional canonical system. [24]
3. This system can then be quantised in a standard way and leads to the Tomonaga-Schrödinger equation
$$i\hbar\frac{\delta\Psi}{\delta T(x)} = \widehat{H}_T(x)\Psi \tag{5.2.4}$$

One of the real advantages of this approach is the relative absence of any ambiguity about what the final quantum theory is supposed to mean. The state-vectors give the probability distributions of the results of measurements on the variables that remain after the "gauge" variables have been removed. In particular, there should be no difficulty in imposing a genuine Hilbert space structure on the quantum states.

[24] Strictly speaking, it is first necessary to handle the Diff(Σ) constraints $\mathcal{C}(\xi) = 0$. This can be done by choosing three functionals $S^a(x;\gamma,\pi]$ to serve as coordinates on Σ, or a hybrid scheme can be used in which the spatial constraints are imposed in the form $\widehat{\mathcal{C}}(\xi)\Psi = 0$.

5.2.3 A Minisuperspace Model

Unfortunately, a number of problems arise with this scheme which can be illustrated with the aid of the simple superspace model introduced at the end of §4 (Blyth and Isham 1975). We will choose the cosmological case $k = 1$ where Σ is a three-sphere. Note that there is no explicit x-dependence in this model since, in effect, we have already chosen the Diff(Σ) gauge $N^a(x) = 0$. Thus the question of a multi-fingered time does not arise.

One natural choice for a global "cosmological" time is the radius a of the universe. Thus, according to the discussion above, the reduced Hamiltonian for this system can be found by solving the constraint

$$\frac{-\pi_a^2}{24a} - 6a + \frac{\pi_\phi^2}{2a^3} + a^3 V(\phi) = 0 \tag{5.2.5}$$

for the variable π_a. Hence

$$H_a = \pm\sqrt{24}\big(- 6t^2 + \frac{\pi_\phi^2}{2t^2} + t^4 V(\phi)\big)^{\frac{1}{2}}. \tag{5.2.6}$$

Another possible choice is the "conjugate" variable $t := \pi_a$ which (in the simple case $V(\phi) = 0$) leads to the Hamiltonian

$$H_{\pi_a} = \pm\left(\frac{-\frac{1}{12}t^2 \pm (\frac{1}{144}t^4 + 48\pi_\phi^2)^{\frac{1}{2}}}{24}\right)^{\frac{1}{2}}, \tag{5.2.7}$$

while a natural "matter" definition is $t = \phi$ which yields

$$H_\phi = \pm\big(\frac{a^2\pi_a^2}{12} + 12a^4 - 2a^6 V(t)\big)^{\frac{1}{2}}. \tag{5.2.8}$$

A number of general problems can already been seen from these simple examples.

1. The constraint equation is typically an algebraic equation for the momentum π_T conjugate to the time variable, and therefore has more than one solution. In effect, we are factorising the classical constraint in the form $C_\perp = (\pi_T - \alpha_1)\dots(\pi_T - \alpha_r)$ and then choosing just one of the roots α_i to be the reduced Hamiltonian. This choice is partly determined by various physical requirements: for example, the reduced Hamiltonian should be a positive, hermitian operator. However, there may still be some residual ambiguity in the procedure.
2. The reduced Hamiltonian can be a very non-linear function of the canonical operators and severe operator-ordering problems occur.

3. The reduced Hamiltonian is time dependent. This has several effects.
   a) A time-dependent Hamiltonian means that energy can be fed into, or taken out of, the quantum system. In a normal physical situation this arises because the system is not closed, and then the time-dependence of the Hamiltonian is determined by the environment. However, a compact three-manifold Σ (the "universe") has no external environment, and so the time-dependence is perhaps a little odd.
   b) If $H(t)$ is time-dependent, the Schrödinger equation

$$i\hbar \frac{d\psi_t}{dt} = \widehat{H}(t)\psi_t \tag{5.2.9}$$

   does not lead to the simple second order equation

$$-\hbar^2 \frac{d^2\psi_t}{dt^2} = (\widehat{H}(t))^2\psi_t \tag{5.2.10}$$

   since there is an extra term involving $\frac{d}{dt}\widehat{H}(t)$. In particular, this means that the Schrödinger equations above do *not* imply the Wheeler-DeWitt equation associated with the constraint (5.2.5). This is not necessarily a bad thing but it does show the inequivalence of the different approaches to the canonical quantisation of gravity.
3. There is no natural choice for the time variable and no reason to suppose that different choices will give equivalent quantum theories. This contrasts strongly with the situation in conventional quantum field theory where different global Lorentz frames lead to unitarily equivalent canonical quantum theories.
4. The square roots can be handled as operators if the object whose square root is being taken is hermitian and positive. However, in general this will only be true for a limited range of t (which depends on $V(\phi)$). It might still be possible to take the square-root outside this range but the result will certainly not be hermitian, resulting in a time evolution operator that is non-unitary.

This last problem has been analysed in depth in a series of papers by Hajicek (1986, 1988, 1989, 1990a, 1990b) which explore the general question of the construction of time functions on the phase space of a constrained system including, for example, the existence of solutions to (5.2.3).

The failure of time functions to be global seems to be a generic property of systems of this type but it is not necessarily a disaster. Much work has been done in recent years on quantum theories of the creation of the universe and, in any such theory, something peculiar must necessarily happen to time near the origination event. For example, in the Hartle-Hawking (1983) and Vilenkin schemes (1988) there is a sense in which time becomes imaginary, and the non-unitarity of the canonical picture may sometimes be an analogue of this effect.

5.3 Time and the Wheeler-DeWitt Equation

5.3.1 Internal Time and the Meaning of $\Psi[\gamma]$

For the reasons above most of the discussion of time in canonical quantum gravity has been within the framework in which all $12\infty^3$ variables $(\gamma_{ab}(x), \pi^{cd}(x))$ are quantised and with the constraints imposed on the state vectors as in (4.4.12-13). In the representation on functionals $\Psi[\gamma]$ this leads to the Wheeler-DeWitt equation , and the problem is to understand the relation of this equation to the concept of time.

An appropriate starting point is to consider the physical meaning of a state-functional Ψ. The operators $\widehat{\gamma}_{ab}(x)$ and $\widehat{\pi}^{cd}(x)$ are defined as in (4.4.16–17) in order that they be *hermitian* with respect to the inner product defined using the "measure" [25] $\mathcal{D}[\gamma]$ on the space Riem(Σ) of all Riemannian metrics on Σ (for our present purposes, the technical problems mentioned in §4.4.5 that are associated with this assumption can be ignored). Similarly, in discussing whether or not the constraint operators $\widehat{\mathcal{C}}_\perp(x)$ and $\widehat{\mathcal{C}}_a(x)$ are, or should be, hermitian, it is the inner product

$$\langle\Psi_1|\Psi_2\rangle := \int_{\mathrm{Riem}(\Sigma)} \Psi_1^*[\gamma]\Psi_2[\gamma]\,\mathcal{D}[\gamma] \tag{5.3.1}$$

on $L^2(\mathrm{Riem}(\Sigma), \mathcal{D}[\gamma])$ which is assumed.

In analogy with normal canonical quantum field theory, this suggests an interpretation of Ψ in which, if a measurement is made of the metric-tensor on the three-manifold Σ, [26] the probability of finding the result in a measureable subset B of Riem(Σ) is

$$\mathrm{Prob}(\gamma \in B; \Psi) := \int_B |\Psi[\gamma]|^2\,\mathcal{D}[\gamma]. \tag{5.3.2}$$

---

[25] The notation $\mathcal{D}[\gamma]$ refers to the non-existing Lebesgue measure on Riem(Σ). I will keep the symbol $d\mu[\gamma]$ for a genuine measure.

[26] Of course, it is not really feasible to measure the gravitational field *everywhere* on Σ, since that is the entire universe! However, (5.3.2) is still meaningful if a trace is taken over the fields on the regions of Σ where no measurements are made. This produces an effective density matrix for the state in the restricted region where measurements *are* made.

If a matter field ϕ is present then the state vector is of the form $\Psi[\gamma, \phi]$, with an analogous interpretation.

This view of $|\Psi[\gamma]|^2$ as a probability density for three-metrics may seem plausible and, indeed, an interpretation of this type has been used a number of times, especially in the context of quantum cosmology. However, as mentioned in §4.4.5, the counting of degrees of freedom is wrong.

A metric tensor γ_{ab} corresponds to $6\infty^3$ degrees of freedom, *i.e.*, six degrees of freedom per spatial point in Σ. Three of these are associated with the action of Diff(Σ) and can be removed by an explicit gauge-fixing or by imposing the constraints $(\widehat{\mathcal{C}}_a(\xi)\Psi)[\gamma] = 0$ as in (4.4.18). In the latter case, the state vector is defined on superspace Riem(Σ)/Diff(Σ), which we will denote by writing $\Psi({}^{(3)}\mathcal{G})$. The inner product (5.3.1) must be replaced by a measure on Riem(Σ)/Diff(Σ) which, following the example of Yang-Mills theory, can be done in a variety of ways. The state-functional $\Psi({}^{(3)}\mathcal{G})$ then has the interpretation of a probability amplitude for three-*geometries*,*i.e.*, equivalence classes of metric tensors with respect to the action of Diff(Σ).

However, the counting is still not correct since a three-geometry ${}^{(3)}\mathcal{G}$ has $6\infty^3 - 3\infty^3 = 3\infty^3$ degrees of freedom whereas the true gravitational system should have a count of $2\infty^3$. The reason for this discrepency can be seen from two perspectives:

1. We still have to impose the constraints $\widehat{\mathcal{C}}_\perp(x)\Psi = 0$, *i.e.*, solve the Wheeler-DeWitt equation. This will remove a further $1\infty^3$ degrees of freedom, leaving $3\infty^3 - 1\infty^3 = 2\infty^3$ as required.
2. From our earlier discussion, we know that the concept of 'time' has to be extracted from the system in some way. In the present context this means that $1\infty^3$ of the $3\infty^3$ degrees of freedom are associated with a multi-fingered time $T(x;\gamma]$ which, once identified, will leave the correct count of $2\infty^3$.

The close relation between these two positions can been seen clearly with the aid of our minisuperspace model. The first view says we must impose the constraint leading to the Wheeler-DeWitt equation (4.4.40). However, the second view says that some (in this case, one) of the configuration variables should be identified with 'time'. For example, if Ω is chosen for this purpose (we recall that $\Omega := \log a$ where a is the radius of the three-sphere) then the Wheeler-DeWitt equation (4.4.40) is to be interpreted as a partial differential equation which describes how the ***physical*** mode ϕ evolves as a function of the time variable Ω.

It is now clear what is wrong physically with the inner product (5.3.1) and the associated interpretation (5.3.2). In the minisuperspace model, the analogue of (5.3.1) is the usual inner product on the Hilbert space $L^2(\mathbb{R}^2, d\Omega\, d\phi)$, which means we are integrating over the *time* variable as well as the physical configuration variable ϕ, and this is something that is not done in conventional quantum theory. Thus the Hilbert space structure (5.3.1) is physically inappropriate because it includes an integral over an internal multi-fingered time.

5.3.2 The Analogue of the Klein-Gordon Equation

The moral of all this is clear. The physical Hilbert space structure should not be imposed on arbitrary functionals on $\mathrm{Riem}(\Sigma)/\mathrm{Diff}(\Sigma)$ but only on ***solutions*** to the Wheeler-DeWitt equation, and in such a way as to make explicit the role of internal time.

The standard procedure at this point is to explore the analogy between the Wheeler-DeWitt equation (4.4.22) and the Klein-Gordon equation of a particle moving in a curved space with an arbitrary, time-dependent potential (DeWitt 1967a). The validity of this analogy can be seen especially clearly in the simple minisuperspace model (4.4.40).

The key observation is that the DeWitt metric (4.4.23) on $\mathrm{Riem}(\Sigma)$ has a hyperbolic character in which, roughly speaking, the conformal modes of the metric play the role of time-like directions, *i.e.*, the transformation $\gamma_{ab}(x) \to F(x)\gamma_{ab}(x)$, $F(x) > 0$, is a "time-like" displacement in $\mathrm{Riem}(\Sigma)$. This suggests defining the formal scalar product

$$\langle\Psi_1|\Psi_2\rangle := i\prod_x \int_\Sigma d\Sigma^{ab}(x)\Psi_1^*[\gamma]\left(\mathcal{G}_{abcd}(\gamma(x))\frac{\overrightarrow{\delta}}{\delta\gamma_{cd}(x)} - \frac{\overleftarrow{\delta}}{\delta\gamma_{cd}(x)}\mathcal{G}_{abcd}(\gamma(x))\right)\Psi_2[\gamma] \tag{5.3.3}$$

where $d\Sigma^{ab}(x)$ is the directed surface element in $\mathrm{Riem}(\Sigma)$ at the point $x \in \Sigma$.

Of course, considerable care would be needed to make this expression rigorous. For example:

1. it necessary to take account of the $\mathrm{Diff}(\Sigma)$-invariance and then project the inner product down to $\mathrm{Riem}(\Sigma)/\mathrm{Diff}(\Sigma)$;

2. the precise form of the scalar product depends on how the operator-ordering problem in the Wheeler-DeWitt equation is solved.

However, the essential idea is clear. In particular, (5.5.3) has the important property that it is invariant under deformations of the "spatial" hypersurface in Riem(Σ). This is the multi-fingered time analogue of the requirement in the normal Klein-Gordon equation that the scalar product be time independent.

In the minisuperspace example (4.4.40) the Wheeler-DeWitt equation can be simplified by multiplying [27] both sides by $e^{3\Omega}$ to give

$$\left(\hbar^2\left(\frac{1}{24}\frac{\partial^2}{\partial\Omega^2}-\frac{1}{2}\frac{\partial^2}{\partial\phi^2}\right)-6ke^{4\Omega}+e^{6\Omega}V(\phi)\right)\psi(\Omega,\phi)=0. \tag{5.3.4}$$

The associated scalar product is simply

$$\langle\psi_1|\psi_2\rangle = i\int_{\Omega=\mathrm{const}} d\phi\,\left(\psi_1^*\frac{\partial\psi_2}{\partial\Omega}-\psi_2\frac{\partial\psi_1^*}{\partial\Omega}\right) \tag{5.3.5}$$

which is conserved in Ω-time by virtue of (5.3.4).

Unfortunately, the right hand side of (5.3.3) cannot serve as a genuine Hilbert space inner product because it is not positive definite. There are two aspects to this:

1. The scalar product vanishes for real functions Ψ. This is not a trivial property since the Wheeler-DeWitt equation is a *real* functional differential equation and therefore introduces no intrinsic coupling between the real and imaginary parts of Ψ. This is fundamentally different from the situation with a genuine time-dependent Schrödinger equation where the explicit appearance of the complex number i enforces such a coupling. This has lead some authors to question the applicability of the inner product (5.3.3) in quantum gravity (Barbour and Smolin 1988; Barbour 1990).
2. Complex solutions Ψ to the Wheeler-DeWitt equation can exist for which $\langle\Psi|\Psi\rangle < 0$.

In the normal Klein-Gordon equation the latter problem is solved by dividing the states into those that have positive frequencies and those whose frequencies are negative, and then restricting the physical space to be just the former. This strategy also works in the

---

[27] This is a contentious step. It is true that a constraint equation $\widehat{C}\Psi = 0$ is not formally affected by multiplying on the left by any invertible operator, but this "renormalisation" of $\widehat{C}$ affects the total constraint algebra, and the implications of this need to be considered at some point. Of course, it also affects the hermiticity properties of the original constraint.

simple minisuperspace model (5.3.5) but, unfortunately, no such division is possible in the general case.

This problem arises already for a relativistic particle moving in a curved spacetime with a metric $g_{\mu\nu}$ and an arbitrary time-dependent potential V. A division into positive and negative frequencies can only be made if $g_{\mu\nu}$ possesses a time-like conformal Killing vector which scales V in an appropriate way (*e.g.*, Kuchar 1981a, 1991). Unfortunately, Kuchar (1981b) showed that Riem(Σ) admits no such vector, and therefore there is no possibility of defining physical states to be some sort of analogue of the positive frequency solutions of the normal Klein-Gordon equation.

5.3.3 Third Quantisation

There have been several different reactions to the failure of the Klein-Gordon approach to the Wheeler-DeWitt equation. In the case of a relativistic particle with an external spacetime-dependent metric and/or potential, the failure to isolate positive-frequency solutions is connected with a breakdown of the one-particle interpretation of the theory. The standard resolution is to second-quantise the system by turning the Klein-Gordon wave function into a quantum field.

It has been suggested several times that a similar process might be needed in quantum gravity with $\Psi[\gamma]$ becoming an operator in some bigger Hilbert space. This procedure is usually called "third" quantisation since the original Wheeler-DeWitt equation is already the result of a quantum field theory (*e.g.*, Coleman 1988, Giddings and Strominger 1988, McGuigan 1988,1989). However, it is unclear what this means, or if the problem of time can really be solved in this way.

1. The approach to third quantisation that is closest to conventional quantum field theory is to construct a Fock space whose "one-particle" sector is associated with the functionals $\Psi[\gamma]$. The problem is how to interpret the states that are tensor products of these basic vectors. In the case of particles, if $|x_1\rangle$ and $|x_2\rangle$ are one-particle states corresponding to a particle localised at points x_1 and x_2 respectively, then the tensor product $|x_1\rangle\,|x_2\rangle$ describes a pair of particles, both of which move in the *same* physical three-space.

   However, if $|\gamma_1\rangle$ and $|\gamma_2\rangle$ are metric states with $\Psi[\gamma]=\langle\gamma|\Psi\rangle$, it is not clear to what the product state $|\gamma_1\rangle\,|\gamma_2\rangle$ refers. The simplest thing might be to say that γ_1 and

γ_2 are both metrics on the same space Σ, but this has no obvious geometrical or physical interpretation. [28]

2. A more meaningful geometrical interpretation is to say that γ_1 and γ_2 are metrics on *different* copies of Σ or, equivalently, a single metric on the disjoint union of two copies of Σ. However this also raises a number of problems.
   a) The transition from a state $|\gamma\rangle$ to a state $|\gamma_1\rangle\,|\gamma_2\rangle$ corresponds to a topology change in which Σ bifurcates into two copies of itself. But this is unlikely to be compatible with the Wheeler-DeWitt equation.
   b) A normal Fock space construction involves Bose statistics but it is not clear physically what it means to say that $|\gamma_1,\gamma_2\rangle$ is symmetric in γ_1 and γ_2. The two copies of Σ are disjoint and presumably no causal connection can be made between them. So what is the operational significance of a Bose structure?
   c) Once the original space Σ has been allowed to bifurcate into a pair of copies of itself it seems logical to extend the topology change to include an *arbitrary* final three-manifold. Thus the scope of the theory is increased enormously.
   d) It is anyway difficult to see how the bifurcation of space helps with the problem of time. Presumably the Wheeler-DeWitt equation will continue to hold in each disconnected piece of the universe, and the problem of internal time reappears in each.

5.3.4 The Use of Extrinsic Time

It could be argued that the problem with the Klein-Gordon approach is that too much weignt is being given to the hyperbolic nature of the Wheeler-DeWitt equation as the source of the concept of time. In §5.2 we have already considered the possibility of a classical time function $T(x;\gamma,\pi]$ that is defined on the full *phase* space of the system, and perhaps this could also be of assistance in discussing the operator constraints $\widehat{C}_\perp(x)\Psi=0$.

The model of parametrised particle dynamics

The central idea is often illustrated with the aid of a simple model of parametrised dynamics. Consider a particle with an action principle

[28] Private communication with Karel Kuchar.

$$S[q] = \int \mathcal{L}(q, \frac{dq}{dT})\, dT \tag{5.3.6}$$

where T is standard Newtonian time. Now let τ be an arbitrary parameter on the path along which the particle moves, subject only to the condition that $T(\tau)$ is a smooth, monotonically increasing function of τ. Furthermore, append $T(\tau)$ to the set of dynamical variables so that the action principle (5.3.6) becomes

$$S[q, T] = \int \mathcal{L}(q, \frac{1}{\dot{T}}\frac{dq}{d\tau})\dot{T}\, d\tau \tag{5.3.7}$$

where the new time parameter is now τ and $\dot{T} := dT/d\tau$.

This action is invariant under redefinitions of time $\tau \mapsto \tau'(\tau)$ and can therefore serve as a simple model of the diffeomorphism invariance of general relativity. In particular, a full canonical analysis (*e.g.*, Kuchar 1981, Unruh and Wald 1989) shows that the variable π_T conjugate to T satisfies the constraint

$$C := \pi_T + H(q, \pi_q) = 0. \tag{5.3.8}$$

Furthermore, the entire set of equations of motion, including (5.3.8), can be derived from the first-order action

$$S[q, \pi_q, T, \pi_T, N] = \int \left(\pi_\alpha \frac{dq^\alpha}{d\tau} - N(\tau)C\right) d\tau \tag{5.3.9}$$

where $N(\tau)$ is a Lagrange multiplier added to enforce the constraint (5.3.8) and the q^α variables are the original configuration variables q in (5.3.6) plus the variable T. The analogy with the gravitational action (4.3.24) is obvious.

The Dirac quantisation of this system is straightforward. If the classical quantity π_T is represented by the operator $-i\hbar\frac{d}{dT}$ then the equation $\widehat{C}\psi = 0$ becomes

$$i\hbar\frac{d\psi}{dT} = \widehat{H}\psi \tag{5.3.10}$$

which is just the conventional time-dependent Schrödinger equation for the original variables q. This illustrates particularly clearly the problem of trying to relate the Hilbert space structure on the original space—on which all the operators are defined—to the inner product on the physical states.

The canonical transformation

The ideal situation would be if the constraints $C_\perp(x)$ in (4.3.23) were already in the form (5.3.8), *i.e.*, ***linear*** in the momentum which is conjugate to the time variable.

Unfortunately, the momentum variables in (4.3.23) have a manifestly quadratic form and so the constraints are not of the desired type. Of course, this is why the Wheeler-DeWitt equation resembles a Klein-Gordon equation.

This means that a "good" time variable (in the sense of (5.3.8) cannot be found merely by restricting the time function T to be a functional $T(x;\gamma]$ of the metric alone. The idea of the "extrinsic" time approach is to extend this to include functionals of π by performing a canonical transformation

$$\left(\gamma_{ab}(x), \pi^{cd}(x)\right) \rightarrow \left(T(x), \pi_T(x); \gamma_A(x), \pi^B(x)\right) \tag{5.3.11}$$

to new variables in which the longitudinal constraint has the form

$$\pi_T(x) + \mathcal{H}(x; T, \gamma_A, \pi^B] = 0 \tag{5.3.12}$$

without the appearance of the difficult square roots, or more general many-valued solutions, which we saw earlier. A Dirac quantisation of this system will lead to the many-fingered time Schrödinger equation

$$i\hbar\frac{\delta\Psi}{\delta T(x)} = \widehat{\mathcal{H}}(x)\Psi. \tag{5.3.13}$$

The York time

This technique works well for linearised gravity (Arnowitt, Deser and Misner 1962) but it is very difficult to find such a canonical transformation in the full theory. The best developed decomposition of variables of this type is the York time in which the choice of time variable is essentially (York 1972)

$$T(x;\gamma\pi] := \frac{2}{3}(\det\gamma(x))^{-\frac{1}{2}}\pi_a{}^a(x) \tag{5.3.14}$$

$$-\pi_T(x;\gamma\pi] := \left(\det\gamma(x)\right)^{\frac{1}{2}} > 0 \tag{5.3.15}$$

and the rest of the canonical variables are unchanged (the full analysis also includes a discussion of the spatial variables).

Defining $\Phi := (-\pi_T)^{\frac{1}{6}} = (\det\gamma)^{\frac{1}{12}}$, the longitudinal constraint becomes

$$\kappa^2\left(\frac{1}{8}\pi^{ab}\pi_{ab}\Phi^{-7} - \frac{3}{64}T^2\Phi^5\right) - \frac{1}{\kappa^2}\left(\frac{1}{8}R(\gamma)\Phi - \nabla_\gamma\Phi\right) = 0 \tag{5.3.16}$$

where ∇_γ is the Laplacian operator constructed from the metric γ. This is a non-linear, elliptic, partial differential equation for Φ which, in principle, can be solved for Φ (and hence for $\pi_T = -(\det\gamma)^{\frac{1}{2}})$ as a functional $-\mathcal{H}$ of the remaining canonical variables. [29] Thus the longitudinal constraints can be written in the form

$$\pi_T(x) + \mathcal{H}(x; T, \gamma, \pi] = 0 \tag{5.3.17}$$

which leads to a many-fingered time Schrödinger equation. The fact that the transformation to the York time is a genuine canonical map was shown by Isenberg and Marsden (1984).

The main problems with this scheme are as follows.

1. The elliptic PDE in (5.3.16) is very non-linear and hence the relation between the solution Φ and the parameters in the equation is only implicit. It is quite unclear if, or how, such a function can be translated into a quantum operator. In particular:
   a) It is important that the Hamiltonian in (5.3.17) should be a hermitian operator but this requires the solution Φ to be expressed as a hermitian function of the operators $\widehat{\gamma}_{ab}$ and $\widehat{\pi}^{cd}$. The associated operator-ordering problem is likely to be quite horrendous.
   b) It may also be very difficult to ensure that $\widehat{\mathcal{H}}$ is a positive operator.
2. Different classical canonical transformations that reduce the constraint to a quasi-linear form like (5.3.12) are likely to produce inequivalent quantum theories.

5.3.5 Quantum Clocks and Material Reference Systems

We have seen that it is difficult to produce a satisfactory definition of time using only the canonical variables of the gravitational field. However, in practice, time is not measured in this way: real clocks are made of matter with definite properties—an observation that has generated a recent flurry of interest in the idea of "quantum clocks" with the hope that they may lead to a more tractable interpretation of the Wheeler-DeWitt equation.

In a sense, the idea of matter clocks has already been implicit in what has been said so far. For example, one of the dynamical variables in our simple minisuperspace model is a homogeneous mode of a scalar field ϕ, and this can be used to define time. Indeed, (5.2.8) is the Hamiltonian obtained from the choice $t = \phi$ for a classical time

---

[29] This is true for a compact manifold Σ provided one avoids the maximal slices where $\pi_a{}^a = 0$ and other odd pathologies.

variable. However, a "quantum clock" does not mean an arbitrary ollection of particle or matter field variables, but rather a device whose self-interaction and coupling to the gravitational field are deliberately optimised to serve as a measure of time. The important question is the extent to which the ability to measure (or, more precisely, define) time using one of these systems is compatible with its realisation as a real object; for example, it must have a positive energy.

Two different types of system have been considered. The first is a clock that can only measure time at a single point in space along its worldline. The second is a cloud of clocks that fills the space Σ and which can therefore provide a global measure of time (and position in space too). This latter idea has its origin in the old classical notion of a reference fluid and was first applied to quantum gravity in a major way by DeWitt (1963).

The work of Kuchar *et al*

There have been a number of discussions of quantum clocks and related topics but I will concentrate here on some particularly interesting recent work by Kuchar. The key idea is to use a coordinate condition on the spacetime manifold but to treat it as something that is physically realisable rather than just being part of the background structure. For example, Kuchar and Torre (1990) use a Gaussian reference frame in the following way.

1. The Gaussian coordinate conditions are $g^{00}(y) = -1$, $g^{0a}(y) = 0$ but a more covariant-looking expression can be obtained by introducing a foliation $X : \Sigma \times \mathbb{R} \to \mathcal{M}$ with an inverse $X^{-1} : \mathcal{M} \to \Sigma \times \mathbb{R}$ that, as in (4.3.2), can be written in the form $X^{-1}(y) = (S(y), T(y))$ where $S : \mathcal{M} \to \Sigma$ and $T : \mathcal{M} \to \mathbb{R}$. Then the Gaussian condition on g with respect to this foliation has the form

$$g^{\mu\nu}(y)T,_{\mu}(y)T,_{\nu}(y) = -1 \tag{5.3.18}$$

$$g^{\mu\nu}(y)T,_{\mu}(y)S^{a},_{\nu}(y) = 0 \tag{5.3.19}$$

   where $a = 1, 2, 3$ denotes any coordinate system on Σ (the condition (5.3.19) is independent of the choice).

2. The conditions (5.3.18-19) are now imposed on the system with the aid of Lagrange multipliers M and M_a, $a = 1, 2, 3$. The extra term in the action is

$$S_F[g,M,M] := \int_{\mathcal{M}} \left(-\frac{1}{2}M\,(g^{\mu\nu}T_{,\mu}T_{,\nu}+1) + M_a g^{\mu\nu}T_{,\mu}S^a{}_{,\nu}\right)(-\det g)^{\frac{1}{2}}\, d^4y. \tag{5.3.20}$$

3. Although (5.3.20) is covariantly defined, the Diff($\mathcal{M}$) invariance is lost since the functions T and S^a are fixed, external, objects in the theory, and the only dynamical variable is $g_{\mu\nu}$. However, invariance can be restored by changing the status of T and S^a to ***bona fide*** dynamical variables that must be varied in the action. This is analogous to the way in which reparametrisation invariance is added to (5.3.6) by the introduction of the function $T(\tau)$.
4. These four extra variables T, S^a can be interpreted as describing a material system—the "Gaussian reference fluid"—that interacts with the gravitational field and has its own energy-momentum tensor, which turns out to be that of a heat conducting fluid.
5. A complete canonical analysis shows that the final longitudinal constraint can be written in the form

$$\Pi_A(x) + \mathcal{H}_A(x; T, \gamma, \pi] = 0 \tag{5.3.21}$$

where $\Pi_A(x)$, $A = 0\ldots3$ are the momenta conjugate to the variables $T^A(x)$ (with $T^a \equiv S^a$). Note that the four quantities T^A are now functions on Σ and can be interpreted as describing an ***embedding*** of Σ in the spacetime manifold $\Sigma \times \mathbb{R}$.

The crucial property of (5.3.21) is that it is ***linear*** in the Π_A variables. Hence Dirac constraint quantisation gives the first-order functional differential equation

$$i\hbar\frac{\delta\Psi}{\delta T^A(x)}[\gamma, T] = \mathcal{H}(x; T, \gamma, -i\hbar\frac{\delta}{\delta\gamma}]\Psi[\gamma, T] \tag{5.3.22}$$

in the embedding variables T^A. The first-order form of this equation means there is no problem in interpreting $\Psi[\gamma, T]$ as a probability density for finding the metric γ on the hypersurface in $\Sigma \times \mathbb{R}$ specified by the four embedding variables T^A.

What goes wrong?

This work of Kuchar and Torre is very elegant and can be extended in a variety of ways, including a treatment (Kuchar 1991) of the "unimodular-gauge" approach to the problem of time suggested by Henneaux and Teitelboim (1989), Unruh and Wald (1989) and Brown and York (1989).

Equation (5.3.22) has the great virtue of leading to what is probably the most plausible-sounding interpretation so far of the various approaches to understanding the Wheeler-DeWitt equation. Unfortunately, it also suffers from a significant defect: the energy-momentum tensor of the embedding variables fails to satisfy the famous energy-conditions of general relativity and therefore the system cannot be regarded as physically realistic.

This seems to be an endemic property of all attempts to produce realistic models for physical reference systems. For example, Unruh and Wald (1989) show that any quantum clock whose Hamiltonian is positive must necessarily sometimes "run backwards", and a similar conclusion was obtained by Hartle [30] in his investigations into the problem of time in the context of a path-integral approach to quantum gravity.

Thus it seems that the concept of 'time' may only be an approximate one—a proposition that has been suggested by a number of workers in a variety of different contexts. For example, it forms an important ingredient in Vilenkin's approach to quantum cosmology (Vilenkin 1988, 1989) as well as in Rovelli's recent work on material reference systems (Rovelli 1990, 1991). However, as was emphasised earlier, the thesis that the concept of time is not exact has many implications for the general applicability of conventional quantum ideas to quantum gravity. It is therefore appropriate to discuss next what is one of the most sustained developments of this position: the so-called "WKB" resolution of the problem of time.

5.3.6 The WKB Approach to the Problem of Time

The WKB Solutions of the Wheeler-DeWitt equation

The main idea is that the notion of 'time' is only meaningful in a quantum state that has some semi-classical component which can serve as its measure. Thus

1. time is an *approximate*, semi-classical concept;
2. the definition of time depends on the quantum *state* of the system.

In particular, time would have no meaning in a quantum cosmology in which the universe never emerges into a semi-classical regime. Thus this approach pays some deference to the general idea that time is part of the "classical background" assumed in the Copenhagen interpretation of quantum theory.

The starting point is the WKB technique for obtaining an approximate solution to the Wheeler-DeWitt equation (see Singh and Padmanabhan (1989) for a comprehen-

[30] Hartle's work is comprehensive and wide-ranging and deserves far more than the cursory reference given here. It is only lack of space that stops me from trying to give a proper account of it.

sive review). This involves looking for a solution in the form

$$\Psi[\gamma] = A[\gamma]e^{\frac{i}{\hbar}S[\gamma]} \tag{5.3.23}$$

where $A[\gamma]$ is a "slowly varying" function of γ in the sense that

$$\left|\frac{\delta A}{\delta\gamma_{ab}}\right| \ll \left|A\frac{\delta S}{\delta\gamma_{ab}}\right|. \tag{5.3.24}$$

Inserting (5.3.23-24) into the Wheeler-DeWitt equation shows that the phase S satisfies the Hamilton-Jacobi equation [31]

$$-\kappa^2\mathcal{G}_{abcd}(\gamma(x))\frac{\delta S[\gamma]}{\delta\gamma_{ac}(x)}\frac{\delta S[\gamma]}{\delta\gamma_{bd}(x)} - \frac{(\det\gamma(x))^{\frac{1}{2}}}{\kappa^2}R(\gamma(x)) = 0 \tag{5.3.25}$$

of classical general relativity. The amplitude factor A satisfies the "conservation law"

$$\mathcal{G}_{abcd}(\gamma(x))\frac{\delta}{\delta\gamma_{ac}(x)}\left(A^2[\gamma]\frac{\delta S[\gamma]}{\delta\gamma_{bd}(x)}\right) = 0. \tag{5.3.26}$$

The concept of time

The recent surge of interest in the application of WKB methods to the problem of time began with an important paper by Banks (1985). A particularly useful discussion in the context of quantum cosmology is Halliwell (1987, 1990). The basic idea is a type of Born-Oppenheimer approach and has been developed further in papers by Zeh (1986, 1988), Brout *et al* (1987,1990), Englert 1989, Kiefer and Singh (1990), and others.

Banks considered a system of matter fields coupled to gravity such that the Wheeler-DeWitt equation for the combined system can be written as (*cf* (4.4.22))

$$-\hbar^2\kappa^2\mathcal{G}_{abcd}(\gamma)\frac{\delta^2\Psi}{\delta\gamma_{ac}\delta\gamma_{bd}}[\gamma] - \left(\frac{(det\gamma)^{\frac{1}{2}}}{\kappa^2}R(\gamma) - \kappa^2\widehat{\mathcal{H}}_m\right)\Psi[\gamma] = 0 \tag{5.3.27}$$

where $\mathcal{H}_m$ is the Hamiltonian density for the matter.

The next step is to find solutions to (5.3.27) of the special form

$$\Psi[\gamma,\phi] = A[\gamma]e^{iS[\gamma]}\Phi[\gamma,\phi] \tag{5.3.28}$$

where A and S are as above and ϕ denotes the collection of matter variables. Inserting this ansatz into the Wheeler-DeWitt equation (5.3.27) and keeping just the lowest order terms in the expansion gives the equation

---

[31] We are essentially performing an expansion in powers of $\hbar\kappa^2 \simeq L_P$

$$-2i\hbar\mathcal{G}_{abcd}(\gamma(x))\frac{\delta S[\gamma]}{\delta\gamma_{ac}(x)}\frac{\delta\Phi[\gamma,\phi]}{\delta\gamma_{bd}(x)}+(\widehat{\mathcal{H}}_m(x)\Phi)[\gamma,\phi]=0. \tag{5.3.29}$$

The key step now is to find a functional $T(x;\gamma]$ such that

$$2\mathcal{G}_{abcd}(\gamma(x))\frac{\delta S[\gamma]}{\delta\gamma_{ac}(x)}\frac{\delta T(x';\gamma]}{\delta\gamma_{bd}(x)}=\delta(x,x'). \tag{5.3.30}$$

Then, if Φ depends on γ only through T, (5.3.29) becomes the multi-fingered time Schrödinger equation

$$i\hbar\frac{\delta\Phi}{\delta T(x)}=\widehat{\mathcal{H}}_m(x)\Phi, \tag{5.3.31}$$

which is the desired result.

There are various things to note about this construction.

1. The equation (5.3.30) shows clearly how the definition of the time function $T(x;\gamma]$ depends on the quantum state via the choice of the solution S to the Hamiltonian-Jacobi equation (5.3.25).
2. Only the gravitational field is treated in this semiclassical way. The matter fields are fully quantised, although the probability associated with (5.3.31) is only conserved to the same order in the L_P expansion to which (5.3.31) is valid.
3. Padmanabhan (1990) and Greensite (1990a,b,c) have suggested an extension of the idea above in which time is defined with respect to the phase of *any* solution of the Wheeler-DeWitt equation, not just one that is in WKB form. Their construction can also be regarded as a result of requiring quantum gravity to satisfy the Ehrenfest principle.
4. If S is a real solution to the Hamilton-Jacobi equation (5.3.25), the function $\Psi = Ae^{\frac{i}{\hbar}S}$ oscillates rapidly along paths in superspace. However, imaginary S solutions can also exist, and these produce an exponential type of behaviour. This is often interpreted as indicating a Riemannian rather than Lorentzian spacetime picture and has been much studied in quantum cosmology with the idea that the universe "tunnels" from an imaginary-time region (Halliwell 1987, 1990).

Problems with the WKB approach

The WKB approach to the definition of time is an interesting one, but it raises a number of difficult questions and problems. For example:

1. There is no evidence that the defining equation (5.3.30) has global solutions on superspace. Indeed, Hajicek's analysis of a minisuperspace model suggests otherwise (Hajicek 1986).

2. The simple WKB approximation breaks down at the turning points of S and a more careful treatment is needed near such regions.
3. The elegant first-order form of (5.3.29) is completely lost at the next order in the WKB approximation. This makes it difficult to assess the proper status of (5.3.29) and to understand how the physics of time changes as one gets nearer to the Planck regime.
4. The ansatz (5.3.28) is only one of a very large number of possible types of solution to the Wheeler-DeWitt equation. Why should it have such a preferred status?
5. Similarly, why should one not consider superpositions of WKB solutions to the Wheeler-DeWitt equation? Indeed, the original discussions of WKB solutions to the Wheeler-DeWitt equation assumed that the aim was to construct a coherent superposition of such solutions to produce a quantum state that approximates a single classical spacetime manifold (Gerlach 1969). However, if the state vector is a sum

$$\Psi[\gamma] := \sum_i A_i[\gamma] e^{\frac{i}{\hbar} S_i[\gamma]} \tag{5.3.32}$$

then each solution S_i of the Hamilton-Jacobi equation will lead to its *own* definition of time! The ensuing difficulty is reminiscent of the Schrödinger cat problem in ordinary quantum theory where a single value for a macroscopic property has to be extracted from a quantum state that is a linear superposition of eigenstates.
6. The alternative is to keep to just a single WKB function, but this also raises several difficulties.
   a) As we remarked earlier in the context of the Klein-Gordon inner product (5.3.13), there is something a little odd about choosing a complex solution Ψ to a functional differential equation that has only real coefficients. As emphasised by Barbour and Smolin (1988) and Barbour (1990) it means that a correlation between the real and imaginary parts of Ψ is imposed in some way. Quantum cosmologists respond that this is precisely what is to be expected in a theory that describes the quantum creation of the universe via the medium of a unique solution to the Wheeler-DeWitt equation. The Vilenkin (1988) scheme does indeed produce a solution that is naturally complex. On the other hand, the Hartle-Hawking ansatz appears to lead to a *real* solution of the equation,

although this issue is clouded by ambiguities in deciding what the ansatz really means.

b) With the aid of Wigner functions and other techniques, Halliwell (1987,1990) interprets a single WKB solution as describing a whole *family* of classical space-times. But this again raises the "Schrödinger cat" problem in the guise of having to decide how any specific spacetime arises.

7. If time is only a semi-classical concept then the notion of probability is also likely to be valid only in an approximate sense. This throws doubt on the utility of quantum theory in general.

6. Quantum Cosmology

6.1 General Remarks

The idea of quantum cosmology was introduced by DeWitt in his fundamental work on the canonical quantisation of gravity (DeWitt 1967a). The subject was developed in a series of papers by Misner (*e.g.*, Misner 1969a,b,c) and generated considerable interest until around the middle of the 1970s. Activity then waned for a while but picked up strongly following the work of Hartle and Hawking on constructing a unique wave function for the universe (Hawking 1982b, Hartle and Hawking 1983), and the parallel approaches by Vilenkin (1982)—summarised in Vilenkin (1988)— and Linde (1984, 1989).

The techniques adopted in the second phase of activity (1982 to the present) differ in several respects from those used in the first phase (1967-1975) although there is much in common, especially in regard to the conceptual problems that arise. However, it is important to appreciate that the goals of the present programme are somewhat different from those of the earlier period. The original workers in quantum cosmology were mainly concerned with understanding the effects of quantum theory on the spacetime singularity that appears in the description of the big-bang given by classical general relativity. Thus one of the primary aims was to replace the deficient classical description with a picture that is physically and mathematically meaningful.

This remains one of the goals of the modern programme but it is supplemented with a desire to produce a theory that predicts a *unique* quantum state of the universe. This is far more ambitious than merely providing a believable picture of the big-bang. It is also potentially open to observational refutation or confirmation.

The subject of quantum cosmology generates a number of conceptual issues over and above those that are present in quantum gravity in general. I will only give a broad sketch of the subject here and refer to the literature for details on the assorted attempts to unravel the various strands of thought. An excellent recent review article is Halliwell (1990a) who has also performed a great service by compiling a very comprehensive bibliography of papers on quantum cosmology (Halliwell 1990b).

6.2 Problems With the Copenhagen Interpretation of Quantum Theory

Any attempt to apply quantum theory to the universe as a whole is bound to encounter severe problems with the conventional "Copenhagen" interpretation of quantum theory. Of course, this itself has been the subject of many lively debates in the past and it is still a topic that generates considerable interest. The problems that arise in the interpretation of quantum theory can usefully be categorised under four headings:

1. the role of measurement;
2. the collapse of the state vector;
3. the meaning of 'probability' as applied in quantum theory ;
4. quantum entanglement (the Einstein-Podolsky-Rosen paradox, Bell's inequalities, *etc*).

These topics are closely interlinked and can be visualised as the four vertices of a tetrahegon, each vertex being closely tied to the remaining three. They raise particularly difficult questions in the context of quantum cosmology, as we can see by considering first the meaning of 'probability'.

6.2.1 The Meaning of Probability

The concept of probability is used in several different ways in theoretical physics, and each poses peculiar problems for quantum cosmology. Let us consider them in turn.

The statistical interpretation

For a physicist, a natural approach to probability is the "statistical" interpretation in terms of relative frequencies of repeated experiments. But this raises obvious difficulties for quantum cosmology. For example, if such a theory predicts that, with probability 0.87, I will be typing these notes on January 27, 1991, what does the 0.87 mean? And is this prediction to be viewed as a great success for the theory? The statistical interpretation says merely that if the experiment is repeated many times, then in 87% of the cases I will indeed be found typing the notes. But how is the experiment to be repeated?

Opinions are sharply divided about the meaning of such a prediction. We are faced here with the old problem of how a probabilistic statement is to be understood if it is intended to apply to a single event. It might perhaps be possible to employ a subjective interpretation involving a "degree of belief" but, in practice, the usual tendency is towards some implementation of the idea of relative frequencies. In the case of quantum cosmology, one school of thought maintains that there really is a large ensemble of universes to which the usual statistical interpretation can apply. Various origins have been suggested for such an ensemble. For example:

1. The universe may continually expand and contract giving rise to an endless series of big-bangs and big-crunches. Each cycle is to be thought of as a member of the statistical ensemble.
2. The initial big-bang may have been such that many parts of the physical universe emerged in a causally disconnected way (although still part of a single, spatially-connected universe). The probabilistic statements of quantum cosmology then apply to the statistical distributions of properties in these separate regions.
3. The physical universe may be spatially *disconnected*, as a result perhaps of some quantum topology operation near the big-bang. The statistical predictions could then be applied to these separate "universes".

Note that one common feature of all these suggestions is the extreme difficulty of *testing* them!

The epistemic interpretation

In classical physics, the most common meaning attached to probability is an epistemic one. Thus it is assumed that probabilities measure the extent to which an (idealised) observer *knows* the precise details of the system. Our ignorance of these details arises from the great complexity of the system (for example, a box of gas) but the underlying assumption is that the uncertainty in the actual behaviour can be made arbitrarily small—at least in principle—with the aid of more precise measurements (or a more detailed theoretical description) of the *possessed* properties of the constituents of the system.

However, unless hidden variables are posited (and these bring problems of their own), the situation in quantum theory is very different. It is generally no longer consistent to talk of individual entities 'possessing' a value for an observable: it is more as if the property concerned is *latent* in the system until such time as an observation "brings it into being". As John Bell has put it, "beables" are replaced by "observables", and the verb "to be" becomes "to be measured".

Needless to say, the situation is even worse in quantum cosmology. For who is the cosmic observer whose "knowledge" is reflected in the wave function of the universe?

Probability for a single system

It is clear that what is needed is an interpretation of probability that is meaningful for a single event and which therefore has a chance of being applied to the universe in its entirety.

One possibility is to stipulate that the only predictions which have any real meaning are those that affirm something with probability one (or zero). This is consistent with the observation that the only time in ordinary quantum theory when it is unequivocably meaningful to say that an individual system *possesses* a value for a quantity is if a measurement of the corresponding observable is guaranteed to yield that value with probability one. However, this is the exception rather than the rule, and to stick to just probability-one statements is to ignore most of the theory.

On the other hand, there are some intriguing "meta-theorems" that purport to show that if the actual universe is sufficiently complex to admit many physical copies of some system, then one of the probability-one statements is that measurements made on this *physical* ensemble will reproduce the statistical distributions of conventional quantum physics (Finkelstein 1963, Hartle 1968, Geroch 1984). This is comforting in its way but it does little to help in understanding the content of quantum cosmology as it applies to the universe as a whole.

6.2.2 The Role of Measurements

One of the problems with the conventional interpretation of quantum theory is the instrumentalist approach that it takes to the relation between theory and reality. In particular, it posits a radical dualism between a system and an observer, the role of the latter being two-fold:

1. the observer *prepares* states of the quantum system;
2. he/she/it then makes *measurements* of some observable pertaining to the system.

The emphasis on measurements (*i.e.*, observables rather than beables) is due mainly to the perceived lack of any quantum-theoretical analogue of the microstates of classical statistical physics. The gap which is thereby placed between the system being observed, and that which is making the observation, contrasts sharply with the situation in classical physics where all aspects of the physical world are deemed to be describable by the same set of physical laws. It poses the obvious problems of (i) when is an interaction between two systems to count as a measurement by one of a property of the other?, and (ii) what happens if an attempt is made to restore a degree of unity by describing the measurement process in quantum mechanical terms? There is no universally accepted answer to either of these questions.

In the context of quantum cosmology such a dualistic picture is particularly problematical. For example:

1. One again we encounter the question of who, or what, is the cosmic observer?
2. The act of measurement is often associated with a *reduction* of the state vector into an eigenvector of the observable measured. This poses no problems for a frequentist interpretation of probability (one simply selects the subsensemble on which the particular value has been obtained) but it raises real difficulties for a more realist, "single-system" interpretation such as we might like to employ in quantum cosmology. There are two quite different reactions to this situation.
   a) The first is to say that the state vector never collapses in quantum cosmology. Indeed, it is difficult to see how such an operation can be consistent with the Wheeler-DeWitt equation. This means that the notion in conventional quantum theory of state vector collapse is deemed to be merely a phenomenological way of describing what really happens in terms of an idealised separation into a system plus an observer. The no-collapse position is almost always associated

with some version of the infamous "many-worlds" interpretation of quantum theory.

b) The more radical view is to say that not only does the state-vector collapse, but this is actually *caused* by the effects of gravity. This is the non-linear picture of quantum theory espoused by Penrose (1986, 1987) to which we referred earlier (see also Karolyhazy *et al* 1986).

3. In conventional quantum mechanics, measurements are made at a fixed time. How is this idea to be reconciled with the many difficulties that arise in quantum gravity when trying to develop an exact concept of time (§5)?

6.3 Where Does a Unique Quantum State Come From?

There are several approaches that might be adopted in an attempt to construct a *unique* quantum state for the universe. For example:

1. Perhaps the post Planck-length quantum state is essentially *independent* of what happened before: a situation that is typical of the so-called "chaotic cosmologies" (Linde 1989). It is difficult to believe that this can be true at the level of fine detail, but it might work for some of the more coarse-grained features of the cosmos.
2. The requirements of *mathematical consistency* may be so restrictive that they single out just one quantum state. Such a view is implicit in DeWitt's seminal discussion of the canonical quantum gravity programme (DeWitt 1967a).
3. It may be possible to select a unique solution to the Wheeler-DeWitt equation by imposing some natural conditions on the values of $\Psi[\gamma]$ at the *boundary* of superspace Riem(Σ)/Diff(Σ). For example, DeWitt's suggestion was that Ψ should be forced to vanish there. Note that what is hoped for is that a relatively *simple* condition on the boundary of Riem(Σ)/Diff(Σ) may lead to a state vector that can describe the very *complex* world we see around us today.

   Vilenkin's scheme is of this type (Vilenkin 1988, 1989). He starts by dividing the boundary of superspace into a "singular part"—which corresponds to the limit of three-geometries that are the slices of a spacetime with a singularity, and which converge towards that singularity—and a regular part, which corresponds to the limit of three-geometries that are the slices of a four-manifold with no singular limit (for example, the equitorial slices of a four-sphere that tend towards a pole on the sphere). Vilenkin requires that the solution to the Wheeler-DeWitt equation must be such that, at the singular boundary, it carries flux *out* of superspace. This suggestion has many attractive features. However, there are several difficulties:

a) The flux is measured using the Klein-Gordon inner product (5.3.3). In the present context, this gives a positive probability density only in the WKB approximation. Thus the prescription as stated may only work at a semi-classical level. A more inclusive version is not known.

b) In a few simple minisuperspace models the prescription does give a unique wave function. However, it is by no means clear that this would apply in the complete theory or, indeed, even in a superspace model with a larger number of degrees of freedom.

4. There may be a simple/natural *ansatz* for constructing a specific solution to the Wheeler-DeWitt equation. An example is the Hartle-Hawking scheme which we discussed briefly in the context of the Euclidean functional integral (4.4.29). Several points should be made about this idea.

   a) It is difficult to give the Hartle-Hawking ansatz a precise mathematical meaning although it is certainly very elegant in a rather formal sense. In practice, it tends to be used in a semi-classical way with the functional integral (4.4.29) being approximated by the value of $e^{-S_E[g]}$ at a classical solution g to the Riemannian field equations subject to the Hartle-Hartle boundary conditions.

   b) More detailed calculations suggest that the value of Ψ calculated using (4.4.29) may depend on the *contour* in the complexified Riem($\mathcal{M}$)-space that must be chosen to make the integral convergent (Halliwell and Hartle 1990).

   c) These problems might be overcome to some extent if the Hartle-Hawking ansatz could be rewritten in terms of boundary conditions on superspace. Unfortunately, this has not yet been possible in any definitive way.

In assessing any of these schemes it should not be forgotten that none of the traditional approaches to tackling the Wheeler-DeWitt equation can surmount the fierce operator divergences that arise. The recent developments in solving the Wheeler-DeWitt equation using Ashtekar's new variables are therefore of particular interest and could have important implications for the future development of quantum cosmology.

6.4 How Does the Classical Universe Emerge?

One of the major problems for any QCU theory is to explain how the world we see around us "emerged" from the primordial quantum soup of the creation region. In particular, the theory needs to explain why it is that (provided we do not probe too deeply) the universe has such a strikingly classical structure.

A unique state Ψ produced by a QCU theory needs to satisfy a number of requirements that culminate in the classical world we see around us.

1. A definition of 'time' must be provided (perhaps using the state Ψ in some way) so as to give meaning to the idea of the "beginning" of the universe and the creation of space and/or time.
2. The initial pure state must have properties that make it look like a mixed, thermal state shortly after the big-bang.
3. If the theory is to agree with modern ideas on early universe physics the state must lead to a period of inflationary expansion. This could happen in two different ways:
   a) An inflationary expansion could be a direct prediction of the QCU theory which generated the state Ψ.
   b) It could happen in a more "phenomenological" way if the QCU theory is blended into one of the, now-standard, ways of producing a quantum-induced inflationary phase.
4. When interpreted in terms of an appropriate definition of 'time', the state vector must have the property that the system that it now describes evolves according to an essentially classical set of dynamical equations.

It is a highly non-trivial technical task to produce a QCU theory that can satisfy these requirements. However, there is also the crucial conceptual problem of understanding what is really *meant* when we talk about "the" classical universe.

It is certainly possible in principle to construct wave-packet solutions $\Psi[\gamma]$ to the Wheeler-DeWitt equation that are "peaked" around a single spacetime in the sense that the support of Ψ is primarily on the set of all spacelike slices of a single Lorentzian metric on a manifold $\mathcal{M} \simeq \Sigma \times \mathbb{R}$. However, the QCU schemes that have been invented so far (*e.g.*, Hartle-Hawking or Vilenkin) produce states whose WKB approximation is a more like a single e^{iS}, or a finite sum of such, rather than a true wave packet.

And, as was mentioned in §5, state-vectors of this type seem to correspond to a whole *family* of classical spacetimes.

This is the problem of Schrödinger's cat writ large: "How can a single macroscopic world emerge from a state vector that looks like a superposition of states corresponding to different macroscopic possibilities?" This is often translated into the question of how a pure state can turn into what is, in effect, a mixed state so that the interference between the different "branches" is lost.

This problem has been extensively discussed in the context of the "measurement problem" of conventional quantum theory, and the view that has been repeatedly presented in one form or another is that the phase interference between the different branches of the superposition is essentially wiped out by the effects of the systems's environment. In recent years this idea has been developed to particulary good effect by Zurek (1981, 1982, 1991), and spirited attempts have been made to develop an analogous concept in the context of quantum cosmology.

The problem of course is that the universe does not have an environment, and therefore any "decoherence" that takes place must be understood purely in terms of the internal structure of the theory; in effect, some of the variables must serve to decohere the rest (*e.g.,* Halliwell 1989, Gell-Mann and Hartle 1990, Hartle 1990). A number of ideas have been developed in the cited references and the whole subject is most intriguing. However, one cannot help but feel a little uneasy about what is being done.

Firstly, there is still the question of what 'probability' is supposed to mean, even if the replacement of the pure state by a mixed state means that the probability under consideration is of a more classical type.

Secondly, decoherence is only an approximate concept. For example, in minisuperspace models that have been studied, the gravitational modes of the system are described using the WKB approximation. This may work in a purely phenomenological sense but it leaves a big question mark about what the state vector Ψ "really" means, especially in the deep quantum regime near the creation event itself.

Thus—as is so often the case in quantum gravity—we seem to end where we began, with the unequivocal statement:

Physics at the Planck length is something of a mystery!

ACKNOWLEDGEMENTS

In recent years I have benefited from discussions with Carlo Rovelli, Lee Smolin and, most particularly, Karel Kuchar. I am very grateful to them all.

References

1 E. Alvarez: Phys. Lett. **201B**, 73 1988
2 E. Alvarez: Rev. Mod. Phys. **61**, 561 (1989)
3 E. Alvarez, J. Cspedes and E. Verdaquer: (1990) CERN preprint
4 D. Amati and C. Klimcik: Phys. Letts. **B219**, 443 (1989)
5 D. Amati D, M. Ciafaloni M and G. Veneziano: Phys. Letts. **197**, 81 (1987)
6 D. Amati, M. Ciafaloni and G. Veneziano: Int. Jour. Theor. Phys. **A3**, 1615 (1988)
7 D. Amati, M. Ciafaloni and G. Veneziano: Phys. Letts. **216**, 41 (1989)
8 I. Antoniadis, C. Bachas, J. Ellis and D.V. Nanopoulos: Phys. Letts. **B211**, 393 (1988)
9 I. Antoniadis, C. Bachas C, J. Ellis and D.V. Nanopoulos: Nucl. Phys. **B328**, 117 (1989)
10 R. Arnowitt, S. Deser and C.W. Misner: Gravitation: An Introduction to Current Research (1962) in ed. S. Witten (Wiley, New York). The dynamics of general relativity
11 A. Ashtekar: Phys. Rev. Letts. **57**, 2244 (1986)
12 A. Ashtekar: Phys. Rev. **36**, 1587 (1987)
13 A. Ashtekar , T. Jacobson and L. Smolin: Comm. Math. Phys. **115**, 631 (1988)
14 T. Banks: Nucl. Phys. **B249**, 332 (1985)
15 U. Bannier: Comm. Math. Phys. **118**, 163 (1988)
16 A.O. Barvinsky: Phys. Letts. **B195**, 344 (1986)
17 A.O.Barvinsky and V.N. Ponomariov: Phys. Letts. **B167**, 289 (1986)
18 P.G. Bergmann and A. Komar: Int. Jour. Theor. Phys. **5**, 15 (1972)
19 W. Blyth and C.J. Isham: Phys. Rev. **D11**, 768 (1975)
20 N. Bohr and L. Rosenfeld: **128**, (1933) Kgl. Danek Vidensk. Selsk, Math.-fys. Medd. On the question of the measurability of the electromagnetic field strength
21 D.G. Boulware and S. Deser: Ann. Phys. (N.Y.) **89**, 193 (1975)
22 R. Brout, G. Horowitz and D. Wiel: Phys. Letts. **B192**, 318(1987)
23 R. Brout and G. Venturi: Phys. Ref. **D39**, 2436 (1989)
24 J.D. Brown and J.W. York: Phys. Rev. **D40**, 3312 (1989)
25 P. Candelas, G. Horowitz, A. Strominger and E. Witten: Nucl. Phys. **B258**, 46 (1985)
26 R. Capovilla, T. Jacobson and J. Dell: Phys. Letts. **63**, 2325 (1989)
27 S. Coleman: Nucl. Phys. **B310**, 643 (1988)
28 B.S. DeWitt: in Gravitation: An Introduction to Current Research (1963) ed. Witten L. (Wiley, New York) The quantization of geometry
29 B.S. DeWitt: Dynamical Theory of Groups and Fields (Gordon and Breach, New York) (1965)
30 B.S. DeWitt: Phys. Rev. **160**, 1113 (1967a)
31 B.S DeWitt: Phys. Rev. **162**,1195 (1967b)
32 B.S. DeWitt: Phys. Rev. **162**, 1239 (1967c)
33 Dirac PAM. Lectures on Quantum Mechanics. Academic Press, New York (1965)
34 J. Earman and J. Norton: (1986) Preprint Pittsburgh University Philosophy Department

35 K. Eppley and E. Hannah: Found. Phys. **7**, 51 (1977)
36 F. Englert: Phys. Letts. **B228**, 111 (1989)
37 M. Fabbrichesi and G. Veneziano: Phys. Letts. **233**, 135 (1989)
38 RP. Feynman: Acta Phys. Pol. **24**, 697 (1963)
39 Finkelstein D. Trans. N.Y. Acad. Sci. **25**, 621 (1963)
40 E.S. Fradkin and A.A. Tseytlin: Phys. Letts. **104B**, 377 (1981)
41 E.S. Fradkin and A.A. Tseytlin: Nucl. Phys. **B201**, 469 (1982)
42 K. Fredenhagen and R. Haag: Comm. Math. **108**, 91 (1987)
43 J.L. Friedman and I. Jack: Phys. Rev. **D37**, 3495 (1988)
44 J.L. Friedman and R. Sorkin: Phys. Rev. Letts. **44**, 1100 (1980)
45 S.A. Fulling: Phys. Rev. **D42**, 4282 (1990)
46 M. Gell-Mann and J.B. Hartle: (1990) in Proceedings of the Santa Fe Institute Workshop on Complexity, Entropy, and the Physics of Information, May 1989. (in press) and in Proceedings of the 3rd International Symposium on The Foundations of Quantum Mechanics in the Light of New Technology, Tokyo 1989, Quantum mechanics in the light of quantum cosmology
47 U.H. Gerlach: Phys. Rev. **117**, 1929 (1969)
48 R. Geroch: Nous. **18**, 617 (1984)
49 S. Giddings and A. Strominger: Nucl. Phys. **B321**, 481 (1988)
50 M.H. Goroff and A. Sagnotti: Phys. Letts. **160B**, 81 (1985)
51 M.H. Goroff and A. Sagnotti: Nucl. Phys. **B226**, 709 (1986)
52 J. Greensite: Nucl. Phys. **B342**, 409 (1990a)
53 J. Greensite: Preprint SFSU-TH-90/1 (1990b)
54 J. Greensite: Preprint NBI-HE-90-38.(1990c)
55 J. Greensite: PreprintNBI-HE-90-71.(1990d)
56 D.J. Gross and P.F. Mende: Phys. Letts. **197B**, 129 (1987)
57 D.J. Gross and PF. Mende: Nucl. Phys. **B303**, 407 (1988)
58 S.N. Gupta: Phys. Rev. **96**, 1683 (1954)
59 R. Güven: Phys. Letts. **191**, 275 (1987)
60 P. Hajicek: Phys. Rev.**D34**, 1040 (1986)
61 P. Hajicek: Phys. Rev.**D38**, 3639 (1988)
62 P. Hajicek: Jour. Math. Phys.**20**, 2488 (1989)
63 P. Hajicek: Class. Quant. Grav. **7**, 861 (1990a)
64 P. Hajicek: Class. Quant. Grav. **7**, 871 (1990b)
65 P. Hajicek and K. Kuchar: Phys. Rev. **D41**, 1091 (1990a)
66 P. Hajicek and K. Kuchar: Jour. Math. Phys. **31**, 172 (1990b)
67 J.J. Halliwell: Phys. Rev. **D36**, 3626 (1987)
68 J.J. Halliwell: Phys. Rev. **D38**, 2468 (1988)
69 J.J. Halliwell: Phys. Rev. **D39**, 2912 (1989)
70 J.J. Halliwell: in Proceedings of the Jerusalem Winter School on Quantum Cosmology and Baby Universes. (1990a) ed. T. Piran (In press). Introductory lectures on quantum cosmology
71 J.J. Halliwell: Int. Jour. Mod. Phys. **A5**, 2473 (1990b)
72 J.J. Halliwell: in Proceedings of the Osgood Hill Conference on Conceptual Problems in Quantum Gravity. (1991), eds. A. Ashtekar and J. Stachel (Boston, Birkhauser) The Wheeler-DeWitt equation and the path integral in minisuperspace quantum cosmology
73 J.J. Halliwell and J.B. Hartle: Phys. Rev. **D41**, 1815 (1990)
74 J.B. Hartle: Amer. Jour. Phys. **36**, 704 (1968)
75 J.B. Hartle: Phys. Rev. **D37**, 2818 (1988a)
76 J.B. Hartle: Phys. Rev. **D37**, 2985 (1988b)
77 J.B. Hartle: Preprint UCSBT90-31. To appear in Proceedings of the Jerusalem Winter School on Quantum Cosmology and Baby Universes ed. T. Piran. The quantum mechanics of cosmology. (1990)
78 J.B. Hartle and S.W. Hawking: Phys. Rev **D28**, 2960 (1983)
79 S.W. Hawking: Nucl. Phys. **B144**, 349 (1978)
80 S.W. Hawking: in General Relativity: An Einstein Centenary Survey. (1979) , eds. S.W. Hawking and W. Israel (Cambridge University Press, Cambridge). The path-integral approach to quantum gravity
81 S.W. Hawking: Comm. Math. Phys. **87**, 395 (1982a)
82 S.W. Hawking: in Astrophysical Cosmology. (1982b), eds. H.A. Brück and M.S. Longair (Pontifica Academia Scientarium, Vatican City). The boundary conditions of the universe
83 S.W. Hawking: Nucl. Phys. **B244**, 135(1984)

84 S.W. Hawking: Phys. Rev. **D37**, 904 (1988)
85 S.W: Hawking and R. LaFlamme: Phys. Letts. **B209**, 39 (1988)
86 G.T. Horowitz: Phys. Rev. **D21**, 1445 (1980)
87 G.T. Horowitz: in Proceedings of the 12th International Conference on General Relativity and Gravitation; Boulder, Colorado. String theory as a quantum theory of gravity. (1989)
88 G.T. Horowitz: Preprint UCSBTH-90-51. (1990a)
89 G.T. Horowitz: in Proceedings of the Strings '90 Conference held at Texas A&M, 1990. (In press). Singularities in string theory. (1990b)
90 G.T. Horowitz and A.R. Steif: Phys. Rev. Letts. **64**, 260 (1990a)
91 G.T. Horowitz and A.R. Steif: (1990b) Preprint UCSB-TH-90-8
92 G.T. Horowitz and A.R. Steif: (1990c) Preprint UCSB-TH-90-8
93 G.T. Horowitz and RM. Wald: Phys. Rev. **D17**, 414 (1978)
94 J. Isenberg and J.E. Marsden: Jour. Geom. and Phys. Phys. Rev. **1**, 85 (1984)
95 C.J. Isham: Phys. Letts. **106B**, 188 (1981)
96 C.J. Isham: (1984) in Relativity, Groups and Topology II: Les Houches, (1983) eds. B.S. DeWitt and R. Stora (North-Holland, Amsterdam). Topological and global aspects of quantum theory
97 C.J. Isham: in Superstrings and Supergravity: Proceedings of the 28th Scottish Universities Summer School in Physics (1985) eds. A.T. Davies and D.G. Sutherland (SUSSP publications, 1986). Aspects of quantum gravity
98 C.J. Isham: (1987) in General Relativity and Gravitation: Proceedings of the 11th International Conference on General Relativity and Gravitation ed M. MacCallum (Cambridge University Press, Cambridge). Quantum gravity—GRG11 review talk
99 C.J. Isham: Class. Quant. Grav. **6**, 1509 (1989)
100 C.J. Isham: (1990) in Proceedings of the Advanced Summer Institute on Physics, Geometry and Topology, 1989, eds. G. Kunstatter and P. Lee. An introduction to general topology and quantum topology
101 C.J. Isham: (1991) in Proceedings of the 1988 Osgood Hill Conference on Conceptual Problems of Quantum Gravity, eds. A. Ashtekar and J. Stachel (Boston, Birkhauser). Canonical groups and the quantization of geometry and topology
102 C.J. Isham and A.C. Kakas: Class. Quant. Grav. **1**, 621 (1984a)
103 C.J. Isham and A.C. Kakas: Class. Quant. Grav. **1**, 633 (1984b)
104 C.J. Isham, Y. Kubyshin and P. Renteln: Class. Quant. Grav. **7**, 1053 (1990)
105 C.J. Isham, R. Penrose and D.W. Sciama: (Clarendon Press, Oxford).(1975)
106 C.J. Isham, R. Penrose and D.W. Sciama: Quantum Gravity 2: A Second Oxford Symposium. (Clarendon Press, Oxford). (1981)
107 T. Jacobson and L. Smolin: Nucl. Phys. **B299**, 295 (1988a)
108 T. Jacobson T and L. Smolin: Class. Quant. Grav. **5**, 583 (1988b)
109 F. Krolyhzy, A. Frenkel and B. Lukcs: (1986) in Quantum Concepts in Space and Time, eds. R. Penrose and C.J. Isham (Clarendon Press, Oxford). On the possible role of gravity in the reduction of the state vector
110 T.W.B. Kibble: (1981) in Quantum Gravity 2: A Second Oxford Symposium, eds. Isham CJ, Penrose R and Sciama DW. (Clarendon Press, Oxford). Is a semi-classical theory of gravity viable?
111 T.W.B. Kibble and S. Randjbar-Daemi: Jour. Phys. **A13**, 141 (1980)
112 C. Kiefer and T.P. Singh: (1990) Preprint
113 A. Komar: Phys.Rev. **D19**, 2908 (1979a)
114 A. Komar: Phys. Rev. **D20**, 830 (1979b)
115 K. Konishi, G. Paffuti and P. Provero: Phys. Letts. **234**, 276 (1990)
116 K. Kuchar: Jour. Math. Phys. **13**, 768 (1972)
117 K. Kuchar: Jour. Math. Phys. **17**, 177 (1976a)
118 K. Kuchar: Journ. Math. Phys. **17**, 792 (1976b)
119 K. Kuchar: Jour. Math. Phys. **17**, 801 (1976c)
120 K. Kuchar: Jour. Math.Phys. **18**, 1589 (1977)
121 K. Kuchar: (1981a) in Quantum Gravity 2: A Second Oxford Symposium, eds. C.J. Isham, R. Penrose and D.W. Sciama. (Clarendon Press, Oxford). Canonical methods of quantization
122 K. Kuchar: Jour. Math. Phys. **22**, 2640 (1981b)
123 K. Kuchar: Phys. Rev.**D34**, 3031 (1986a)
124 K. Kuchar: Phys.Rev. **D34**, 3044 (1986b)
125 K. Kuchar: Phys. Rev. **D35**, 596 (1987)

126 K. Kuchar: (1991) in Proceedings of the Osgood Hill Conference on Conceptual Problems in Quantum Gravity, eds. A. Ashtekar and J. Stachel. (Boston, Birkhauser) The problem of time in canonical quantization of relativistic systems
127 J. Klauder and E.W. Aslaksen: Phys. Rev. **D2**, 272 (1970)
128 J. Klauder: (1970) in Relativity, eds. M.S. Carmeli, S.L. Flicker and L. Witten (Plenum, New York). Soluble models of quantum gravitation
129 J. Klauder: Ann. Phys. **117**, 19 (1979)
130 A. Linde: Nuov. Cim. **39**, 401 (1984)
131 A. Linde: (1989) Particle Physics and Inflationary Cosmology. (Academic Press, Boston)
132 N. Linden and M. Perry: (1991) Preprint DAMTP R90/29
133 M. McGuigan: Phys. Rev. **D38**, 3031 (1988)
134 M. McGuigan: Phys. Rev. **D39**, 2229 (1989)
135 C.W. Misner: Phys. Rev. **186**, 1319(1969a)
136 C.W. Misner: Phys. Rev. **186**, 1328(1969b)
137 C.W. Misner: Phys. Rev. Letts. **22**,1071 (1969c)
138 C. Moller: Les theories relativistes de la gravitation (1962) (CNRS, Paris)
139 T. Padmanabhan: Gen. Rel. Grav. **17**, 215 (1985a)
140 T. Padmanabhan: Ann. Phys. **165**, 38 (1985b)
141 T. Padmanabhan: Quant. Grav. **4**, L107 (1987)
142 T. Padmanabhan: Jour. Phys. **35**, L199 (1990)
143 D.N. Page and C.D. Geilker: Phys. Rev. Letts. **47**, 979 (1981) See also the ensuing correspondence. Phys. Rev. Letts. **48**, 520
144 W. Pauli: Theory of Relativity. (1967) (Pergamon Press, London)
145 R. Penrose: (1975) in Quantum Gravity: An Oxford Symposium, eds. C.J. Isham, R. Penrose and D.W. Sciama. (Clarendon Press, Oxford). Twistor theory, its aims and achievements
146 R. Penrose: (1977) in Magic Without Magic ed. J. Klauder (W.H. Freeman, Reading). On the nature of quantum geometry
147 Penrose R. (1986) in Quantum Concepts in Space and Time, eds. R. Penrose and C.J. Isham (Clarendon Press, Oxford). Gravity and state vector reduction
148 R. Penrose: (1987) in Three Hundred Years of Gravitation, eds. S.W. Hawking and W. Israel (Cambridge University Press, Cambridge). Newton, quantum theory and reality
149 A. Peres: Nuov. Cim. **26**, 53 (1962)
150 M. Pilati: Phys. Rev. **D26**, 2645 (1982)
151 M. Pilati: Phys. Rev. **D28**, 729 (1983)
152 S. Randjbar-Daemi, B.S. Kay and T.W.B. Kibble: Phys. Rev. Letts. **91B**, 417 (1980)
153 L. Rosenfeld: Nucl. Phys. **40**, 353 (1963)
154 C. Rovelli: Preprint SISSA 101/89/ep
155 C. Rovelli: (1990) Preprint PITT-90-11
156 C. Rovelli: Phys. Rev. (1991) inpress
157 C. Rovelli and L. Smolin: Nucl. Phys. **B331**, 80 (1990)
158 J. Scherk and J. Schwarz: Nucl.Phys. **B81**, 118 (1974)
159 A. Sen: Phys. Lett. **119B**, 89(1982)
160 J.Z. Simon: (1990) Preprint UCSBTH-90-43
161 T.P. Singh and T. Padmanabhan: Ann. Phys. **196**, 296 (1989)
162 L. Smolin: (1991) in Proceedings of the Osgood Hill Conference on Conceptual Problems in Quantum Gravity eds. A. Ashtekar and J. Stachel (Boston, Birkhauser). Space and time in the quantum universe.
163 J. Stachel: (1987) in General Relativity and Gravitation: Proceedings of the 11th International Conference on General Relativity and Gravitation ed M. MacCallum (Cambridge University Press, Cambridge). How Einstein discovered general relativity: a historical tale with some contemporary morals
164 K. Stelle: Phys. Rev. **D16**, 953 (1977)
165 W.M. Suen: Phys. Rev. Letts. **62**, 2217 (1989a)
166 W.M. Suen: Phys. Rev. **D40**, 315 (1989b)
167 C. Teitelboim: Phys. Rev. **D25**, 3159 (1982)
168 C. Teitelboim: Phys. Rev. **D28**, 297 (1983a)
169 C. Teitelboim: Phys. Rev. **D28**, 310 (1983b)
170 E.T: Tomboulis: Phys. Rev. Letts. **52**, 1173 (1984)
171 W. Unruh and R. Wald: Phys. Rev. **D40**, 2598 (1989)
172 G. Veneziano: (1989a) Preprint CERN-TH.5366/89

173 G. Veneziano: (1989a)Preprint CERN-TH.5581/89
174 A. Vilenkin: Phys. Letts. B117, 25 (1982)
175 A. Vilenkin: Phys. Rev. D37, 888 (1988)
176 A. Vilenkin: Phys. Rev. D39, 1116 (1989)
177 R. Wald: Phys.Rev. D33, 3613 (1986)
178 S. Weinberg: Phys. Rev. B135, 1049 (1964)
179 S. Weinberg: Phys. Rev. B138, 988 (1965)
180 S. Weinberg: (1979) in General Relativity—An Einstein Centenary Survey eds. S.W. Hawking and W. Israel (Cambridge University Press, Cambridge). Ultraviolet divergences in the quantum theories of gravitation
181 J. Wheeler: (1964) in Relativity, Groups and Topologyeds. C. DeWitt and B. DeWitt (Blackie and Sons, London). Geometrodynamics and the issue of the final state
182 J. Wheeler: (1968) in Battelle Rencontres 1967, eds. C. DeWitt and J.A. Wheeler (Benjamin, New York) Superspace and the nature of quantum geometrodynamics
183 E. Witten: Comm. Math. Phys. 92, 455 (1984)
184 R.P. Woodward : (1989) Brown University preprint HET-723
185 T. Yoneya: Prog. Theor. Phys 51, 1907 (1974)
186 J.W. York: Jour. Math. Phys. 13, 125 (1972)
187 H.D. Zeh: Phys. Letts. A116, 9 (1986)
188 H.D. Zeh: Phys. Letts. A116, 9 (1986)
189 H.D. Zeh: Phys. Letts. A126, 311 (1988)
190 W.H. Zurek: Phys. Rev. D24, 1516 (1981)
191 W.H. Zurek: Phys.Rev. D26, 1862 (1982)
192 W.H. Zurek: (1991) in Proceedings of the Osgood Hill Conference on Conceptual Problems in Quantum Gravity eds. A. Ashtekar and J. Stachel(Boston, Birkhauser) Quantum measurements and the environment induced transition from quantum to classical

Two-Dimensional Gravities and Supergravities as Integrable Systems

H. Nicolai

II. Institut für Theoretische Physik,
Universität Hamburg,
Luruper Chaussee 149, 2000 Hamburg 50, F.R.G.

1. Introduction

The study of exact solutions of Einstein's field equations in general relativity has a long history (for a comprehensive account of the subject, see [1]. A feature common to all such solutions is their independence of certain coordinates, or, in more technical terms, the existence of one or more Killing vectors. This independence is, first of all, important from a practical point of view because one would like to simplify the structure of the equations so as to make them amenable to explicit solution. In a more general perspective, the omission of two of the coordinates can be regarded as a special case of the Kaluza Klein program [2]. To study the solutions with two Killing vectors in a systematic fashion, one considers a Kaluza Klein type reduction of Einstein's theory to two dimensions, where the gravitational degrees of freedom are split into a "matter sector" describing the two physical helicity degrees of freedom and the remaining unphysical (non-propagating) degrees of freedom, in particular the conformal factor. The complications inherent in Einstein's theory are best illustrated by the long time it took to realize that the existence of exact solutions in this reduction is not just a fortuitous and accidental property of the theory, but intimately related to the existence of hidden symmetries and of certain infinitesimal transformations, which connect different solutions and which can be used to construct new solutions from old ones [3]. An important breakthrough occurred in 1970, when Geroch was able to demonstrate that this set of transformations is, in fact, infinite-dimensional [4] (the corresponding group is nowadays referred to as the "Geroch group"). This was the first indication that Einstein's theory is actually integrable after the reduction to two dimensions. The idea was further developed and elaborated by the general relativists in the years following this discovery, with special emphasis on the search for "solution-generating procedures" [1,5-10]. This program has met with considerable success, especially with regard to the so-called stationary axisymmetric solutions possessing one space-like and one time-like Killing vector. The integrability of Einstein's equations was explicitly demonstrated with the construction of a linear system (or Lax pair) for the gravitational field equations by Belinskii and Sakharov [7] and Maison [8]. Thus, it became possible to apply techniques such as the inverse scattering method from the theory of integrable systems to Einstein's theory. There are, however, important differences between Einstein's theory and ordinary integrable systems: for instance, solutions of the gravitational field equations define their own domain of validity, whereas the space-time supporting the solution is a priori given for flat space integrable systems. This peculiar feature is reflected in the x-dependence of the spectral parameter entering the linear system [7,8,10]. Solutions describing colliding gravitational plane waves with two space-like Killing vectors have come into focus only much more recently [11], and the application of the solution-generating methods described above to this type of solutions is still in its infancy [12,13].

Quite independently of these developments, particle physicists had become interested in Kaluza Klein theories and higher dimensional supergravities in the late seventies in their attempts to formulate a unified theory of the fundamental interactions (see e.g. [14]). One of the most remarkable developments in this context was the discovery of "hidden symmetries" in dimensionally reduced supergravities, culminating in the discovery of a hidden E_7 invariance in the maximally extended $N = 8$ supergravity theory [15]. Apparently, B. Julia was the first to realize the connection between these developments and the work of the general relativists [16]. He emphasized the importance of group theoretical concepts for the investigation of the structural properties of dimensionally reduced gravity and supergravity theories and was led to conjecture that the emergence of infinite-dimensional symmetries in the reduction to two dimensions is a generic phenomenon. He was further able to show quite explicitly that the Geroch group in infinitesimal form is nothing but the affine Kac Moody algebra $A_1^{(1)}$ and to demonstrate the presence of a central term, which had gone unnoticed by the general relativists, and which acts as a shift operator on the conformal factor [17]. The underlying group theoretic structure and the connection with the σ-models encountered in particle physics were further elucidated by Breitenlohner and Maison [10,18]. They discovered a beautiful cocycle formula explicitly solving the equation of motion for the conformal factor. Through this work it has become clear that one can think of the space of solutions as some kind of infinite-dimensional coset space, which is analogous to the finite-dimensional coset spaces of gauge-equivalent field configurations in ordinary σ-models. It is, of course, crucial here that many concepts can be generalized from the finite-dimensional models to the infinite-dimensional case.

As for the purely bosonic models, these developments have been amply documented, and I will therefore try to avoid reduplication of the existing literature in these lectures by concentrating on those aspects which have not been given so much attention until now, and by shifting the emphasis from the systematics of solution-generating procedures to the discussion of group theoretical aspects and the search for yet bigger symmetries. This point of view is of more concern to the particle physicist and string theorist (and was already adopted in [16]). Nonetheless, I will also describe some concrete applications of the solution-generating procedures in section 3.4, where a special colliding plane wave solution of the Ferrari-Ibanez type [12] is reconstructed via the monodromy matrix. Special attention will be given to the fermionic models in these lectures. Although many results from the bosonic theory carry over to supergravity (such as the existence of a linear system yielding the supergravitational field equations [19,20]), there remain some puzzling questions which will be discussed at the end of these lectures. There is some evidence (to be presented in section 4.3) that there is more to the addition of fermions than just the popular recipe of turning every item into a super-item. For instance, the bosonic sector of the maximally extended $N = 16$ theory in three dimensions [21] is described by an $E_8/SO(16)$ coset space non-linear σ-model, the two-dimensional reduction of which possesses a non-linearly realized E_9 symmetry independently of whether there are fermionic degrees of freedom or not. Contrary to one's naive expectations, the symmetry of the fully supersymmetric theory cannot be some ordinary supergroup, as E_8 has no supergroup extension [22]! Nonetheless, the symmetry must be enlarged in some way as there appears to be "no room" for the fermionic solutions in the bosonic coset spaces. The possibility that the required extension leads to hyperbolic algebras [23], which was already raised in [24], is discussed in section 4.3.

The search for more symmetries in the context of low dimensional supergravities could also provide some much needed hints for the formulation of a proper theory of quantum gravity, which is likely to be as much a problem in finding the correct underlying symmetry as in finding the correct conceptual framework. Let me briefly explain why it is that the study of $d = 2$ or even $d = 1$ reductions should afford any insight into this difficult problem. One reason for this expectation is what I would like to refer to (somewhat loosely) as the "string magic", by which an essentially trivial two-dimensional field theory is converted into a complicated higher-dimensional theory that contains (super)gravity as a tiny subsector. This is accomplished by re-interpreting the physical states of the theory on the two-dimensional world sheet as the one-particle states of a quantum field theory in a higher-dimensional target space. This basic idea has been realized so far only in the context of conformally invariant theories, but has not been tried for the kind of theories we are about to discuss. Let us suppose that this is actually possible. Then, the Geroch group, which previously acted upon the space of solutions of a classical field equation, becomes a symmetry on the Hilbert space of physical quantum states defined through the imposition of the canonical constraints. Re-interpreting this space as a space of one-particle states of a higher-dimensional theory, one realizes that the quantum analog of the Geroch group now acts as a spectrum generating symmetry: it transforms the physical states into one another and possibly generates all of them from some groundstate (if it acts transitively). This spectrum generating algebra is reminiscent of an analogous spectrum generating algebra in string theory, which consists of the physical vertex operators [25] (a vertex operator is called physical if it weakly commutes with the Virasoro constraints). These similarities will be further illuminated in section 2.3. The point

I would like to emphasize here is that the Geroch group has been converted from an abstract group acting on some equally abstract space of solutions into a "space-time symmetry" in this fashion. Since this symmetry constitutes a huge extension of the ordinary space-time symmetries, it may well be related to the elusive fundamental symmetry that string theorists have been searching for in vain until now.

There are also some interesting geometrical aspects which will be discussed in chapter 4. In the formulation of $d = 2$ (super)gravity as an integrable system described in chapter 3, the basic object of interest is a matrix $\widehat{\mathcal{V}}$ depending not only on the space-time coordinates but also on a so-called spectral parameter t. This matrix appears in the linear system (cf. section 3.3), and contains all the information about the equations of motion, including the fermionic ones in the case of supergravity. It may be viewed as an infinite-dimensional generalization of the vielbein in gravity. This analogy is suggested by the fact that the symmetry groups acting on this "∞-bein" are in one-to-one correspondence with the groups acting on the ordinary vielbein: there is an infinite group G^∞ (the Geroch group) acting rigidly on $\widehat{\mathcal{V}}$ from the left and another H^∞ acting as a gauge group from the right. These groups contain the transverse subgroups of the corresponding finite-dimensional groups, namely $SL(D-2, \mathbb{R})$ and the transverse Lorentz group $SO(D-2)$. (This is the reason why I referred to the quantum Geroch group as a "space-time symmetry" above). The construction of even bigger symmetry groups involving "boost generators" will most probably require the further reduction to one dimension and the introduction of fermions.

Since the reduction to one dimension may be of interest in its own right, some pertinent results concerning maximal supergravity in one dimension are collected in the appendix. A novel feature is the proposal to perform the reduction with respect to one of the light-cone coordinates x^+ or x^-, rather than x^0 or x^1. The potentially most important reason in favor of such a chiral (or "unidexterous") reduction is the possibility that it may give rise to a new kind of dual model, as I will briefly explain at the end of the appendix.

2. Preliminaries

This chapter summarizes some well-known results about the hidden symmetries of Einstein's theory in three space-time dimensions. As this subject has already been extensively dealt with in the literature [1], the presentation will be cursory and mainly serve to set up the notation. The results for $N = 2$ supergravity in three dimensions are also given, because this is the simplest supersymmetric model, where the emergence of infinite-dimensional symmetries in the reduction to two dimensions can be studied. The main reason for the emphasis on three-dimensional Lagrangians here is that, in contradistinction to other approaches, I will always take the three-dimensional non-linear σ model as the point of departure, thus bypassing the labor of explicitly working out the Kaluza Klein reduction from higher dimensions. Some salient features of the canonical treatment of hidden symmetries, which has been largely ignored in the relativists' literature, will be briefly discussed in section 2.3. (For further details concerning canonical gravity the reader is advised to consult the standard references and Isham's lectures in this volume). The relevance of hidden symmetries for the construction of non-trivial observables in canonical gravity is pointed out.

2.1 The Reduction of Einstein's Theory to $d = 3$

The simplest example of a hidden symmetry is obtained when one reduces Einstein's theory from $d = 4$ to $d = 3$. After the reduction and a duality redefinition of the Kaluza Klein vector field, the theory possesses a hidden $SL(2,\mathbb{R})$ symmetry, the two physical gravitational degrees of freedom being parametrized by the $SL(2,\mathbb{R})/SO(2)$ coset space. This fact is, of course, implicit in the earlier works on the subject. The presentation here follows [16,17]; the general case (e.g. Maxwell-Einstein) is discussed in [26], where further references can be found. The dimensional reduction simply amounts to dropping the dependence on the fourth coordinate for all fields ϕ, viz.

$$\phi(x^0, x^1, x^2, x^3) \longrightarrow \phi(x^0, x^1, x^2) \tag{2.1.1}$$

This applies in particular for the vierbein, for which, in addition, we make partial use of the local $SO(1,3)$ invariance invariance to fix a triangular gauge, viz.

$$E_M^A = \begin{pmatrix} \Delta^{-1/2} e_m^a & B_m \Delta^{1/2} \\ 0 & \Delta^{1/2} \end{pmatrix} \tag{2.1.2}$$

where the indices on the left hand side refer to four dimensions, while $m, ... = 0,1,2$ and $a, ... = 0,1,2$ on the right hand side refer to three dimensions, and the extra factors of Δ have been inserted for convenience. Writing out Einstein's action in terms of

$$F_{mn} = \partial_m B_n - \partial_n B_m \tag{2.1.3}$$

we obtain

$$\begin{aligned} \mathcal{L} &= -\tfrac{1}{4} E R(E) \\ &= -\tfrac{1}{4} e R^{(3)}(e) + \tfrac{1}{8} e g^{mn} \Delta^{-1} \partial_m \Delta \Delta^{-1} \partial_n \Delta - \\ &\quad - \tfrac{1}{16} e \Delta^2 g^{mp} g^{nq} F_{mn} F_{pq} \end{aligned} \tag{2.1.4}$$

The part involving e_m^a describes pure gravity in three dimensions and contains no propagating degrees of freedom, whereas the physical degrees of freedom (the two helicity states of the graviton) reside in the remainder of the Lagrangian. In the equations of motion, one can perform a duality redefinition

$$\Delta^2 F_{mn} = \epsilon_{mnp} \partial^p B \tag{2.1.5}$$

The integrability of this equation is guaranteed (on-shell) by Maxwell's equation for the vector field B_m. The resulting equations of motion can be equivalently obtained from the σ-model Lagrangian

$$\mathcal{L} = -\tfrac{1}{4}eR^{(3)}(e) + \tfrac{1}{8}eg^{mn}\Delta^{-2}\left(\partial_m B\partial_n B + \partial_m \Delta\partial_n \Delta\right) \tag{2.1.6}$$

or, in terms of the complex field $Z \equiv B + i\Delta$,

$$\mathcal{L} = -\tfrac{1}{4}eR^{(3)}(e) + \tfrac{1}{2}eg^{mn}\frac{\partial_m Z\partial_n \bar{Z}}{\left(Z-\bar{Z}\right)^2} \tag{2.1.7}$$

(The superscript (3) will henceforth be dropped.) Note that the field Z is related to the so-called Ernst potential $\mathcal{E}$ via $Z = i\bar{\mathcal{E}}$ (see section 3.2). It is easy to check invariance of (2.1.7) under the $SL(2,\mathbb{R})$ transformations

$$Z \to Z' = \frac{aZ+b}{cZ+d} \quad ; \quad \begin{pmatrix} a & b \\ c & d \end{pmatrix} \in SL(2,\mathbb{R}) \tag{2.1.8}$$

This is the simplest example of a non-linearly realized symmetry acting on the space of physical field configurations. In the following section, (2.1.7) will be shown to be just a special case of the more general construction.

A more sophisticated example is the reduction of $N=1$ supergravity [27,28] from four to three dimensions. The gravitational sector is, of course, exactly as above, but we now have in addition a four component Majorana vector spinor ψ_M (the gravitino), which upon reduction to three dimensions splits into two gravitinos ψ_m^I and two matter fermions χ^I. The extra internal index $I=1,2$ is necessary because spinors in three dimensions have only two components, so every four-dimensional spinor gives rise to *two* three-dimensional spinors transforming irreducibly under the Lorentz group $SO(1,2)$. In complete analogy with the dreibein in pure gravity the three-dimensional vector spinors ψ_m^I carry no longer any propagating degrees of freedom, while the two physical degrees of freedom (the helicity $\frac{3}{2}$ states of the gravitino) reside in the "matter fermions" χ^I. To work out the reduction with all the necessary field redefinitions is a bit more laborious than in pure gravity (readers interested in the details are invited to have a look at [15,29] where a more complicated example is treated). It is also unnecessary here, for in the following section I will explain the general formalism which will allow us to formulate the models directly in three dimensions by exploiting their symmetry structures.

2.2 Coset Space σ-Models and Supergravity

It can now be shown that the structure of which we just had a first glimpse is actually generic in Kaluza-Klein (super)gravity: there are always hidden symmetries which extend the manifest symmetries obtained in the naive dimensional reduction [15,16,26]. One invariably obtains a non-linear σ-model of the non-compact type based on a coset space G/H with H as the maximally compact subgroup. Of course, the coset space one ends up with depends on which theory one starts with, but once the coset has been identified (for instance by simply counting the scalar degrees of freedom), the model is unalterably fixed. This fact can be exploited by taking the three-dimensional Lagrangian as one's point of departure in analyzing the reduction to two dimensions rather than starting from the higher-dimensional theory. In this section, I will first summarize the general formalism for the bosonic σ-models, which was invented by particle physicists long ago [30] (see also [31] for a more recent treatment) and then present $N = 2$ supergravity in three dimensions, which is the locally supersymmetric extension of (2.1.6). The limitation to the $N = 2$ theory is mainly for pedagogical reasons, as this is the simplest supersymmetric model to be studied in the present context. However, the notational conventions are such that the comparison with the results described in the appendix is quite straightforward.

To formulate the bosonic σ-models in a general way [31], consider a maximally symmetric space G/H with the associated Lie algebra decomposition

$$\mathbf{G} = \mathbf{H} \oplus \mathbf{K} \tag{2.2.1}$$

For the sake of simplicity, we will only consider the case of interest here, where G is non-compact and H is its maximally compact subgroup. The maximal symmetry of the coset space is reflected in the Lie algebra relations

$$[\mathbf{H},\mathbf{H}] \subset \mathbf{H} \quad [\mathbf{H},\mathbf{K}] \subset \mathbf{K} \quad [\mathbf{K},\mathbf{K}] \subset \mathbf{H} \tag{2.2.2}$$

This decomposition is invariant under the symmetric space automorphism

$$\tau(\mathbf{H}) = \mathbf{H} \qquad \tau(\mathbf{K}) = -\mathbf{K} \tag{2.2.3}$$

which can alternatively be formulated in terms of Lie group elements g directly through

$$\tau(g) = \eta^{-1}(g^T)^{-1}\eta \tag{2.2.4}$$

where the matrix η depends on the group G (e.g. $\eta = \mathbf{1}$ for $G = SL(n,\mathbb{R})$).

As already pointed out, our principal example will be the $SL(2,\mathbb{R})/SO(2)$ coset space. In this case, there are the three generators

$$Y^1 = \begin{pmatrix} 1 & 0 \\ 0 & -1 \end{pmatrix} \;,\; Y^2 = \begin{pmatrix} 0 & 1 \\ 1 & 0 \end{pmatrix} \;,\; Y^3 = \begin{pmatrix} 0 & 1 \\ -1 & 0 \end{pmatrix} \tag{2.2.5}$$

Clearly,

$$\mathrm{Tr}\,(Y^1)^2 = \mathrm{Tr}\,(Y^2)^2 = -\mathrm{Tr}\,(Y^3)^2 = 2 \tag{2.2.6}$$

so Y^1 and Y^2 are the non-compact generators, while Y^3 generates the $SO(2)$ subgroup. I will label the coset generators with capital letters $A, B, ... = 1, 2$.

The scalar fields $\mathcal{V}(x)$ of the associated non-linear σ model are now taken to be elements of the group G for all x. To reduce the number of physical degrees of freedom to the desired number, one requires invariance under the transformations

$$\mathcal{V}(x) \rightarrow \mathcal{V}'(x) = g^{-1}\mathcal{V}(x)h(x) \tag{2.2.7}$$

so the local H invariance can be used to remove the unwanted degrees of freedom. For instance, we can pick a "unitary" gauge such that

$$\mathcal{V}(x) = \exp \varphi_0(x) \qquad \varphi_0(x) \in \mathbf{K} \tag{2.2.8}$$

which shows that there are precisely $\dim \mathbf{K}$ physical degrees of freedom. If one acts on this matrix with an arbitrary element $g \in G$, the "unitarity" of the gauge will be lost in general, so a compensating H transformation $h(x; g, \mathcal{V})$ is necessary to restore the gauge. In this way, the symmetry becomes non-linearly realized on the physical fields $\varphi_0(x)$ [30]. Note that if no special gauge is chosen, there are two symmetries acting on $\mathcal{V}$, namely the rigid group G and the local group H. Consequently, one can think of the matrix $\mathcal{V}$ as some kind of vielbein, with G and H replacing the groups $GL(D, \mathbb{R})$ and $SO(1, D-1)$, respectively. This analogy will be encountered again in these lectures.

The next step is, of course, the construction of a Lagrangian with the required symmetries. To this aim, consider the Lie-algebra-valued expression

$$\mathcal{V}^{-1}\partial_m \mathcal{V} = Q_m + P_m \qquad Q_m \in \mathbf{H}\ ,\ P_m \in \mathbf{K} \tag{2.2.9}$$

or, equivalently,

$$\mathcal{V}^{-1} D_m \mathcal{V} \equiv \mathcal{V}^{-1}\left(\partial_m \mathcal{V} - \mathcal{V} Q_m\right) = P_m \tag{2.2.10}$$

which defines the H covariant derivative D_m. It is straightforward to verify that Q_m transforms indeed as a gauge field with respect to the local group H, and that $P'_m = h^{-1} P_m h$. The formula (2.2.7) implies the integrability relations

$$\begin{gathered} \partial_m Q_n - \partial_n Q_m + [Q_m, Q_n] = -[P_m, P_n] \\ D_m P_n - D_n P_m = 0 \end{gathered} \tag{2.2.11}$$

The Lagrangian is given by

$$\mathcal{L}' = +\tfrac{1}{4} e g^{mn} \operatorname{Tr} P_m P_n = \tfrac{1}{2} e g^{mn} P_m^A P_n^A \tag{2.2.12}$$

where the gravitational fields have been included again. The field equations read

$$D_m\left(\sqrt{g} g^{mn} P_n\right) = 0 \tag{2.2.13}$$

To see that this formulation leads to the same equations as the Einstein Lagrangian of the previous section, take $G = SL(2, \mathbb{R})$ with the matrix $\mathcal{V}$ in the triangular gauge

$$\mathcal{V} = \begin{pmatrix} \Delta^{1/2} & B\Delta^{-1/2} \\ 0 & \Delta^{-1/2} \end{pmatrix} \tag{2.2.14}$$

Decomposing the expression

$$\mathcal{V}^{-1}\partial_m \mathcal{V} = \begin{pmatrix} \frac{1}{2}\Delta^{-1}\partial_m \Delta & \Delta^{-1}\partial_m B \\ 0 & -\frac{1}{2}\Delta^{-1}\partial_m \Delta \end{pmatrix} \tag{2.2.15}$$

into its components along $\mathbf{K}$ and $\mathbf{H}$, we obtain

$$P_m^1 = \tfrac{1}{2}\Delta^{-1}\partial_m \Delta\ ,\ P_m^2 = \tfrac{1}{2}\Delta^{-1}\partial_m B\ ,\ Q_m = \tfrac{1}{2}\Delta^{-1}\partial_m B \tag{2.2.16}$$

Substituting (2.2.15) into (2.2.12), we immediately recover the "matter part" of (2.1.6).

Let us now turn to $N = 2$ supergravity. Since the derivation of the results below is standard, I will omit details and only quote the final results, referring the reader to [28] for more information on supergravity technology. In addition to the dreibein e_m^a and the scalar fields Δ and B introduced in (2.2.14), the $N = 2$ theory contains two "matter fermions" χ^I and two gravitinos ψ_m^I $(I, J, ... = 1, 2)$. In the present formulation, these fields are inert under the action of the rigid group G, but do transform under $SO(2)$. There is only one consistent choice of $SO(2)$ charge assignments. For an $SO(2)$ rotation

$$h = \begin{pmatrix} \cos\alpha & \sin\alpha \\ -\sin\alpha & \cos\alpha \end{pmatrix} \tag{2.2.17}$$

we must have

$$\begin{aligned} P_m^1 + iP_m^2 &\to e^{2i\alpha}\left(P_m^1 + iP_m^2\right) \\ \chi^1 + i\chi^2 &\to e^{\frac{3}{2}i\alpha}\left(\chi^1 + i\chi^2\right) \\ \psi_m^1 + i\psi_m^2 &\to e^{\frac{1}{2}i\alpha}\left(\psi_m^1 + i\psi_m^2\right) \end{aligned} \tag{2.2.18}$$

This is in accord with our expectation that the "hidden" $SO(2)$ is just the helicity (or transverse) subgroup of the Lorentz group in four dimensions, with $P_m^1 \pm iP_m^2$ and $\chi^1 \pm i\chi^2$ corresponding to the physical states of helicity ± 2 and $\pm\frac{3}{2}$, respectively (the supersymmetry transformation parameter ϵ^I transforms like ψ_m^I). More general helicity groups are obtained for higher extended theories; for instance, in the maximally extended $N = 16$ theory, we have $H = SO(16)$, which contains the helicity group $SO(9)$ of $D = 11$ supergravity as a subgroup. Given the charge assignments (2.2.18), we can immediately deduce the form of the fully covariant derivatives acting on the spinors

$$\begin{aligned} (D_m\chi)^I &= \left(\partial_m + \tfrac{1}{4}\omega_{mab}\gamma^{ab}\right)\chi^I + \tfrac{3}{2}Q_m\epsilon^{IJ}\chi^J \\ (D_m\epsilon)^I &= \left(\partial_m + \tfrac{1}{4}\omega_{mab}\gamma^{ab}\right)\epsilon^I + \tfrac{1}{2}Q_m\epsilon^{IJ}\epsilon^J \end{aligned} \tag{2.2.19}$$

which shows how the scalars and the spinors couple via the composite gauge field Q_m determined from (2.2.9).

At this point, we could, of course, easily switch to a slightly more economical complex notation, but I prefer not to do so in order to keep the notational uniformity with the $N = 16$ results to be discussed in the appendix. Thus we continue to use real fields in the remainder of this subsection, using indices $I, J, ... = 1, 2$ and $A, B, ... = 1, 2$ for the "spinorial" and "vectorial" representations of $SO(2)$, respectively. The full Lagrangian of $N = 2$ supergravity is[1]

---

1 Readers may wonder why there are explicit quartic spinorial terms in this Lagrangian and the transformation rules below, whereas there are no such terms in the usual first (or 1.5) order formalution of $d = 4, N = 1$ supergravity [27]. The reason for their appearance here is that we are using 1.5 order formalism *only* for the three-dimensional spin connection, whereas second order formalism is used for the $SO(2)$ part of the connection, and that one of the components of the vierbein has been dualized.

$$\mathcal{L} = -\tfrac{1}{4}eR(e) + \tfrac{1}{2}\epsilon^{mnp}\bar{\psi}^I_m D_n \psi^I_p + \tfrac{1}{4}eg^{mn}P^A_m P^A_n - \tfrac{i}{2}e\bar{\chi}^I\gamma^m(D_m\chi)^I -$$
$$- e\, t^A_{IJ}\,\bar{\chi}^I\gamma^m\gamma^n\psi^J_m P^A_n + \tfrac{1}{8}e\left(\bar{\chi}^I\chi^I\right)^2 - \tfrac{1}{6}e\left(\bar{\chi}^I\gamma^a\chi^J\right)^2 +$$
$$+ \tfrac{1}{4}e\bar{\chi}^I\gamma_p\chi^J\left(\bar{\psi}^I_m\gamma^{mnp}\psi^J_n - \bar{\psi}^I_m\gamma^p\psi^{mJ}\right) - \tfrac{1}{8}e\bar{\chi}^I\chi^I\bar{\psi}^J_m\gamma^n\gamma^m\psi^J_n \tag{2.2.20}$$

with the "Clebsch Gordan coefficients" $t^A_{IJ} \equiv \frac{1}{\sqrt{2}}(Y^A)_{IJ}$ (see (2.2.5)). It is invariant under the local supersymmetry variations

$$\begin{aligned}
\delta V^a_m &= i\bar{\epsilon}^I\gamma^a\psi^I_m \\
\delta\psi^I_m &= D_m\epsilon^I + \tfrac{i}{2}\gamma^n\epsilon^J\,\bar{\chi}^I\gamma_{mn}\chi^J \\
\delta\chi^I &= i\gamma^m\epsilon^J\widehat{P}_m\,t^A_{IJ} \\
\mathcal{V}^{-1}\delta\mathcal{V} &= t^A_{IJ}\bar{\epsilon}^I\chi^J
\end{aligned} \tag{2.2.21}$$

where $\widehat{P}^A_m$ denotes the supercovariant extension of P^A_μ

$$\widehat{P}^A_m = P^A_m - t^A_{IJ}\,\bar{\psi}^I_m\chi^J \tag{2.2.22}$$

Later, we will also need the equations of motion for the physical fields, which read

$$D^m\left(P^A_m - t^A_{IJ}\bar{\chi}^I\gamma^n\gamma_m\psi^J_n\right) = \epsilon^{AB}P^B_m\left(\tfrac{3}{2}i\epsilon^{IJ}\bar{\chi}^I\gamma^m\chi^J + \tfrac{1}{2}\epsilon^{mnp}\epsilon^{IJ}\bar{\psi}^I_n\psi^J_p\right) \tag{2.2.23}$$

and

$$-i\gamma^m D_m(\widehat{\omega})\chi^I = t^A_{IJ}\gamma^n\gamma^m\psi^J_n\widehat{P}^A_m + \tfrac{1}{2}\left(\bar{\chi}^J\chi^J\right)\chi^I \tag{2.2.24}$$

where the Fierz-identity $(\bar{\chi}^I\gamma^a\chi^J)^2 = \frac{3}{2}(\bar{\chi}^I\chi^I)^2$ has been used, and $\widehat{\omega}$ is the supercovariant spin connection

$$\widehat{\omega}_{mab}(e,\psi,\chi) = \omega_{mab}(e) + i\bar{\psi}^I_{[a}\gamma_{b]}\psi^I_m + \tfrac{i}{2}\bar{\psi}^I_a\gamma_m\psi^I_b + \tfrac{1}{4}\epsilon_{mab}\bar{\chi}^I\chi^I \tag{2.2.25}$$

The Lagrangian (2.20) is, of course, also invariant under general coordinate transformations in three dimensions, the local Lorentz group $SO(1,2) \cong SL(2,\mathbb{R})$ and $SO(2)$ gauge transformations. It has already been pointed out that the fermions are inert under the rigid group G unless a $SO(2)$ gauge is chosen. If, on the other hand, a gauge is fixed as in (2.2.8), the fermions do transform under G via the induced compensating H rotation $h(x,g;\mathcal{V})$.

2.3 Remarks on the Canonical Formalism

It is well known that the invariance of a Lagrangian under a set of rigid symmetry transformations leads to a conserved charge $\mathcal{Q}$. For the bosonic σ-models discussed in the preceding section the conserved charge associated with the rigid symmetry under G can be easily obtained by integrating the time component of the conserved current

$$\mathcal{J}^m = \mathcal{V}P^m\mathcal{V}^{-1} \tag{2.3.1}$$

over a space-like hypersurface. Note that the vielbein $\mathcal{V}$ converts P_m, which transforms under H and is inert under the rigid symmetry G, into a quantity which is inert under the local symmetry and transforms covariantly under G (cf. (2.2.7))

$$\mathcal{J}^m \to g^{-1}\mathcal{J}^m g \tag{2.3.2}$$

The conserved current exists, of course, also for the locally supersymmetric theory. For instance, the conserved $SL(2,\mathbb{R})$ current for $N=2$ supergravity is given by

$$\begin{aligned}\mathcal{J}^m =&\mathcal{V}Y^A\mathcal{V}^{-1}\left(P^{mA} - \bar{\psi}^I_n\gamma^m\gamma^n\chi^J t^A_{IJ}\right)+\\ &+\mathcal{V}Y^3\mathcal{V}^{-1}\left(\tfrac{1}{4}e^{-1}\epsilon^{mnp}\epsilon^{IJ}\bar{\psi}^I_n\psi^J_p - \tfrac{3}{4}i\epsilon^{IJ}\bar{\chi}^I\gamma^m\chi^J\right)\end{aligned} \tag{2.3.3}$$

The algebra of symmetry transformations generated by the conserved charges and the interplay between rigid and local symmetries is most conveniently analyzed in the canonical framework. Of course, there is no room here to explain the the basics of canonical gravity in all but the scantiest detail. Fortunately, there are several excellent introductory articles available (see e.g. Isham's contribution in this volume or [32]), to which we can refer the reader for more details. At this point, I would also like to alert the reader to [33] where the issues raised here are discussed in much more detail for maximally extended supergravity in the framework of Ashtekar's formalism.

Canonical gravity is substantially different from other canonical theories in that the Hamiltonian becomes a constraint which, in the classical theory, singles out a physical submanifold in the phase space of general relativity [34]. This constraint is a consequence of the invariance of Einstein's theory under reparametrizations of the time coordinate. In the quantum theory, the constraint must be imposed as an operator constraint on the Hilbert space of wave functionals, and then becomes the famous Wheeler-DeWitt operator $\mathcal{H}_0$ [35]. In addition to $\mathcal{H}_0$, there are the constraints $\mathcal{H}_i$ associated with the invariance under spatial diffeomorphisms, and, for the kind of theories we are concerned with here, the constraints associated with the invariance under H gauge transformations, which carry no dynamical content, however. In a locally supersymmetric theory, there is also a supersymmetry constraint $\mathcal{S}$, which generates all the other constraints upon commutation and thus carries both dynamical and kinematic information. Because $\{\mathcal{S},\mathcal{S}\} = \mathcal{H}_0 + \ldots$, the equation $\mathcal{S}\Psi = 0$ can be regarded as the square root of the Wheeler-DeWitt equation in the same way as the Dirac equation is the square root of the Klein Gordon equation.

A central (and unsolved) problem in canonical gravity is the construction of genuine observables. An observable in the sense of Dirac is any functional $\mathcal{O}$ which does not vanish on the physical subspace generated by the canonical constraints and which weakly commutes with all the constraints. Thus, one demands

$$\{\phi, \mathcal{O}\} \approx 0 \tag{2.3.4}$$

where ϕ collectively denotes all the constraint generators. Since, in contrast to ordinary canonical theories, there is no a priori "time" in canonical gravity, (2.3.4) replaces the statement of conservation. The difficult part here is, of course, to get $\mathcal{O}$ to commute with the Hamiltonian constraint, whereas the kinematic constraints are comparatively easy to satisfy. We now see that conserved charges of the above type are good candidates for observables. All one has to show is that $\mathcal{O} = \mathcal{Q}$ weakly commutes with all the constraints according to (2.3.4). In the case at hand, it is rather easy to check that (2.3.4) is indeed satisfied for the charge $\mathcal{Q}$ obtained from (2.3.1). This shows that, in the presence of matter, it is not at all difficult to construct genuine observables, if the matter sector has some rigid invariance group. There is, however, one difference between symmetries introduced "by hand" and the hidden symmetries discussed here: the latter originate from gravity itself. In other words, even in pure gravity, it is possible to construct observables provided one reduces the theory by one (or more) dimensions.[2]

In principle, these statements carry over to the quantum theory. Of course, ordering problems and ill-defined operator products will vitiate many arguments, but let us assume for the moment that these difficulties can be overcome (this is one motivation for considering extended supergravity rather than pure gravity!). Then equation (2.3.4) schematically translates into

$$[\phi, \mathcal{Q}] = f(...)\phi \tag{2.3.5}$$

where $f(...)$ denotes a functional of the fields. This means that from any given solution Ψ of the Wheeler-DeWitt equation, one can generate another according to $\Psi' = \mathcal{Q}\Psi$ [38,33]. Consequently, the solutions of the Wheeler-DeWitt equation form multiplets of the rigid symmetry group G (which are, in fact, infinite-dimensional owing to the non-compactness of G, if the representation is unitary). Although there is so far no concrete calculation exemplifying this idea, it is clear that what we have here is nothing but the quantum analog of the solution-generating procedures for the classical field equations. The argument also shows that the widely acclaimed uniqueness of the wave function of the universe is merely a consequence of overly restrictive assumptions about the boundary conditions imposed on Ψ. In any physically "reasonable" theory, and certainly in any theory originating from higher dimensions, there will be hidden symmetries leading to conserved charges, and the uniqueness of the wave function Ψ will be lost. It is amusing to speculate about the cosmological interpretation of the conserved quantum numbers.

Ultimately, we are interested in the reduction to two dimensions. There now remain only two constraints from the gravitational sector, which when written in the form

[2] This was independently noted in [36], where Ashtekar's version of $d = 4$ gravity [37] is directly reduced to $d = 2$. Their $SL(2,\mathbb{R})$ group is, however, not the same as the one generated by the charge $\mathcal{Q}$ and corresponds to the "Matzner-Misner $SL(2,\mathbb{R})$" rather than the "Ehlers $SL(2,\mathbb{R})$" generated by $\mathcal{Q}$ above. It is also not obvious to me whether the loop group identified in [36] is related to the Geroch group, as the latter arises through the interplay of the Ehlers *and* the Matzner-Misner groups.

$$T_{\pm\pm} = \mathcal{H}_0 \pm \mathcal{H}_1 \approx 0 \tag{2.3.6}$$

are easily recognizable as the Virasoro constraints well known from string theory [39]. (For chiral reductions to *one* dimension of the type considered in the appendix, only one of the constraints (2.3.6) survives.) Despite the obvious similarities with string theory, there are also some differences. A glance ahead at (3.2.7) shows that, in contrast to (critical) string theory, the properly defined energy momentum tensor $T_{\pm\pm}$ receives a contribution not only from the matter degrees of freedom, but also from the conformal factor, and that the role of the unphysical target space coordinates $X^{\pm}$ in string theory is now played by certain privileged components of the dreibein[3]. The Wheeler-DeWitt equation then amounts to the familiar requirement that the physical states should be annihilated by $T_{\pm\pm}$. The really new feature here is the presence of an infinite-dimensional rigid symmetry G^{∞} (to be described in chapter 4), which acts non-trivially on the space of physical states and has no analog in string theory. To what extent then can we generalize the above considerations to this case? Clearly, a first desideratum is the construction of a suitable conserved Kac Moody current $\widehat{\mathcal{J}}_{\pm}$ corresponding to (2.3.1) and (2.3.3). A candidate will be presented in section 3.5. The conserved charges constructed from it are, however, non-local, and the proper canonical treatment has met with some technical difficulties already in much simpler models [40,41]. Nonetheless, it is possible to show that $\{T_{\pm\pm}, \mathcal{Q}\} \approx 0$, at least for classical theory, although the full algebra of conserved charges remains to be evaluated.

3. From $d = 3$ to $d = 2$

It is the main purpose of this chapter to explain why Einstein's theory as well as more general (supersymmetric) models become integrable after the reduction to two dimensions. At the heart of the construction is the linear system (3.3.2) and its generalizations. The emphasis is more on the practical side in the following sections, and I will give an explicit example illustrating how the method works for colliding plane wave solutions. The supersymmetric case is discussed in section 3.5.

3.1 General Remarks

To reduce the theory to two dimensions, we can choose to drop the dependence on either the time-like coordinate x^0 or on the space-like coordinate x^2. In more technical terms, the corresponding solutions will then have one time-like and one space-like or two space-like Killing vectors, respectively. The so-called stationary axisymmetric solutions belong to the first class of solutions; well-known examples are the Schwarzschild and Kerr-Newman solutions. Of course,

[3] See the appendix for a further discussion, if you find these remarks too cryptic.

many solutions of the stationary axisymmetric type are known [1], but not all of them are physical (as they possess either naked singularities or need stiff "rods" for stability). The second class of solutions describes colliding plane waves. The first non-trivial example of such a solution was discovered in 1970, and more complicated solutions have been found since [11,12]. The two cases require a very different treatment: the study of stationary axisymmetric solutions leads to an elliptic boundary value problem while, in the second case, one has to analyze a hyperbolic partial differential equation. The structure of solutions is also very different; for instance, in the stationary axisymmetric case, the singularities are located "on the axis", whereas in the second case, they occur "after the collision".

While the solution generating procedures have been extensively applied to the stationary axisymmetric case, it is only recently that the attention has shifted to the study of colliding plane wave solutions in this framework [13]. It is for this reason, that I will concentrate here on the plane wave case. A new result presented in these lectures is the reconstruction of some recently found colliding plane wave solution via the group theoretical method. These solutions are usually constructed piecewise, by first solving the equations in the interaction region where the collision takes place. Once a solution has been found it is often straightforward to extend it from the lower boundary of the interaction region into two separated regions with pure radiation; in between the incoming waves the space-time is Minkowski (the known solutions can, however, not be continued into the future as they exhibit singularities on the upper boundary of the interaction region, see the figure). It is noteworthy that the group theoretical method yields the pertinent information (i.e. the solution) where it is most interesting, namely in the interaction region.

3.2 Once More: Einstein Gravity

As already indicated in the foregoing section, we will take the three-dimensional σ models as our point of departure, starting with the Lagrangian (2.2.12) coupled to gravity (or the corresponding supergravity Lagrangian, see section 3.5), rather than reducing the theory directly from $d = 4$ to $d = 2$. Thus we proceed from

$$\mathcal{L} = -\tfrac{1}{4}eR(e) + \tfrac{1}{4}eg^{mn}\,\mathrm{Tr}\,P_m P_n \tag{3.2.1}$$

As in section 2.1, we can make use of the local Lorentz group to simplify the dreibein and bring it into the form

$$e_m^{\ a} = \begin{pmatrix} \lambda\delta_\mu^\alpha & \rho B_\mu \\ 0 & \rho \end{pmatrix} \tag{3.2.2}$$

where Greek indices refer to two dimensions. In contrast to the vector B_m in (2.1.2), the two-dimensional Kaluza Klein vector B_μ carries no physical degrees of freedom any more, but is an auxiliary field which can be dropped with impunity here (however, it gives rise to higher order fermionic terms in supergravity, see section 3.5). So, effectively,

$$e_m^a = \begin{pmatrix} \lambda\delta_\mu^\alpha & 0 \\ 0 & \rho \end{pmatrix} \tag{3.2.3}$$

from which we can immediately compute the components of the three-metric and other quantities of interest. At this point, we can drop the distinction between flat and curved indices $\alpha, \ldots$ and $\mu, \ldots$. Furthermore, derivatives such as D_μ in (3.2.5) below are only covariantized with respect to H, whereas the Lorentz covariantizations have been written out and are properly accounted for by the extra factors of ρ (and the rescalings of fermions by a factor $\lambda^{\frac{1}{2}}$ in supergravity) in all equations referring to two dimensions. It is well known that the form of this gauge is preserved under the conformal reparametrizations

$$\delta x^+ = \xi_-(x^+) \qquad \delta x^- = \xi_+(x^-) \tag{3.2.4}$$

with the light-cone coordinates $x^\pm \equiv \frac{1}{\sqrt{2}}(x^0 \pm x^1)$. Substituting the gauge (3.2.3) into the scalar field equation (2.2.13), we arrive at

$$\rho^{-1} D^\mu (\rho P_\mu) = 0 \tag{3.2.5}$$

Readers should recall that D_μ is the H-covariant derivative. The dependence of this equation on ρ is all that remains of three-dimensional gravity.

The remaining equations, which follow from higher dimensions, are the equations for the field ρ and for the conformal factor λ. ρ is a free field in two dimensions which can be solved for in terms of two arbitrary functions (left-movers and right-movers)

$$\rho(x) = \rho_+(x^+) + \rho_-(x^-) \tag{3.2.6}$$

In the Euclidean case, where the fields depend on two space-like coordinates (x^1, x^2) or on the complex coordinates $(z, \bar{z})$ where $z = \frac{1}{\sqrt{2}}(x^1 + ix^2)$, ρ would be the sum of a holomorphic and an anti-holomorphic part. There remains some freedom in the choice of these functions as we still have the conformal transformations (3.2.4) at our disposal[4]. In the Euclidean case, we can make use of them to identify ρ with one of the coordinates (so-called "Weyl canonical coordinates", see [1,10]. For plane-wave type solutions, another choice is usually preferred, see section 3.4.

The equation of motion for the conformal factor reads (for later convenience, I employ light-cone notation)

$$\rho^{-1}\partial_\pm\rho\, \lambda^{-1}\partial_\pm\lambda = P_\pm^A P_\pm^A + \tfrac{1}{2}\rho^{-1}\partial_\pm^2\rho \tag{3.2.7}$$

Note that the second term on the right hand side can be absorbed into λ through the replacement $\lambda \to \lambda(\partial_+\rho)^{-\frac{1}{2}}(\partial_-\rho)^{-\frac{1}{2}}$, and that this equation determines λ only up to a constant factor (this corresponds to the freedom in the

---

[4] Similarities with string theory [39] were already mentioned before. The field ρ corresponds to the target space coordinate X^+, which can be identified with one of the world-sheet coordinates by a conformal coordinate transformation of the type (3.2.4). The conformal factor λ, on the other hand, resembles the string coordinate X^- in that both are dependent fields which can be solved for in terms of the physical fields (the transverse degrees of freedom) by means of the Virasoro constraints.

normalization of the first term in (3.2.9) below). Observe also that this equation has no analog in flat space theories and this, together with the presence of the field ρ makes an enormous difference. For instance, we cannot simply put $\rho = const$, as this would imply the vanishing of the right hand side of (3.2.7), which by the positivity of the Killing metric on the subalgebra $\mathbf{K}$ would imply $P_\pm = 0$ and leave us only with the trivial solution $\mathcal{V} = const$ (modulo H gauge transformations). It is also evident from (3.2.7) that the theory would be conformally invariant without the extra field ρ, as the conformal factor λ would then completely decouple from the rest of the theory.

It is instructive to work out the equation of motion (3.2.5) explicitly for the $G/H = SL(2,\mathbb{R})/SO(2)$ coset space. To do so, we again employ the triangular gauge (2.2.14) together with the explicit expressions for P_μ and Q_μ in (2.2.16). After a little algebra one arrives at the result

$$\Delta\partial_\mu(\rho\partial^\mu\mathcal{E}) = \rho\partial_\mu\mathcal{E}\partial^\mu\mathcal{E} \tag{3.2.8}$$

in terms of the complex potential $\mathcal{E} \equiv \Delta + iB$. This is just the well known Ernst equation [42]. Solving Einstein's equations is now simply a matter of choosing the appropriate $\rho(x)$, finding a solution to the non-linear partial differential equation (3.2.8) and finally determining the conformal factor λ by integration of (3.2.7). For the colliding plane wave solutions, one distinguishes waves with collinear polarization, where $B = 0$, or, equivalently, $g_{23} = 0$, and waves with non-collinear polarization. For collinearly polarized waves, the non-linear Ernst equation can be reduced to a *linear* partial differential equation through the replacement $\Delta = \exp\psi$ [43]. For later reference, let us also quote the expression for the four-dimensional line element in the case of collinearly polarized waves, which reads

$$ds^2 = 2\Delta^{-1}\lambda\, dx^+ dx^- - \Delta^{-1}\rho^2(dx^2)^2 - \Delta(dx^3)^2 \tag{3.2.9}$$

where Δ, ρ and λ depend only on x^+ and x^-.

3.3 The Linear System

The integrability of the non-linear equation of motion (3.2.5) is reflected in the existence of a linear system (or Lax pair). This means that there is a set of *linear* differential equations, whose compatibility conditions yield just the non-linear equations that one tries to solve. Such linear systems are well known in flat space integrable models (such as the KdV system or the KP hierarchy), so let us first recall how this method works for flat space σ-models and ignore the conformal factor λ, putting $\rho = const$. To formulate the linear system one must introduce a so-called spectral parameter t as an extra variable and replace $\mathcal{V}(x)$ by a matrix $\widehat{\mathcal{V}}$ which also depends on t [44] (For the reduction to a Euclidean space, one must replace $t \to it$, cf. [10])

$$\mathcal{V}(x^0, x^1) \to \widehat{\mathcal{V}}(x^0, x^1; t) \tag{3.3.1}$$

The linear system is now given by

$$\widehat{\mathcal{V}}^{-1}\partial_\mu\widehat{\mathcal{V}} = Q_\mu + \frac{1+t^2}{1-t^2}P_\mu + \frac{2t}{1-t^2}\epsilon_{\mu\nu}P^\nu \tag{3.3.2}$$

To prove this assertion, one must check the integrability condition

$$\partial_\mu\left(\widehat{\mathcal{V}}^{-1}\partial_\nu\widehat{\mathcal{V}}\right) - \partial_\nu\left(\widehat{\mathcal{V}}^{-1}\partial_\mu\widehat{\mathcal{V}}\right) + \left[\widehat{\mathcal{V}}^{-1}\partial_\mu\widehat{\mathcal{V}}\,,\,\widehat{\mathcal{V}}^{-1}\partial_\nu\widehat{\mathcal{V}}\right] = 0 \tag{3.3.3}$$

It is an elementary exercise to verify that (3.3.3) is equivalent to the integrability relations (2.2.11) *and* the scalar equation of motion (3.2.5) with $\rho = 1$ (Note that Pohlmeyer's original linear system involves the matrix $\mathcal{U} \equiv \widehat{\mathcal{V}}\mathcal{V}^{-1}$). Once this equation is solved, one can retrieve the original matrix $\mathcal{V}$ by setting $t = 0$, i.e. $\mathcal{V} = \widehat{\mathcal{V}}(t = 0)$; the equations of motion will then be satisfied automatically. In practice, it is, of course, not so easy to solve (3.3.3), and the solutions of the flat space σ-models have not been completely classified despite the existence of the linear system [45].

Restoring the ρ-dependence in (3.2.5), we see that a constant spectral parameter will no longer do, and that some modification is necessary. The trick is to make the spectral parameter x-dependent [7,8,10]. It takes a little more work to check that the proper ρ-dependence is recovered if t obeys the differential equation[5]

$$\partial_\mu t = -\tfrac{1}{2}\epsilon_{\mu\nu}\partial^\nu\left(\rho\left(t + \frac{1}{t}\right)\right) \tag{3.3.4}$$

This first order equation can be explicitly solved with the result

$$t(x;w) = \frac{\sqrt{w+\rho_+(x^+)} - \sqrt{w-\rho_-(x^-)}}{\sqrt{w+\rho_+(x^+)} + \sqrt{w-\rho_-(x^-)}} \tag{3.3.5}$$

where w is an integration constant. An equivalent expression is

$$\frac{1-t}{1+t} = \sqrt{\frac{w-\rho_-(x^-)}{w+\rho_+(x^+)}} \tag{3.3.6}$$

For fixed x, the function $t(x;w)$ lives on a two-sheeted Riemann surface over the complex w-plane, with an x-dependent cut extending from $\rho_-(x^-)$ to $\rho_+(x^+)$. The integration constant w can be regarded as an alternative spectral parameter. Evidently, the linear system could also be formulated in terms of w, but at the expense of introducing x-dependent poles and cuts in the linear system. This phenomenon appears to have no analog for flat space integrable models.

---

5 A useful alternative version of this formula is

$$t^{-1}\partial_+ t = \rho^{-1}\partial_+\rho\frac{1-t}{1+t} \quad , \quad t^{-1}\partial_- t = \rho^{-1}\partial_-\rho\frac{1+t}{1-t}$$

As we will see in chapter 4, both t and w play an essential role in the investigation of the underlying group theoretical structure. Below we will also make use of the inverse relation

$$w = \frac{1}{4}\rho\left(t + \frac{1}{t}\right) - \frac{1}{2}(\rho_+ - \rho_-) \tag{3.3.7}$$

which shows, incidentally, that the values $t = 0$ and $t = \infty$ correspond to $w = \infty$ on the different sheets.

Let us pause to contemplate the significance of the replacement (3.3.1). Obviously, a spectral parameter is required if one wants to enlarge the finite Lie group to its affine extension, and the appearance of t in (3.3.1) fits nicely with this expectation. There is now an infinite hierarchy of fields, as one can see by expanding $\widehat{\mathcal{V}}$ in t. For convenience, let us go to a generalized "triangular gauge" defined by the requirement that $\widehat{\mathcal{V}}$ should be regular at $t = 0$ [17,10], or

$$\widehat{\mathcal{V}}(x;t) = \exp \sum_{n=0}^{\infty} t^n \varphi_n(x) \tag{3.3.8}$$

Substitution into the linear system shows that the higher potentials are all related to the original field φ_0 containing the physical degrees of freedom in a non-linear and non-local fashion; in addition, φ_0 must obey its equation of motion. This is clearly illustrated by the lowest order result

$$\partial_\mu \varphi_1 = \epsilon_{\mu\nu}\partial^\nu \varphi_0 + \ldots \tag{3.3.9}$$

For the special choice $G/H = SL(2,\mathbb{R})/SO(2)$ this infinity of new fields in just the infinite hierarchy of "dual potentials" first introduced by Kinnersley and Chitre in their study of the Geroch group [5]. Note also that t-dependent transformations will mix the fields in this hierarchy, and therefore constitute an infinite-dimensional generalization of "duality symmetries" of Maxwell's equations in empty space.

Another important feature of the linear system (3.3.2) is its invariance under a generalization of the symmetric space automorphism (2.2.3) [10]. Define (for $\eta = 1$)

$$\tau^\infty \widehat{\mathcal{V}}(t) \equiv \left(\widehat{\mathcal{V}}^T\right)^{-1}\left(\frac{1}{t}\right) \tag{3.3.10}$$

or, in terms of the Lie algebra,

$$Q_\mu \to Q_\mu \quad P_\mu \to -P_\mu \quad t \to \frac{1}{t} \tag{3.3.11}$$

It is straightforward to verify that

$$\tau^\infty\left(\widehat{\mathcal{V}}^{-1}\partial_\mu \widehat{\mathcal{V}}\right) = \widehat{\mathcal{V}}^{-1}\partial_\mu \widehat{\mathcal{V}} \tag{3.3.12}$$

a property which does not hold for (2.2.9) if we replace τ^∞ by τ.

3.4 An Example: Recovering Colliding Plane Wave Solutions

At this point we can already convince ourselves that the results obtained so far can actually be used to construct non-trivial solutions of Einstein's equation, and I will here discuss a specific example to demonstrate the usefulness of the method. Of central importance for the reconstruction of solutions is the monodromy matrix [10], which is defined as follows.

$$\mathcal{M} = \widehat{\mathcal{V}}\tau^{\infty}\widehat{\mathcal{V}}^{-1} = \widehat{\mathcal{V}}(x;t)\widehat{\mathcal{V}}^{T}\left(x;\frac{1}{t}\right) \tag{3.4.1}$$

A short calculation reveals that

$$\partial_{\mu}\mathcal{M} = \widehat{\mathcal{V}}\left(\widehat{\mathcal{V}}^{-1}\partial_{\mu}\widehat{\mathcal{V}} - \tau^{\infty}\left(\widehat{\mathcal{V}}^{-1}\partial_{\mu}\widehat{\mathcal{V}}\right)\right)\tau^{\infty}\widehat{\mathcal{V}}^{-1} = 0 \tag{3.4.2}$$

where the result (3.3.12) was used. Consequently, $\mathcal{M}$ can only depend on w. The solution generating procedure now consists in choosing a matrix $\mathcal{M}(w)$ and finding a factorization as in (3.4.1). In order to be able to extract the components of the metric tensor, the matrix $\widehat{\mathcal{V}}$ should be chosen in a triangular gauge, so that in (3.4.1) one factor is analytic in a neighborhood of $t = 0$ while the other is analytic in a neighborhood of $t = \infty$, such that the overlap of the respective domains of analyticity is an annular region. This is a variant of the famous Riemann-Hilbert problem. Observe that the x-dependence of t, which may have seemed a nuisance in the analysis of the linear system, is absolutely essential here because it is precisely the x-dependence of t in the factorization (3.4.1) which generates the x-dependence of $\widehat{\mathcal{V}}$ and thereby of $\mathcal{V}$ and the metric. In fact, it is obvious for this reason that this factorization method would not work for a flat space integrable theory.

To facilitate the comparison with the standard literature, I will in this subsection use the notation

$$u \equiv x^{+} \quad , \quad v \equiv x^{-} \tag{3.4.3}$$

for the light-cone coordinates. Furthermore, the remaining conformal invariance is entirely fixed by choosing the coordinates such that

$$\rho_{+}(u) = \tfrac{1}{2}(1-2u^2) \ , \ \rho_{-}(v) = \tfrac{1}{2}(1-2v^2) \ \Rightarrow \ \rho(u,v) = 1 - u^2 - v^2 \tag{3.4.4}$$

The interaction region, where the waves collide, is $0 < u,v < 1, \rho(u,v) > 0$. In the region "before the collision" the solution is obtained by defining $g(u,v) \equiv g(0,v)$ for $-\infty < u \leq 0; 0 < v < 1$ and $g(u,v) \equiv g(u,0)$ for $-\infty < v \leq 0; 0 < u < 1$. Between the incoming waves, the space-time is Minkowski ($g(u,v) = g(0,0)$). Although the solution is continous across the boundary by construction, the matching introduces certain discontinuities in the higher derivatives (gravitational shock wave). Furthermore, the solution

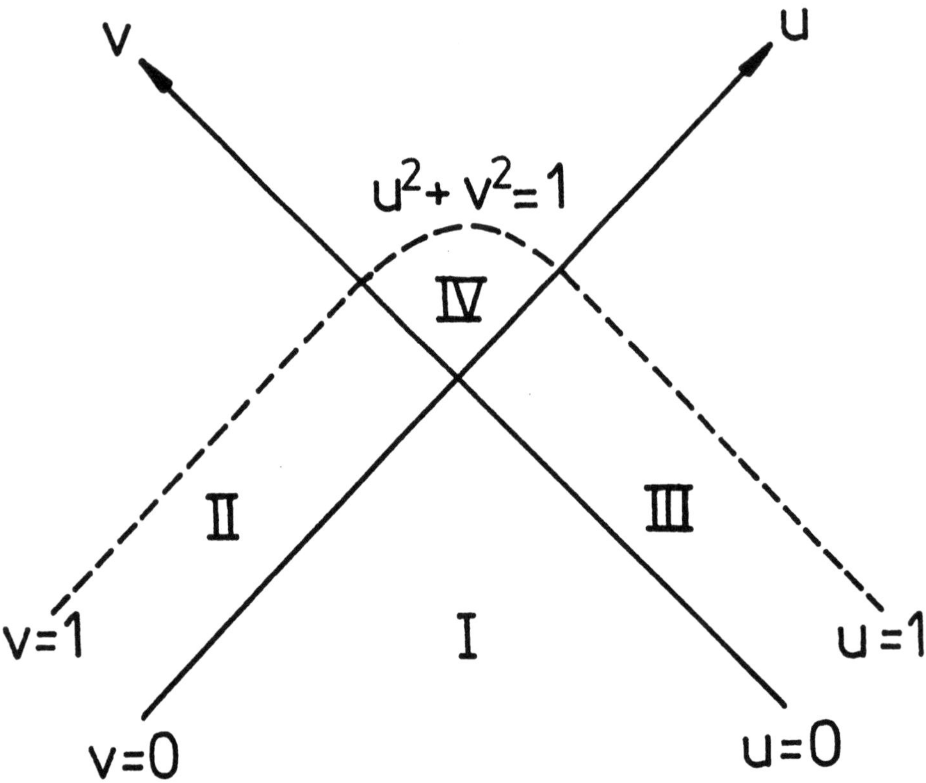

Fig. 1. Space-time for a colliding plane-wave solution. The metric is flat (Minkowski) in (I), and corresponds to pure gravitational radiation in regions (II) and (III). In the interaction region (IV) it is given by the solution (3.4.14) below. The dashed line indicates a curvature (Weyl tensor) singularity [11-13].

cannot be extended beyond $\rho(u,v) = 0$ [11-13]. The situation is illustrated in the figure.

We now consider the simplest non-trivial monodromy matrix

$$\mathcal{M}(w) = \begin{pmatrix} \frac{w_0-w}{w_0+w} & 0 \\ 0 & \frac{w_0+w}{w_0-w} \end{pmatrix} \tag{3.4.5}$$

for the special value $w_0 = \frac{1}{2}$ as an example[6]. Next we use (3.3.7) to derive

$$w - w_0 = -\frac{\rho}{4t_0}(t-t_0)\left(\frac{1}{t} - t_0\right) \tag{3.4.6}$$

with $t_0(u,v) \equiv t(u,v;w=w_0)$. Substituting (3.4.6) into (3.4.5), and defining the functions (see (3.3.5))

---

6 For a Euclidean metric, essentially the same monodromy matrix yields the Schwarzschild solution with mass $m = w_0$ in Weyl canonical coordinates [18]. The relative minus sign is due to the fact that, for the solution (3.4.9) below, $\widehat{\mathcal{V}}(t=0) = \mathcal{V}$ and $\widehat{\mathcal{V}}(t=\infty) = -\tau\mathcal{V}$, so $\mathcal{M}(\infty) = -1$.

$$t_1(u,v) \equiv t(u,v;w=\tfrac{1}{2}) = \frac{\sqrt{1-u^2}-v}{\sqrt{1-u^2}+v} > 0 \qquad (3.4.7)$$

and

$$t_2(u,v) \equiv t(u,v;w=-\tfrac{1}{2}) = -\frac{\sqrt{1-v^2}+u}{\sqrt{1-v^2}-u} < 0 \qquad (3.4.8)$$

(the inequalities hold in the interaction region), we obtain the desired factorization in the form (3.4.1) with

$$\widehat{\mathcal{V}}(u,v;t) = \begin{pmatrix} \sqrt{-\frac{t_2}{t_1}\frac{t-t_1}{t-t_2}} & 0 \\ 0 & \sqrt{-\frac{t_1}{t_2}\frac{t-t_2}{t-t_1}} \end{pmatrix} \qquad (3.4.9)$$

Putting $t = 0$, we can directly read off the result for Δ by virtue of (2.2.14)

$$\Delta = -\frac{t_1}{t_2} = \frac{1-\xi}{1+\xi} > 0 \qquad (3.4.10)$$

where, on the right hand side, I have introduced the oblate spherical coordinates, widely used in the literature:

$$\begin{aligned} \xi &\equiv u\sqrt{1-v^2} + v\sqrt{1-u^2} \\ \eta &\equiv u\sqrt{1-v^2} - v\sqrt{1-u^2} \end{aligned} \qquad (3.4.11)$$

To determine the conformal factor, we integrate the equation (3.2.7), using

$$P_+^1 = \tfrac{1}{2}\rho^{-1}\partial_+\rho\left(\frac{1-t_1}{1+t_1} - \frac{1-t_2}{1+t_2}\right) \qquad (3.4.12)$$

Some further calculation shows that

$$\lambda^2 = 8uv\frac{(1-t_1t_2)^2}{(1-t_1^2)(1-t_2^2)} \qquad (3.4.13)$$

where the undetermined overall factor has been chosen for convenience. In section 4.2, I will re-derive this result directly from the Breitenlohner-Maison cocycle formula. Collecting all contributions, reexpressing them in terms of the coordinates (3.4.11) and substituting the result into (3.2.9) yields the metric

$$ds^2 = (1+\xi)^2\left(\frac{d\xi^2}{1-\xi^2} - \frac{d\eta^2}{1-\eta^2}\right) - \rho^2\frac{1-\xi}{1+\xi}(dx^2)^2 - \frac{1+\xi}{1-\xi}(dx^3)^2 \qquad (3.4.14)$$

This is just one of the Ferrari Ibanez metrics[7]. It is quite obvious now that more complicated monodromy matrices with off-diagonal terms in (3.4.7) will lead to colliding plane wave solutions with non-collinear polarization of the type discussed in [13].

To conclude this section I will sketch how the factorization works in the general case, given a "reasonable" form of the monodromy matrix (I here follow

[7] Equation (2.1) of the first paper in [13] for $n = -1$, to be completely precise.

the derivation given in [46]). For simplicity, we again assume absence of higher order poles in the monodromy matrix

$$\mathcal{M}(w) = Y + \sum_{i=1}^{N} \frac{A_i}{w - w_i} \tag{3.4.15}$$

where $Y = \mathcal{M}(\infty)$ and the A_i are constant symmetric but otherwise arbitrary matrices. Defining $t_i \equiv t(x; w_i)$, we have

$$\frac{1}{w - w_i} = \nu_i \left(\frac{t_i}{t - t_i} + \frac{1}{1 - t_i t} \right) \tag{3.4.16}$$

with the functions ν_i

$$\nu_i(x) \equiv -4\rho^{-1} \frac{t_i}{1 - t_i^2} \tag{3.4.17}$$

Inspired by the explicit result (3.4.9), we make the following ansatz for $\widehat{\mathcal{V}}$

$$\widehat{\mathcal{V}}(t) = B_0 + \sum_i \frac{B_i}{t - t_i} \tag{3.4.18}$$

where the functions $B_0(x)$ and $B_i(x)$ are to be determined. Once the solution of the factorization problem has been brought into the form (3.4.18), we can immediately read off the result for $\mathcal{V}(x)$, which is

$$\mathcal{V} = B_0 - \sum_i \frac{B_i}{t_i} \tag{3.4.19}$$

Inserting the ansatz (3.4.18) into (3.4.1), we obtain the following matrix equations after a little algebra

$$\begin{aligned} Y &= B_0 \mathcal{V}^T \\ \nu_i t_i A_i &= B_i \left(\mathcal{V}^T + \sum_j \frac{B_j^T}{t_j} \frac{1}{1 - t_i t_j} \right) \\ \nu_i t_i A_i &= \left(B_0 + \sum_j B_j \frac{t_i}{1 - t_i t_j} \right) B_i^T \end{aligned} \tag{3.4.20}$$

Assuming that these equations can be solved, we have therefore succeeded in reducing the problem of solving a non-linear partial differential equation to a purely algebraic problem in this way! Finally, we have to make sure that acting with a derivative on $\widehat{\mathcal{V}}$, we do get the right pole structure for the linear system. After a little further calculation (not forgetting the derivatives on t and t_i !), we find

$$\begin{aligned}\partial_+\widehat{\mathcal{V}} =&\partial_+B_0 + \sum_i \frac{1}{t-t_i}\left(\partial_+B_i + \rho^{-1}\partial_+\rho B_i\frac{t_i}{1+t_i}\right) - \\ &- \frac{1-t}{1+t}\rho^{-1}\partial_+\rho\sum_i \frac{1}{t-t_i}\frac{B_i}{1+t_i}\end{aligned} \tag{3.4.21}$$

which displays the desired structure explicitly.

3.5 Extension to Supergravity

The foregoing considerations can be generalized to supergravity theories [19]. In this section, I will give the results for $N = 2$ supergravity, making use of the results in section 2.2, and also present the Kac-Moody generalization of the current (2.3.3), which can be cast into a very simple and model-independent form. As before much labor is saved if we start from the three dimensional Lagrangian (2.2.20) rather than $N = 1$ supergravity in four dimensions. Again it proves convenient to fix certain gauges. For the dreibein, we make the same choice as in (3.2.2). In addition, we invoke local supersymmetry (2.2.21) to put

$$\psi_a^I = (\gamma_\alpha\psi^I, \psi_2^I) \tag{3.5.1}$$

(the index a is a *flat* Lorentz index in three dimensions). Thus, in two dimensions, we have the correspondence

$$\lambda \leftrightarrow \psi^I \quad , \quad \rho \leftrightarrow \psi_2^I \tag{3.5.2}$$

It is important that, just as in the bosonic theory, there is a residual invariance which preserves the form of this gauge. In addition to the conformal reparametrizations (3.2.4), we now have the residual superconformal transformations with a parameter ϵ^I subject to the constraint

$$\gamma^\nu\gamma^\mu D_\nu\epsilon^I = 0 \tag{3.5.3}$$

This residual invariance can be used to put $\psi_2^I = 0$ in the equations of motion (however, it is also possible to keep the gravitinos by *not* fixing this remaining gauge, see [20]). A further simplification of the equations of motion of the physical fields occurs when one passes to second order formalism and eliminates the auxiliary vector B_μ in (3.2.2) through its equation of motion [19]. Defining $\epsilon_{\mu\nu}F \equiv \lambda^{-1}\rho(\partial_\mu B_\nu - \partial_\nu B_\mu)$ and rescaling the fermions by a factor $\lambda^{\frac{1}{2}}$, a calculation completely analagous to the one leading to (2.1.4) reveals that the field F appears only in the following combination

$$\mathcal{L}' = \tfrac{1}{8}\rho F^2 - \tfrac{1}{4}\rho F\left(\bar{\psi}^I\psi^I + \tfrac{1}{2}\bar{\chi}^I\chi^I\right) \tag{3.5.4}$$

Elimination of F therefore produces the following extra quartic terms

$$\mathcal{L}' = -\tfrac{1}{8}\rho\left(\bar{\psi}^I\psi^I\right)^2 - \tfrac{1}{8}\rho\bar{\psi}^I\psi^I\bar{\chi}^J\chi^J - \tfrac{1}{32}\rho\left(\bar{\chi}^I\chi^I\right)^2 \tag{3.5.5}$$

to which we must add the terms originating from the second order spin connection

$$\mathcal{L}'' = \tfrac{1}{8}\rho\left(\bar{\psi}^I\psi^I\right)^2 + \tfrac{1}{8}\rho\bar{\psi}^I\psi^I\,\bar{\chi}^J\chi^J - \tfrac{3}{32}\rho\left(\bar{\chi}^I\chi^I\right)^2 \tag{3.5.6}$$

Adding up these contributions, we see that the remaining gravitino components ψ^I decouple, since no further contributions involving ψ^I arise from the extra terms in (2.2.23) if we put $\psi_2^I = 0$. Taking into account all the $(\bar{\chi}\chi)^2$ terms and rescaling the fermion fields by $\lambda^{\frac{1}{4}}$, we arrive at the equations of motion

$$\begin{aligned} \rho^{-1}D^\mu\left(\rho P_\mu^1\right) &= 3iP_\mu^2\,\bar{\chi}^1\gamma^\mu\chi^2 \\ \rho^{-1}D^\mu\left(\rho P_\mu^2\right) &= -\,3iP_\mu^2\,\bar{\chi}^1\gamma^\mu\chi^2 \end{aligned} \tag{3.5.7}$$

and

$$\begin{aligned} -i\rho^{-\frac{1}{2}}\gamma^\mu D_\mu\left(\rho^{\frac{1}{2}}\chi^1\right) &= \left(\bar{\chi}^2\chi^2\right)\chi^1 \\ -i\rho^{-\frac{1}{2}}\gamma^\mu D_\mu\left(\rho^{\frac{1}{2}}\chi^2\right) &= \left(\bar{\chi}^1\chi^1\right)\chi^2 \end{aligned} \tag{3.5.8}$$

These equations constitute the supersymmetric generalization of the Ernst equation (3.2.8).

The bosonic linear system (3.3.2) admits the following generalization for $N = 2$ supergravity

$$\begin{aligned} \widehat{\mathcal{V}}^{-1}\partial_\mu\widehat{\mathcal{V}} &= \left(\frac{1+t^2}{1-t^2}P_\mu^A + \frac{2t}{1-t^2}\epsilon_{\mu\nu}P^{\nu A}\right)Y^A \\ &+ \left(Q_\mu - \frac{2t^2}{(1-t^2)^2}3i\bar{\chi}^1\gamma_\mu\chi^2 - \frac{t(1+t^2)}{(1-t^2)^2}3i\bar{\chi}^1\gamma^3\gamma_\mu\chi^2\right)Y^3 \end{aligned} \tag{3.5.9}$$

To prove this assertion, the integrability relation (3.3.3) must again be checked; the explicit calculations are now considerably more tedious than in the bosonic case due to the extra contributions arising from the derivatives on the spectral parameter and the quartic spinor terms. The main feature distinguishing (3.5.9) from the bosonic result (3.3.2) is the appearance of higher order poles in the spectral parameter t (even higher order poles appear when the gravitinos are retained [20]). For $N = 2$, some information is lost in the linear system because the cubic terms on the right hand side of (3.5.8) as well as the coupling between χ and the gauge field Q_μ drop out because the subgroup $SO(2)$ is abelian. This is no longer true for non-abelian H. Apart from this, the on-shell information of $N = 2$ supergravity is now entirely contained in the matrix $\widehat{\mathcal{V}}$. It is noteworthy that the linear system with the fermions remains invariant under the generalized symmetric space automorphism τ^∞ of (3.3.12).

There is also an analog of the conserved current (2.3.3) [38]. It is

$$\begin{aligned} \widehat{\mathcal{J}}_\mu = \rho K\widehat{\mathcal{V}}\Bigg(&\left(\frac{1+t^2}{1-t^2}P_\mu^A + \frac{2t}{1-t^2}\epsilon_{\mu\nu}P^{\nu A}\right)Y^A - \\ &- \left(\frac{1+6t^2+t^4}{(1-t^2)^2}\tfrac{3}{2}i\bar{\chi}^1\gamma_\mu\chi^2 + \frac{4t(1+t^2)}{(1-t^2)^2}\tfrac{3}{2}i\bar{\chi}^1\gamma^3\gamma_\mu\chi^2\right)Y^3\Bigg)\widehat{\mathcal{V}}^{-1} \end{aligned} \tag{3.5.10}$$

where the factor K satisfies the differential equation

$$K^{-1}\partial_\mu K = \frac{4}{1-t^2}\epsilon_{\mu\nu}\partial^\nu t \tag{3.5.11}$$

and is explicitly given by[8]

$$\rho K = -8\rho^{-1}\frac{t^2}{(1-t^2)^2} = \frac{2}{1-t^2}\frac{\partial t}{\partial w} \tag{3.5.12}$$

The somewhat unexpected prefactor ρK ensures that the divergence $\widehat{\mathcal{J}}_\mu$ is indeed proportional to the equations of motion (with the correct factors of ρ!) and therefore vanishes. Readers skeptical that we are on the right track with the above expression for the Kac Moody current should note that in flat space (i.e.rigidly) supersymmetric models, one finds exactly the same expression for the conserved current [47], but *without* the prefactor ρK, since we have $\rho = 1$ and $t = const$ in that case.

There is a very nice simplification of the result (3.5.10). To derive it, re-write (3.5.10) as follows

$$\widehat{\mathcal{J}}_\mu = \tfrac{1}{2}\rho K(1-t^2)\widehat{\mathcal{V}}\left(\frac{\partial}{\partial t}\left(\epsilon_{\mu\nu}\widehat{\mathcal{V}}^{-1}\partial^\nu\widehat{\mathcal{V}}\right)\right)\widehat{\mathcal{V}}^{-1} \tag{3.5.13}$$

Now using

$$\frac{\partial}{\partial t}\left(\widehat{\mathcal{V}}^{-1}\partial_\mu\widehat{\mathcal{V}}\right) = \partial_\mu\left(\widehat{\mathcal{V}}^{-1}\frac{\partial\widehat{\mathcal{V}}}{\partial t}\right) + \left[\widehat{\mathcal{V}}^{-1}\partial_\mu\widehat{\mathcal{V}}, \widehat{\mathcal{V}}^{-1}\frac{\partial\widehat{\mathcal{V}}}{\partial t}\right] + \widehat{\mathcal{V}}^{-1}\left[\frac{\partial}{\partial t}, \partial_\mu\right]\widehat{\mathcal{V}} \tag{3.5.14}$$

and

$$\left[\frac{\partial}{\partial t}, \partial_\mu\right] = -\partial_\mu\left(\frac{\partial w}{\partial t}\right)\frac{\partial}{\partial w} \tag{3.5.15}$$

together with (3.5.12) above, some re-arrangement leads to a much more elegant formula for the Kac Moody current

$$\widehat{\mathcal{J}}_\mu = \epsilon_{\mu\nu}\partial^\nu\left(\frac{\partial\widehat{\mathcal{V}}}{\partial w}\widehat{\mathcal{V}}^{-1}\right) \tag{3.5.16}$$

In this form the result is completely general and model-independent, and remains valid in particular when gravitinos are included. We will come back to it in section 4.2 when discussing the central term.

Finally, a remark on the status of fermionic solutions is in order. To date, no such solutions of the supersymmetric Ernst equation (3.5.7) and (3.5.8) describing colliding supergravitational plane waves are known, but there is no reason of principle why such solutions should not exist. This is strongly suggested by the existence of fermionic solutions extending the extreme Reissner-Nordstrom solution [48]. There remains, of course, the question of how to interpret Grassmann valued solutions. Their main importance in the present context derives

[8] This simple solution was overlooked in [38], where it was erroneously claimed that K is a non-local function of t.

from the new insights one might hope to gain into the structure of the linear system (3.5.9).

4. Group Theoretical Structure

As already emphasized in the introduction, one of the main motivations for studying the reduction of gravity and supergravity theories to low dimensions is the search for a better understanding of the underlying symmetry structures. In this section, I will describe the emergence of infinite-dimensional symmetries of the affine Kac-Moody type for the purely bosonic theory. Although the work of the last few years has provided us with a fairly satisfactory understanding in that case, there have been some surprising developments such as the discovery of the central extension [17,10] and of a Virasoro algebra acting on the space of solutions [49,18]. These will be reviewed in sections 4.2 and 4.3, respectively. A new result is the formula (4.2.18) for the variation of the conformal factor. While the first two sections of this chapter deal mostly with the bosonic theory, the last section is devoted to a discussion of the supersymmetric theory, where some perplexing questions remain. These are in part related to the fact that, until now, no fermionic analog of the infinite hierarchy of dual potentials in (3.3.7) has been found: the chiral fermions behave like eigenstates under the duality rotations associated with the infinite-dimensional extensions of G (cf. equation (4.2.29) below). This and our lack of understanding of how to treat fermionic solutions in the present framework may be taken as a hint at the existence of yet more symmetries. In fact, by considering the hyperbolic extension of certain affine Dynkin diagrams, B.Julia was led to conjecture the emergence of hyperbolic symmetries in the reduction to one dimension, and in particular of an E_{10} symmetry in the maximally extended $N = 16$ supergravity theory [24]. In section 4.3, I will argue that the appearance of such algebras is related to the introduction of fermionic degrees of freedom. The suggestive relation between hyperbolic Lie algebras and locally supersymmetric theories is most impressively exemplified by the fact that the "maximally extended" simply laced hyperbolic Lie algebras are extensions of E_8 and $D_8 = SO(16)$ [23], which are precisely the groups that appear in the maximally extended $N = 16$ supergravity theory.

4.1 Infinite Dimensional Coset Spaces

In section 2.2, some pertinent results for finite-dimensional σ-models were briefly reviewed. We would now like to generalize these results to the infinite-dimensional case. The appearance of the spectral parameter t in section 3.3 was the first evidence for the emergence of an affine Kac-Moody symmetry. The situation is somewhat complicated by the x-dependence of t, and in section 3.3 we also noted the existence of another spectral parameter w, which does not depend on x, and which is, in principle, interchangeable with t. As already pointed out before, *both* these parameters are essential for the generalizations of the various concepts introduced in section 3.3. To see why this is so, recall that the Kac Moody group G^∞, which extends (and contains) G must act rigidly. Therefore the elements of this group must depend on w rather than t. On the other hand, the infinite-dimensional generalization τ^∞ of the symmetric space automorphism introduced in (3.3.10) involves the parameter t, and therefore automatically brings in an x-dependence. This shows already that we must introduce matrices which depend on both x and w, if we want to define an infinite-dimensional extension of the coset space G/H. We are thus naturally led to consider the matrix $\widehat{\mathcal{V}}(x, t(x, w))$ as an element of G^∞ for each (fixed) x. The actual definition of G^∞ also requires some technical assumptions about the regularity properties of $g(w) \in G^\infty$ such as the requirement that $g(w)$ should be a meromorphic function of w, but I will not further elaborate on this issue here (see [10] for a discussion). The subgroup H^∞ extending H is then defined to consist of all matrices $h(x;t)$ in G^∞ invariant under the generalized automorphism τ^∞, or

$$h(x;t) \in H^\infty \Longleftrightarrow (h(x;t))^{-1} = h\left(x;\frac{1}{t}\right)^T \tag{4.1.1}$$

The Lie algebras associated with G^∞ and H^∞ will be called $\mathbf{G}^\infty$ and $\mathbf{H}^\infty$. An element of $\mathbf{G}^\infty$ is written as

$$\delta g(w) = \sum_{n=-\infty}^{\infty} \delta g_n \, w^n \tag{4.1.2}$$

whereas the general element of $\mathbf{H}^\infty$ has the form

$$\delta h(t) = \delta h_0 + \sum_{n>0} \delta h_n \left(t^n + \frac{1}{t^n}\right) + \sum_{n>0} \delta k_n \left(t^n - \frac{1}{t^n}\right) \tag{4.1.3}$$

with $\delta h_n \in \mathbf{H}$ and $\delta k_n \in \mathbf{K}$. So, roughly speaking, $\mathbf{H}^\infty$ is "half" of the Lie algebra $\mathbf{G}^\infty$.

We are now interested in finding and classifying the solutions $\widehat{\mathcal{V}}(x,t)$ of the linear system (3.3.2). According to (3.3.12), the expression $\widehat{\mathcal{V}}^{-1}\partial_\mu \widehat{\mathcal{V}}$ must be an element of $\mathbf{H}^\infty$, which, by construction, is "pure gauge". This suggests that[9].

$$\widehat{\mathcal{V}}(x,t) = g^{-1}(w)h(x,t) \tag{4.1.4}$$

because we still have the freedom of multiplication by an x-independent but w-dependent matrix from the left, since $\widehat{\mathcal{V}}$ is an element of G^∞ , and not just H^∞. The explicit t-dependence on the right hand side of (3.3.2) imposes a stronger constraint on $\widehat{\mathcal{V}}^{-1}\partial_\mu \widehat{\mathcal{V}}$, since matrices of the form (4.1.3) will not yield

---

9 The question of whether indeed *all* solutions of (3.3.3) can be represented in this way is related to the question of whether the group G^∞ acts transitively, see the discussion below.

the t-dependence prescribed by (3.3.2) in general. In addition, we demand that $\widehat{\mathcal{V}}$ should be regular at $t = 0$, so we can recover $\mathcal{V}$ in the limit $t \to 0$. This requirement is tantamount to the existence of a triangular gauge for $\widehat{\mathcal{V}}$ as in (3.3.8). For arbitrary matrices in (4.1.4), $\widehat{\mathcal{V}}$ will, however, fail to satisfy these requirements, because the replacement of w by t through (3.3.7) introduces poles at $t = 0$. Nonetheless, we can argue that there exists a *special* $h_0(x, t; g)$, implicitly depending on $g(w)$, such that

$$\widehat{\mathcal{V}}_0(x, t) = g^{-1}(w)h_0(x, t; g) \tag{4.1.5}$$

has the desired properties. To see this, let us first determine the monodromy matrix from (4.1.4), which is $\mathcal{M}(w) = (g^T(w)g(w))^{-1}$. If the matrix equations (3.4.20) can be solved, there exists a solution $\widehat{\mathcal{V}}_0$ to the Riemann Hilbert factorization problem, which is of the form (3.4.18) and thus indeed regular at the origin. The H^∞ matrix in (4.1.5) is then simply $h_0 = g(w)\widehat{\mathcal{V}}_0$. Differentiating this solution generates at most first order poles at $t = \pm 1$ by (3.4.21), and this, in turn, guarantees the absence of higher order poles from the linear system (3.3.2). Altogether, these arguments show that, given a suitably regular matrix $g(w)$ in (4.1.3), we can always construct a solution of the linear system with the requisite properties.

These considerations make it highly plausible that the space of solutions to the non-linear equation (3.2.5) can be identified with the coset space G^∞/H^∞. To see this, consider the space of all matrices $\widehat{\mathcal{V}}(x; t)$ with $\widehat{\mathcal{V}}^{-1}\partial_\mu\widehat{\mathcal{V}} \in \mathbf{H}^\infty$, but not necessarily of the special form (3.3.2). On this space, there is a combined action of $G^\infty \times H^\infty$

$$\widehat{\mathcal{V}}(x; t) \to \widehat{\mathcal{V}}'(x; t) = g^{-1}(w)\widehat{\mathcal{V}}(x; t)h(x; t) \tag{4.1.6}$$

Thus the group G^∞ acts rigidly from the left, while H^∞ acts as a gauge group from the right. The infinitesimal counterpart of (4.1.6) is

$$\delta\widehat{\mathcal{V}}(x; t) = -\delta g(w)\,\widehat{\mathcal{V}}(x; t) + \widehat{\mathcal{V}}(x; t)\,\delta h(x; t) \tag{4.1.7}$$

where $\delta g(w)$ and $\delta h(x; t)$ are to be expanded as in (4.1.2) and (4.1.3), but with constant δg_n and x-dependent δh_n and δk_n. Observe that through the elevation of H^∞ to a gauge group, the infinitely many "dual potentials" introduced in (3.3.8) (now with $n \in \mathbb{Z}$) become *independent* gauge degrees of freedom. The transformation law (4.1.6) generalizes (2.2.7) to the infinite-dimensional case. The coset space G^∞/H^∞ can now be defined in complete analogy with the finite-dimensional space G/H: two matrices $\widehat{\mathcal{V}}$ and $\widehat{\mathcal{V}}'$ will be called gauge equivalent if and only if there exists a matrix $h \in H^\infty$ such that $\widehat{\mathcal{V}}' = \widehat{\mathcal{V}}h$. The equivalence classes of such matrices constitute the elements of the coset space G^∞/H^∞. By the above reasoning, each equivalence class contains one particular element $\widehat{\mathcal{V}}_0$, which is analytic at $t = 0$ and from which the matrix $\mathcal{V}$ solving the field equations (3.2.5) can be reconstructed. Therefore, the equivalence classes are in one-to-one correspondence with the solutions of (3.3.2), and thereby of (3.2.5). Conversely, one can proceed from a given solution $\widehat{\mathcal{V}}_0$

and act on it with an arbitrary matrix $g(w) \in G^\infty$. This transformation will in general take us out of the coset to which $\widehat{\mathcal{V}}_0$ belongs, and a compensating H^∞ rotation must be applied to restore the triangular gauge.

In summary, the affine Kac-Moody group G^∞ acts on the space of matrices $\widehat{\mathcal{V}}$ satisfying $\widehat{\mathcal{V}}^{-1}\partial_\mu \widehat{\mathcal{V}} \in \mathbf{H}^\infty$. If we insist that $\widehat{\mathcal{V}}$ be triangular as in (3.3.8), or, equivalently, that $\widehat{\mathcal{V}}^{-1}\partial_\mu \widehat{\mathcal{V}}$ be subject to the more restrictive condition (3.3.2), the symmetry is realized non-linearly. Since we can then extract the solution of the original non-linear field equation from $\widehat{\mathcal{V}}$, there is also a non-linear action of G^∞ on the space of solutions. There is an obvious analogy between $\widehat{\mathcal{V}}$ and the vielbein in general relativity, with G^∞ replacing $GL(D, \mathbb{R})$ and H^∞ the Lorentz group (however, G^∞ and H^∞ contain only their transverse subgroups). For this reason, and because G^∞ is infinite-dimensional, the matrix $\widehat{\mathcal{V}}$ might be appropriately referred to as "∞-bein". This analogy is deficient only insofar as all the elements of G^∞/H^∞ are "on shell" by construction, whereas this is not the case for the finite-dimensional coset spaces of section 2.2.

To be sure, there remain some thorny technical issues. One is the question of whether all the solutions of (3.2.5) can indeed be obtained in this way, or, equivalently, whether the group G^∞ acts transitively on the space of solutions, and all solutions can be "reached" from $\widehat{\mathcal{V}}_0 = \mathbf{1}$ via (4.1.6). For stationary axisymmetric solutions of Einstein's equations, this has been demonstrated modulo some technical assumptions in [50]. For the plane wave solutions discussed in section 3.4, the corresponding result is still lacking[10]. A second, and perhaps more significant indication of our incomplete understanding is the existence of further symmetries. The considerations above did not take into account the conformal factor, a defect that will be remedied in the following section. As a consequence, the space of solutions is actually bigger, and the coset space defined above must be modified. In fact, there are more symmetries at work even in the purely bosonic theory, and in section 4.3, a Virasoro type symmetry on the solution space will be exhibited. There, we will also return to the fermionic theories.

4.2 The Central Extension

It is well known that every affine Kac-Moody algebras admits a central extension [23]. In physical examples, the appearance of such a term is usually linked to quantization (see [51] for a very readable introduction). It is therefore rather surprising that, for the affine algebra $\mathbf{G}^\infty$, a central term is already present in the *classical* theory [17,10]. Moreover, this central term is not a c-number, but acts as a scaling operator on the conformal factor. In this section, a derivation of this result is given which generalizes previous ones and also applies to supergravity theories. Although the final result (4.2.18) is completely general, the proof is only given for the bosonic model in order to avoid cumbersome formulas. Furthermore, I will employ light-cone notation from now on (with

---

[10] At least to the best of the author's knowledge.

the usual conventions $V^{\pm} \equiv \frac{1}{\sqrt{2}}(V^0 \pm V^1)$ for vectors, etc.), since the derivation becomes much more transparent in this way. Details of the calculation are provided only for positive (+) chirality components, since they are the exactly the same for the opposite chirality.

The linear system (3.3.2) will now be used in the form

$$\widehat{\mathcal{V}}^{-1}\partial_+\widehat{\mathcal{V}} = Q_+ + \frac{1-t}{1+t}P_+ \quad , \quad \widehat{\mathcal{V}}^{-1}\partial_-\widehat{\mathcal{V}} = Q_- + \frac{1+t}{1-t}P_- \tag{4.2.1}$$

so the right and left moving components of the linear system have poles at $t=-1$ and $t=1$, respectively. The correspondence between the chirality and the location of the poles in t remains valid for supergravity, as one can easily verify from (3.5.9) and (3.5.10).

The starting point is equation (3.2.7), which determines the conformal factor in terms of the "matter fields" up to an overall factor, and which we now write as

$$\rho^{-1}\partial_{\pm}\rho\, \lambda^{-1}\partial_{\pm}\lambda = \tfrac{1}{2}\mathrm{Tr}\, P_{\pm}P_{\pm} \tag{4.2.2}$$

disregarding the term involving second derivatives of ρ, which can be absorbed into a redefinition of λ, from now on. To analyze the action of G^{∞} on λ, we must first find out how this group acts on the right hand side of (4.2.2). From (4.1.6) and (3.3.2), it is clear that $P_{\pm}$ and $Q_{\pm}$ are inert under the action of G^{∞}, and therefore this group can act only via the induced compensating H^{∞} rotation needed to restore the generalized triangular gauge. Hence, we must examine the compensating generalized duality rotations in a little more detail. It is advantageous to analyze infinitesimal transformations of the form (4.1.2) with a "cutoff" on the number of components, i.e. $\delta g_n = 0$ for $|n|$ sufficiently large. Assuming $\widehat{\mathcal{V}}$ to be regular at $t=0$, we now demand that the the compensating H^{∞} variation in (4.1.7) should be chosen in such a manner that $\delta\widehat{\mathcal{V}}$ is also regular at $t=0$. Inserting the expansion (4.1.2) into (4.1.7) and using (3.3.7), we see that only the *positive* powers of w in (4.2.1) lead to poles in t which must be compensated by an appropriate $\delta h(t)$. If, on the other hand, only negative powers of w appear, we can put $\delta h(t) = 0$ because $w^{-1} = 4\rho^{-1}t + O(t^2)$ for $t \to 0$ from (3.3.7). For this reason, we can restrict attention at this point to variations of the form

$$\delta g(w) = \sum_{n=1}^{N} \delta g_n w^n \tag{4.2.3}$$

For such $\delta g(w)$, we have

$$\widehat{\mathcal{V}}^{-1}\delta g(w)\widehat{\mathcal{V}} = \frac{\alpha_{-n}}{t^n} + \ldots + \frac{\alpha_{-1}}{t} + \alpha_0 + \sum_{\nu=1}^{\infty}\alpha_\nu t^\nu \tag{4.2.4}$$

by (3.3.7); the G-valued matrix functions $\alpha_\nu = \alpha_\nu(x)$ are implicitly defined through this equation. The compensating transformation $\delta h(t)$ needed to remove the pole terms is therefore

$$\delta h(t) = \frac{\alpha_{-n}}{t^n} + \ldots + \frac{\alpha_{-1}}{t} + \tfrac{1}{2}\Big(\alpha_0 + \tau(\alpha_0)\Big) + \tau(\alpha_{-1})t + \ldots + \tau(\alpha_{-n})t^n =$$
$$= \delta h_0 + \sum_{\nu=1}^{N} \delta h_\nu \left(t^\nu + \frac{1}{t^\nu}\right) + \sum_{\nu=1}^{N} \delta k_\nu \left(t^\nu - \frac{1}{t^\nu}\right) \tag{4.2.5}$$

with $\delta h_\nu \equiv \big(\alpha_{-\nu} + \tau(\alpha_{-\nu})\big)$, etc.; for the definition of τ, see (2.2.3). We can express the constraints $\alpha_\nu(x)$, or equivalently on $\delta h_\nu(x)$ and $\delta k_\nu(x)$, which follow from the definition (4.2.4), in a somewhat different form. Differentiating the left hand side of (4.2.4), we get

$$\partial_+ \left(\widehat{\mathcal{V}}^{-1}\delta g(w)\widehat{\mathcal{V}}\right) = \left[\widehat{\mathcal{V}}^{-1}\delta g(w)\widehat{\mathcal{V}}\,,\,\widehat{\mathcal{V}}^{-1}\partial_+\widehat{\mathcal{V}}\right] \tag{4.2.6}$$

Inserting (4.2.1) and the expansion (4.2.4) back into this expression, and comparing this with the result obtained by directly differentiating the right hand side of (4.2.4), it is straightforward to show that

$$D_+\alpha_\nu + D_+\alpha_{\nu-1} + \rho^{-1}\partial_+\rho\Big(\nu\alpha_\nu - (\nu-1)\alpha_{\nu-1}\Big) + \big[P_+, \alpha_\nu - \alpha_{\nu-1}\big] = 0 \tag{4.2.7}$$

where we define $\alpha_{-n-2} \equiv \alpha_{-n-1} \equiv 0$. The variation of Q_+ and P_+ can now be deduced from

$$\delta\left(\widehat{\mathcal{V}}^{-1}\partial_+\widehat{\mathcal{V}}\right) = \partial_+\delta h + \left[\widehat{\mathcal{V}}^{-1}\partial_+\widehat{\mathcal{V}}\,,\,\delta h\right] \tag{4.2.8}$$

It is essential here that the t-dependence of the linear system is preserved in this variation because contributions potentially violating this form of the t-dependence vanish by virtue of (4.2.7). Comparing the coefficients of the remaining terms in (4.2.8) and neglecting the (trivial) term with δh_0, one obtains

$$\delta P_+ = \tfrac{1}{2}D_+\delta k_1 + \tfrac{1}{2}\rho^{-1}\partial_+\rho\,\delta k_1 - \tfrac{1}{2}\left[P_+, \delta h_1\right] \tag{4.2.9}$$

(The result for δQ_+ is similar, but not needed here). Next, we eliminate $D_+\delta k_1$, using (4.2.7) for $\nu = 2$, and repeat this procedure until all derivatives have disappeared (remember that $\delta h_\nu = \delta k_\nu = 0$ for $\nu = n+1, n+2$). The variation of P_+ is thus found to be

$$\delta P_+ = \rho^{-1}\partial_+\rho \sum_{\nu=1}^{n}(-1)^{\nu+1}\nu\delta k_\nu + \left[P_+\,,\,\sum_{\nu=1}^{n}(-1)^\nu\delta h_\nu\right] \tag{4.2.10}$$

Inserting this result into the variation of (4.2.2), we see that the second term in (4.2.10) drops out, so the factor $\rho^{-1}\partial_+\rho$ appears on both sides and cancels out (notice that ρ is not varied). Therefore,

$$\partial_+\left(\lambda^{-1}\delta\lambda\right) = \sum_{\nu=1}^{n}(-1)^{\nu+1}\nu\,\mathrm{Tr}\left(\delta k_\nu P_+\right) \tag{4.2.11}$$

In order to exhibit the right hand side as a derivative, we return to the formula (3.5.10) which, of course, remains valid for the purely bosonic theory if we put $\chi^I = 0$. By use of the various relations given there, the relevant component of the Kac-Moody current can be written as

$$\widehat{\mathcal{J}}_+ = \frac{2}{(1+t)^2}\frac{\partial t}{\partial w}\widehat{\mathcal{V}}P_+\widehat{\mathcal{V}}^{-1} \tag{4.2.12}$$

A short calculation now shows that the result (4.2.11) can be alternatively expressed as a contour integral in two different ways[11]

$$\begin{aligned} \partial_+\left(\lambda^{-1}\delta\lambda\right) &= \frac{1}{4}\oint_{t=-1} dt\frac{\partial w}{\partial t}\operatorname{Tr}\left(\widehat{\mathcal{V}}^{-1}\widehat{\mathcal{J}}_+\widehat{\mathcal{V}}\cdot\delta h(t)\right) = \\ &= -\frac{1}{2}\oint_{t=0} dt\frac{\partial w}{\partial t}\operatorname{Tr}\left(\widehat{\mathcal{V}}^{-1}\widehat{\mathcal{J}}_+\widehat{\mathcal{V}}\cdot\delta h(t)\right) \end{aligned} \tag{4.2.13}$$

Therefore, integrating over the chain $\mathcal{C} \equiv \mathcal{C}_0 + \frac{1}{2}\mathcal{C}_{-1}$, with $\mathcal{C}_{t_0}$ encircling the point t_0 in the complex t-plane once, we obtain the identity

$$\oint_{\mathcal{C}} dt\frac{\partial w}{\partial t}\operatorname{Tr}\left(\widehat{\mathcal{J}}_+\cdot\delta g(w)\right) = \oint_{\mathcal{C}} dt\frac{\partial w}{\partial t}\operatorname{Tr}\left(\widehat{\mathcal{J}}_+(\delta g(w) - \widehat{\mathcal{V}}\delta h(t)\widehat{\mathcal{V}}^{-1})\right) \tag{4.2.14}$$

(for the opposite $(-)$ chirality, one must integrate over $\mathcal{C} \equiv \mathcal{C}_0 + \frac{1}{2}\mathcal{C}_1$). On the right hand side of (4.2.14), we can now deform the contour to a circle around $t = -1$, since the compensating transformation $\delta h(t)$ was designed precisely for the purpose of removing the pole at $t = 0$ (for triangular $\widehat{\mathcal{V}}$). Hence,

$$\oint_{\mathcal{C}} dt\frac{\partial w}{\partial t}\operatorname{Tr}\left(\widehat{\mathcal{J}}_+\cdot\delta g(w)\right) = \frac{1}{2}\oint_{t=-1} dt\frac{\partial w}{\partial t}\operatorname{Tr}\left(\widehat{\mathcal{J}}_+(\delta g(w) - \widehat{\mathcal{V}}\delta h(t)\widehat{\mathcal{V}}^{-1})\right) \tag{4.2.15}$$

Replacing the integral involving $\delta h(t)$ by $\partial_+(\lambda^{-1}\delta\lambda)$ according to (4.2.13) (which is the expression we set out to calculate), we arrive at

$$\oint_{t=0} dt\frac{\partial w}{\partial t}\operatorname{Tr}\left(\widehat{\mathcal{J}}_+\cdot\delta g(w)\right) = -2\partial_+\left(\lambda^{-1}\delta\lambda\right) \tag{4.2.16}$$

In the last step, we convert the integral on the left hand side into an integral over w, remembering that the point $t = 0$ corresponds to $w = \infty$ in the complex w-plane, and invoke formula (3.5.16) in the form

$$\oint_{w=\infty} dw\operatorname{Tr}\left(\widehat{\mathcal{J}}_+\cdot\delta g(w)\right) = -\partial_+\left(\oint_{w=\infty} dw\operatorname{Tr}\left(\frac{\partial\widehat{\mathcal{V}}}{\partial w}\widehat{\mathcal{V}}^{-1}\cdot\delta g(w)\right)\right) \tag{4.2.17}$$

In dropping the derivative operator ∂_+, we must allow for an arbitrary function depending only on x^-. This function is then fixed by repeating the above steps for $\lambda^{-1}\partial_-\lambda$. The final result (and the main formula of this section) is

$$\lambda^{-1}\delta\lambda = \frac{1}{2}\oint_{w=\infty} dw\operatorname{Tr}\left(\frac{\partial\widehat{\mathcal{V}}}{\partial w}\widehat{\mathcal{V}}^{-1}\cdot\delta g(w)\right) \tag{4.2.18}$$

From this formula, we can now recognize the action of the central term if we perform a second variation on $\widehat{\mathcal{V}}$ according to (4.1.7). To pick out the central term, it is sufficient to consider a $\delta g^{(2)}(w)$ of the form

---

11 The definition of the contour integral includes a factor $(2\pi i)^{-1}$.

$$\delta g^{(2)} = \sum_{n=1}^{N} \delta g^{(2)}_{-n} w^{-n} \tag{4.2.19}$$

for which, by the arguments preceding (4.2.3), the compensating H^∞ rotation vanishes. This trick simplifies the calculation of the commutator considerably and leaves us with the result

$$\begin{aligned}\lambda^{-1}[\delta_1,\delta_2]\lambda = &\oint_{w=\infty} dw \,\mathrm{Tr}\left(\delta g^{(1)}(w)\cdot\frac{\partial}{\partial w}\delta g^{(2)}(w)\right) + \\ &+\frac{1}{2}\oint_{w=\infty} dw\,\mathrm{Tr}\left(\frac{\partial\widehat{\mathcal{V}}}{\partial w}\widehat{\mathcal{V}}^{-1}\left[\delta g^{(1)}(w),\delta g^{(2)}(w)\right]\right)\end{aligned} \tag{4.2.20}$$

The first contribution is the central term advertized before, while the second term is just the result expected from the Lie algebra structure of $\mathbf{G}^\infty$. A remarkable feature of this result, already alluded to before, is that the central term is not just a c-number, but a transformation operator. Already after (3.2.7), it was noted that the equations of motion admit constant rescalings of the conformal factor as an extra symmetry. The corresponding generator $\mathcal{Z}$ acts according to

$$\lambda^{-1}\delta\lambda = const \tag{4.2.21}$$

The $\mathbf{G}^\infty$ Lie algebra is therefore characterized by the commutation relations (with self-explanatory notation)

$$[T^a_m, T^b_n] = f^{abc}T^c_{m+n} + m\delta_{m+n,0}\delta^{ab}\mathcal{Z} \tag{4.2.22}$$

The symmetric space automorphism τ^∞ can be generalized to the centrally extended algebra [10]. Using (3.3.10) it is easy to check that (4.2.22) is invariant under

$$\tau^\infty(T^a_m) = \tau(T^a_{-m}) \quad , \quad \tau^\infty(\mathcal{Z}) = -\mathcal{Z} \tag{4.2.23}$$

This means that, in contrast to $\mathbf{G}^\infty$, the τ^∞ invariant subalgebra $\mathbf{H}^\infty$ does not possess a central extension.

The appearance of the operator $\mathcal{Z}$ in the Kac Moody algebra (4.2.20) suggests that the conformal factor λ should be included into an enlarged version of the "∞-bein" $\widehat{\mathcal{V}}$. In accordance with this idea, it was proposed in [10] to consider pairs $(\widehat{\mathcal{V}},\lambda)$ with the multiplication rule

$$(\widehat{\mathcal{V}}_1,\lambda_1)\circ(\widehat{\mathcal{V}}_2,\lambda_2) = (\widehat{\mathcal{V}}_1\widehat{\mathcal{V}}_2\,,\,\lambda_1\lambda_2\exp\Omega(\widehat{\mathcal{V}}_1,\widehat{\mathcal{V}}_2)) \tag{4.2.24}$$

where $\Omega(\widehat{\mathcal{V}}_1,\widehat{\mathcal{V}}_2)$ is a group two-cocycle (an introduction for physicists is given in [52]). This cocycle is only determined up to a coboundary, which can be absorbed into a multiplicative redefinition of λ. The above multiplication rule correctly implements the central extension of the Kac Moody algebra $\mathbf{G}^\infty$ at the level of the Kac Moody group G^∞. Denoting the centrally extended version of G^∞ by G^∞_{ce}, it is clear that the coset space G^∞_{ce}/H^∞ is related to the space of solutions of (3.2.5) *and* (4.2.2) in the same way as G^∞/H^∞ was related to the solutions of (3.2.5) alone. The Geroch group is now properly identified as G^∞_{ce}.

It is perhaps worth emphasizing that through the enlargement of the original "∞-bein" to a new entity also containing the conformal factor, the unification of symmetries, which is observed in the reduction from higher dimensions, for the first time touches upon the unphysical gravitational degrees of freedom (while, incidentally, the last remnant of local Lorentz invariance on the zweibein has disappeared). This indicates that any further unification of symmetries beyond G^{∞} will involve the boost generators.

As pointed out in [10], the automorphism τ^{∞} can generalized to the centrally extended Kac Moody group. The analog of (3.3.10) is

$$\tau^{\infty}(\widehat{\mathcal{V}},\lambda) = (\tau^{\infty}\widehat{\mathcal{V}},\lambda^{-1}) \tag{4.2.25}$$

We can now repeat the argument which was used in (3.4.2) to prove the x-independence of the monodromy matrix and show that

$$\widetilde{\mathcal{M}} \equiv (\widehat{\mathcal{V}},\lambda) \circ \tau^{\infty}(\widehat{\mathcal{V}},\lambda)^{-1} = \left(\widehat{\mathcal{V}}\tau^{\infty}\widehat{\mathcal{V}}^{-1}, \lambda^2 \exp\Omega\left(\widehat{\mathcal{V}},\tau^{\infty}\widehat{\mathcal{V}}^{-1}\right)\right) \tag{4.2.26}$$

is likewise x-independent. Thus, the second entry on the right hand side must be constant, which immediately leads to a beautiful formula for the conformal factor in terms of $\widehat{\mathcal{V}}$ [10], viz.

$$\ln\lambda = -\tfrac{1}{2}\Omega\left(\widehat{\mathcal{V}},\tau^{\infty}\widehat{\mathcal{V}}^{-1}\right) + const \tag{4.2.27}$$

Unfortunately, the practical usefulness of this result is limited by the scarcity of explicit formulas for group cocycles [52]. Nonetheless, there are some special cases where an explicit computation is possible and (4.2.27) can be "tested". One is the colliding plane wave solution reconstructed in section 3.4, for which the matrix $\widehat{\mathcal{V}}$ is diagonal due to the collinear polarization of the waves. By use of the formulas given in appendix B of [10], (4.2.27) reduces to

$$\begin{aligned} \ln\lambda &= -\frac{1}{2}\oint dt\frac{\partial}{\partial t}\ln\left(\sqrt{-\frac{t_1}{t_2}\frac{t-t_1}{t-t_2}}\right)\ln\left(\sqrt{-\frac{t_1}{t_2}\frac{1-tt_1}{1-tt_2}}\right) = \\ &= +\frac{1}{2}\ln\frac{(1-t_1t_2)^2}{(1-t_1^2)(1-t_2^2)} \end{aligned} \tag{4.2.28}$$

This agrees with (3.4.13) after re-insertion of the factor $(\partial_+\rho)^{-\frac{1}{2}}(\partial_-\rho)^{-\frac{1}{2}}$ and adjustment of the integration constant.

I conclude this section with some comments on the supersymmetric theory. As already mentioned, the above derivation of the central term is more general than previous ones in that it also works for supergravity (the method of commuting an Ehlers with a Matzner-Misner transformation and evaluating its effect on the conformal factor [17,10] does not obviously generalize beyond $SL(2,\mathbb{R})$, and is certainly insufficient for supergravity). Remarkably, formula (4.2.18) for the variation of the conformal factor remains valid as it stands, although the intermediate steps are different and more cumbersome due to the extra fermionic terms in (4.2.2) and the appearance of higher order poles in the linear system. Amongst other things, one must work out the action of the compensating duality rotation on the fermions. In a chiral basis (with $\chi_+ \equiv \frac{1}{2}\gamma_-\gamma_+\chi$ and $\chi_- \equiv \frac{1}{2}\gamma_+\gamma_-\chi$, etc.), one finds that the chiral components of χ transform as eigenstates under the action of $\delta h(x;t)$ at $t = \pm 1$, viz.

$$\delta\chi_+ = \delta h(x;-1)\chi_+ \quad , \quad \delta\chi_- = \delta h(x;+1)\chi_- \tag{4.2.29}$$

This is, of course, not an unexpected result in view of the fact that the fermionic analogs of the "dual potentials" in (3.3.7) are still missing.

4.3 Hints of More Symmetries

The foregoing analysis has revealed an impressive array of "hidden symmetries" in Einstein's theory and its various generalizations. Nevertheless, there are still more symmetries even in the purely bosonic theory, and it seems like we have only had a glimpse of the full structure so far! As discovered only very recently, one can realize a Virasoro algebra on the space bosonic solutions [49,18]. It is most simply defined through its action on $\widehat{\mathcal{V}}$

$$\widehat{\mathcal{V}} \rightarrow \delta w \frac{\partial \widehat{\mathcal{V}}}{\partial w} \tag{4.3.1}$$

where $\delta w = f(w)$ with an arbitrary meromorphic function $f(w)$. Expanding $f(w)$ in powers of w, we have the following correspondence with the well-known Virasoro generators

$$L_m : \widehat{\mathcal{V}} \rightarrow -w^{m+1} \frac{\partial \widehat{\mathcal{V}}}{\partial w} \tag{4.3.2}$$

To keep $\widehat{\mathcal{V}}$ in the triangular gauge, a compensating H^∞ rotations is again necessary. However, there is a further subtlety now. To maintain the pole structure of the linear system, a compensating conformal coordinate transformation of the type (3.2.4) is also needed. More precisely, one demands that

$$\delta t = \delta w \frac{\partial t}{\partial w} + \delta x^+ \partial_+ t + \delta x^- \partial_- t \tag{4.3.3}$$

should be regular as $t \rightarrow \pm 1$, so no higher order poles are introduced through the transformation (4.3.1) into the linear system (this compensating transformation is also necessary to take care of the extra contribution $\delta(\rho^{-1}\partial_+\rho) \neq 0$ in the variation of (4.2.2)). It is not difficult to prove that the generators L_m indeed satisfy a Virasoro algebra, and that the commutator of a Virasoro and a G^∞ transformation gives expected result

$$\left[L_m, T_n^a\right] = -nT_{m+n}^a \tag{4.3.4}$$

Finally, the conformal factor also transforms under the Virasoro algebra, and its variation can be determined from (4.2.2) as before.

The presence of the extra symmetry (4.3.1) suggests that the coset space G_{ce}^∞ / H^∞ may not be enough to properly characterize the space of solutions, since it does not take the extra symmetry (4.3.1) into account. Likewise, the subgroup H^∞ should be extended so as to accomodate the conformal coordinate transformations (3.2.4). This, too, may not be enough for the supersymmetric theory because it was shown in [20] that after re-introducing the gravitino components into the linear system (3.5.11), the local superconformal invariance of the theory could be expressed as a Grassmann valued H^∞ gauge transformation (plus a compensating conformal coordinate transformation, see the appendix). All of this hints at the existence of an even larger coset space in terms of an as yet unknown quotient $G^{\infty\infty}/H^{\infty\infty}$. In view of our ignorance about $G^{\infty\infty}$ and $H^{\infty\infty}$, it is advisable to collect the available information. Designating the associated Lie algebras by $\mathbf{G}^{\infty\infty}$ and $\mathbf{H}^{\infty\infty}$, respectively, we know that the set of rigid symmetry generators must contain $\mathbf{G}^\infty, \mathcal{Z}$ and the Virasoro algebra $\mathrm{Vir}(w)$. Thus

$$\mathbf{G}^{\infty\infty} \supset \mathbf{G}^\infty \oplus \mathcal{Z} \oplus \mathrm{Vir}(w) \tag{4.3.5}$$

For the gauge symmetries, we have in addition to $\mathbf{H}^\infty$ the generators of the conformal coordinate transformations (but no central term!), or

$$\mathbf{H}^{\infty\infty} \supset \mathbf{H}^\infty \oplus \mathrm{Vir}(x^+) \oplus \mathrm{Vir}(x^-) \tag{4.3.6}$$

where $\mathrm{Vir}(x^\pm)$ stands for the algebra of conformal coordinate transformations (3.2.4) depending on x^+ or x^-. What could the algebra $\mathbf{G}^{\infty\infty}$ be? A glance at the table of Cartan Kac Moody Lie algebras shows that, beyond the known affine Kac Moody algebras, there remain only the so-called hyperbolic algebras [23,53]. It was in fact this observation which led to the conjecture that such hyperbolic algebras should emerge in the reduction to *one* dimension [24]. In the appendix, I will argue that this last step should involve a *chiral* reduction, so one of the summands $\mathrm{Vir}(x^\pm)$ should be omitted from (4.3.6). Nonetheless, even without actually considering this reduction in detail, we can adduce some evidence for the correctness of the conjecture by taking a closer look at (4.3.5). The point here is that the extra generators $\mathcal{Z}$ and $L_0 \in \mathrm{Vir}(w)$ identified in (4.3.5) are precisely the generators needed to enlarge the Cartan subalgebra of $\mathbf{G}$ to that of its hyperbolic extension! This is most easily understood in the language of string vertex operators. In the vertex operator construction of Kac Moody algebras [51,54], the Cartan subalgebra is always spanned by the zero mode string momentum operators $\oint \varepsilon_\mu \partial X^\mu$. Choosing the polarization vector ε_μ space-like, one gets the Cartan subalgebra of the finite dimensional Lie algebra $\mathbf{G}$. Taking ε_μ to be one of the remaining two light-like vectors, on the other hand, one obtains just the two operators $\mathcal{Z}$ and L_0 above (these operators are called k and d in [51]). Observe that the metric in root space has the required signature $(- + \ldots +)$. In fact, one can embed the *whole* Virasoro algebra $\mathrm{Vir}(w)$ into the hyperbolic extension by considering vertex operators of the form $\oint :\bar{\varepsilon}\partial X \exp(im\varepsilon X):$, where ε_μ and $\bar{\varepsilon}_\mu$ are the two light-like vectors obeying $\varepsilon \cdot \bar{\varepsilon} = 1$, and $m \in \mathbb{Z}$.

Obviously, these considerations do not suffice to really prove that $\mathbf{G}^{\infty\infty}$ is the hyperbolic extension of $\mathbf{G}$, even after reduction to one dimension. Moreover, it appears that the inclusion of fermions is absolutely indispensable if one is to get the full hyperbolic algebra. The reason for this expectation is related to the following puzzle, which has been carefully kept under the rug so far. Up to this point, I have pretended that the addition of fermions is hardly more than a technicality, resulting in higher order poles in the linear system and the like, but not fundamentally altering the structure of the theory. For this reason, we have so far mainly dealt with the bosonic theory, identifying the coset space $G^{\infty}_{ce}/H^{\infty}$ with the *bosonic* solutions. Arguments were given at the end of section 3.4, that these solutions indeed cover the whole coset space. What happens when we switch on the fermionic degrees of freedom? Surely, the space of solutions must be enlarged in one way or another, if the bosonic solutions fill out the coset space already by themselves, leaving "no more room" for the fermions[12]. The folklore line of reasoning might now entice one into believing that cosets should simply be replaced by super-cosets, but here we encounter trouble: once the local supersymmetry has been eliminated by fixing a gauge, no rigid supersymmetry remains that could act on a super coset space. Moreover, in many cases of interest (e.g. $\mathbf{G} = E_8$ and $\mathbf{G}^{\infty} = E_9$), a super-extension of the algebra $\mathbf{G}^{\infty}$ does not exist [22]! There seems to be only one way out of this impasse: the space of combined bosonic and fermionic solutions must be related to the hyperbolic extension of the affine Kac-Moody algebra $\mathbf{G}^{\infty}$. The missing generators on the right hand side of (4.3.5) would then correspond to transformations mixing bosonic and fermionic solutions. These conclusions seem to concur with the analysis of [55], where the the hyperbolic algebra extending $A_1^{(1)}$ was shown to exhibit a graded structure reminiscent of a fermionic Fock space. Could it be that the new "supersymmetry transformations", which have apparently no analog in higher dimensions, are related to the hyperbolic generators associated with negative norm roots, for which no manageable representation has been found so far?

Acknowledgements: I would like to thank the organizers for inviting me to this very pleasant meeting and P. Aichelburg and C. Isham for some interesting discussions related to this work. I am especially grateful to P. Breitenlohner and D. Maison for having contributed to these lectures through many stimulating (and occasionally frustrating) discussions.

Appendix: Maximally Extended Supergravity in One Dimension

This appendix deals with maximally extended ($N = 16$) supergravity in *one* dimension. The notation and conventions are taken from [19-21], which readers are advised to consult for further explanations and details. In addition to the gauge conditions (3.2.2) and (3.5.1), I will employ the light-cone conventions already used in the main body of this article, in particular $\chi_+ \equiv \frac{1}{2}\gamma_-\gamma_+\chi$ and $\chi_- \equiv \frac{1}{2}\gamma_+\gamma_-\chi$ for the fermionic fields (in fact, we could introduce "one-component spinors" at this point, but I will not do so). Although the results below are deficient in that they only represent the very first steps in the search

[12] Nor does the coset space of the bosonic theory become larger after the reduction to one dimension, which again indicates that the reduction of the bosonic theory without fermions will not lead to the desired enhancement of symmetry.

for more symmetries, they might nonetheless be useful at some later stage, and this is my reason for including them in these lectures. Contrary to the prescription followed in chapters 2 and 3, it is proposed here to perform the dimensional reduction by dropping the dependence on one of the light-cone coordinates $x^{\pm}$ rather than x^0 or x^1. In other words, the last step in the dimensional reduction should involve a *chiral* truncation, so instead of (2.1.1)

$$\phi(x^0, x^1) \longrightarrow \phi(x^+) \quad \text{or} \quad \phi(x^-) \tag{A.1}$$

for all fields. One advantage of this procedure is that, unlike in the ordinary dimensional reduction, the equations of motion are *enforced* by this truncation. As a consequence, the algebra of local supersymmetry transformations closes without further ado (i.e. off-shell) on all fields with the exception of ψ^I_{2+}, where a term proportional to the supersymmetry current J^I_+ remains in the commutator.

Imposing (A.1) in the two-dimensional equations of motion immediately implies that certain chiral components vanish. More explicitly, using the notation of [19,20], one has

$$\chi^{\dot{A}}_- = \psi^I_- = \psi^I_{2-} = Q^{IJ}_- = P^A_- = 0 \tag{A.2}$$

As already mentioned, the remaining (+) components of the "matter fields" then satisfy their respective equations of motion automatically. Observe that through (A.2) half of the physical degrees of freedom are eliminated. This would not have been the case for an ordinary reduction, where we would have ended up with $N = 32$ locally supersymmetric quantum mechanics instead. The only equations of motion that must still be imposed are those obtained by varying the traceless components of the two-metric and the gravitino, which were eliminated through the gauge conditions (3.2.2) and (3.5.1), respectively. These are just the generators of the remaining local invariances, namely the supersymmetry current J^I_+ and the energy momentum tensor T_{++}.

The full supersymmetry variations of chiral $N = 16$ supergravity can be derived by truncation of the corresponding transformations in [21] and are characterized by the chiral transformation parameter

$$\epsilon^I = \epsilon^I_-(x^+) \quad \text{i.e.} \quad \epsilon^I_+ = 0 \tag{A.3}$$

In addition to the terms obtained directly from [21], a compensating $SO(1,1)$ rotation necessary to preserve the diagonality of the zweibein must be taken into account. Furthermore, it should be remembered that the gravitino components below are defined from $\psi^I_a = e^m_a \psi^I_m$ (flat indices), and that the second order spin connections in the gravitino variations have been written out explicitly. Keeping track of all these contributions, one arrives at the the supersymmetry variations

$$\rho^{-1}\delta\rho = -\bar{\epsilon}^I_- \psi^I_{2+} \quad , \quad \lambda^{-1}\delta\lambda = i\bar{\epsilon}^I_- \psi^I_+$$
$$\mathcal{V}^{-1}\delta\mathcal{V} = \bar{\epsilon}^I_- \chi^{\dot{A}}_+ \Gamma^I_{A\dot{A}} Y^A \tag{A.4}$$

for the bosons, and

$$\delta\chi_+^{\dot A} = \tfrac{i}{2}\lambda^{-1}\gamma_-\epsilon_-^I\Gamma^I_{A\dot A}\widehat{P}_+^A + \tfrac{i}{2}\chi_+^{\dot A}\cdot\bar\epsilon_-^J\psi_+^J$$

$$\delta\psi_+^I = \tfrac{1}{2}\lambda^{-1}\gamma_-\left(D_+ + \tfrac{1}{2}\lambda^{-1}\partial_+\lambda\right)\epsilon_-^I - \tfrac{3i}{2}\psi_+^I\cdot\bar\epsilon_-^J\psi_+^J + \tfrac{i}{8}\gamma_-\epsilon_-^J\cdot\bar\chi_+\gamma_+\Gamma^{IJ}\chi_+$$

$$\begin{aligned}\delta\psi_{2+}^I = &-\tfrac{i}{2}\lambda^{-1}\rho^{-1}\partial_+\rho\gamma_-\epsilon_-^I + \tfrac{i}{2}\psi_{2+}^I\cdot\bar\epsilon_-^J\psi_+^J + \tfrac{1}{2}\gamma_-\epsilon_-^J\cdot\bar\psi_{2+}^I\gamma_+\psi_{2+}^J + \\ &+\tfrac{i}{2}\gamma_-\epsilon_-^I\cdot\bar\psi_+^J\gamma_+\psi_{2+}^J + \tfrac{1}{4}\gamma_-\epsilon_-^J\cdot\bar\chi_+\gamma_-\Gamma^{IJ}\chi_+\end{aligned} \qquad (A.5)$$

for the fermions with

$$\widehat{P}_+^A \equiv P_+^A - \lambda\Gamma^I_{A\dot A}\bar\chi_+^{\dot A}\gamma_+\psi_+^J \qquad (A.6)$$

On all fields but ψ_{2+}^I, this algebra closes "off shell". With some heavy Fierzing (see the appendix of [21] for the necessary technology), one finds

$$\left[\delta(\epsilon_1),\delta(\epsilon_2)\right] = \delta_{conf}(\xi') + \delta_{SUSY}(\epsilon') + \delta_{SO(16)}(\omega') \qquad (A.7)$$

with the parameters

$$\xi'_- = i\lambda^{-1}\bar\epsilon_{2-}^I\gamma_-\epsilon_{1-}^I$$

$$\epsilon'^I_- = -i\gamma_+\psi_+^I\cdot\bar\epsilon_{2-}^K\gamma_-\epsilon_{1-}^K + \tfrac{i}{2}\gamma_+\psi_+^K\cdot\epsilon_{2-}^{(K}\gamma_-\epsilon_{1-}^{I)}$$

$$\omega'^{IJ} = \xi'_-Q_+^{IJ} + \tfrac{1}{4}\bar\epsilon_{2-}^K\gamma_-\epsilon_{1-}^K\cdot\bar\chi_+\gamma_+\Gamma^{IJ}\chi_+ + \bar\epsilon_{2-}^K\gamma_-\epsilon_{1-}^{[I}\cdot\bar\chi_+\gamma_+\Gamma^{J]K}\chi_+ \qquad (A.8)$$

On the gravitino component ψ_{2+}^I, one gets an extra contribution in the commutator

$$\left[\delta(\epsilon_1),\delta(\epsilon_2)\right]\psi_{2+}^I = \ldots - i\bar\epsilon_{2-}^J\gamma_-\epsilon_{1-}^J\cdot J_+^I + i\bar\epsilon_{2-}^{(I}\gamma_-\epsilon_{1-}^{J)}\cdot J_+^J \qquad (A.9)$$

where the supercurrent J_+^I (the gravitino equation of motion) is given by

$$\begin{aligned}J_+^I =&\lambda^{-1}\rho^{-1}D_+(\rho\psi_{2+}^I) - \tfrac{1}{2}\lambda^{-2}\partial_+\lambda\,\psi_{2+}^I + i\lambda^{-1}\rho^{-1}\partial_+\rho\,\psi_+^I - \lambda^{-1}\Gamma^I_{A\dot A}\chi_+^{\dot A}\widehat{P}_+^A - \\ &- i\bar\psi_+^K\gamma_+\psi_{2+}^K\cdot\psi_+^I - \tfrac{1}{2}\bar\chi_+\gamma_+\Gamma^{IJ}\chi_+\cdot\psi_+^J - \tfrac{i}{4}\bar\chi_+\gamma_+\Gamma^{IJ}\chi_+\cdot\psi_{2+}^J\end{aligned} \qquad (A.10)$$

Thus, on ψ_{2+}^I, the algebra closes only if $J_+^I = 0$. As explained in section 2.3, this is just the condition defining the physical sector in phase space. The above results can be streamlined somewhat by rescaling the fermion fields according to

$$\chi \to \lambda^{\frac{1}{2}}\chi \quad , \quad \psi_+^I \to \lambda^{\frac{1}{2}}\psi_+^I \quad , \quad \psi_{2+}^I \to \lambda^{\frac{1}{2}}\psi_{2+}^I \qquad (A.11)$$

and the supersymmetry transformation parameter according to

$$\epsilon_-^I \to \lambda^{-\frac{1}{2}}\epsilon_-^I \qquad (A.12)$$

(The supersymmetry current is rescaled by a factor $\lambda^{\frac{3}{2}}$.) In the remainder of this appendix I will use rescaled fields.

The $(+)$ component of the linear system for $N = 16$ supergravity takes the form

$$\widehat{\mathcal{V}}^{-1}\partial_+\widehat{\mathcal{V}} = \tfrac{1}{2}\mathcal{Q}_+^{IJ}(t)X^{IJ} + \mathcal{P}_+^A(t)Y^A \qquad (A.13)$$

with

$$
\begin{aligned}
\mathcal{Q}_+^{IJ}(t) = Q_+^{IJ} + &\frac{t}{(1+t)^2}\left(-i\bar{\chi}_+\gamma_+\Gamma^{IJ}\chi_+ + 8\bar{\psi}_{2+}^{[I}\gamma_+\psi_+^{J]}\right) \\
&-16\frac{t^2}{(1+t)^4}i\bar{\psi}_{2+}^I\gamma_+\psi_{2+}^J
\end{aligned}
\tag{A.14}
$$

and

$$
\mathcal{P}_+^A(t) = P_+^A + 2\frac{t(1-t)}{(1+t)^3}i\Gamma_{A\dot{A}}^I\bar{\psi}_{2+}^I\gamma_+\chi_+^{\dot{A}}
\tag{A.15}
$$

Of course, (A.13) loses some of its lustre in one dimension since the compatibility condition (3.3.3) is trivially satisfied for x^--independent $\widehat{\mathcal{V}}$. Nevertheless, the matrix $\widehat{\mathcal{V}}$ remains useful as it describes all the degrees of freedom in a unified manner. In [20], it was pointed pointed out that the local supersymmetry transformations can be re-expressed as a Grassmann-valued H^∞ gauge transformation

$$
\widehat{\mathcal{V}}^{-1}\delta\widehat{\mathcal{V}} = \tfrac{1}{2}S^{IJ}(t)X^{IJ} + S^A(t)Y^A
\tag{A.16}
$$

with

$$
S^{IJ}(t) = -\frac{8t}{(1+t)^2}\bar{\epsilon}_-^{[I}\psi_{2+}^{J]} \quad , \quad S^A(t) = \frac{1-t}{1+t}\Gamma_{A\dot{A}}^I\,\bar{\epsilon}_-^I\chi_+^{\dot{A}}
\tag{A.17}
$$

provided the spectral parameter is also varied according to

$$
t^{-1}\delta t = -\frac{1-t}{1+t}\bar{\epsilon}_-^I\psi_{2+}^I
\tag{A.18}
$$

One finds

$$
\begin{aligned}
\tfrac{1}{2}\delta\mathcal{Q}_+^{IJ}(t)X^{IJ} + \delta\mathcal{P}_+^A(t)Y^A =&\partial_+\left(\widehat{\mathcal{V}}^{-1}\delta\widehat{\mathcal{V}}\right) + \left[\widehat{\mathcal{V}}^{-1}\partial_+\widehat{\mathcal{V}}\,,\,\widehat{\mathcal{V}}^{-1}\delta\widehat{\mathcal{V}}\right] - \\
&-\frac{8t}{(1+t)^2}\bar{\epsilon}_-^I J_+^J\cdot\tfrac{1}{2}X^{IJ}
\end{aligned}
\tag{A.19}
$$

It should come as no surprise that the results obtained by varying the right hand side of (A.13) by use of (A.4) and (A.5), and the one obtained by substituting (A.16) into (A.19) again only agree on the physical subspace $J_+^I = 0$. Agreement is likewise obtained for the cubic fermionic terms in the supersymmetry variations (A.4) and (A.5), as I have explicitly verified (this consistency check was not performed in [20]). As already remarked in [20], the result (A.18) is suggestive of some hierarchy of H^∞ (or $H^{\infty\infty}$?) gauge transformations, of which local supersymmetry would only be a tiny part. Finally, the supersymmetry variation of the spectral parameter (A.18) can be replaced by a conformal transformation with parameter ξ_- defined through

$$
\partial_+\rho\,\xi_- = -\rho\bar{\epsilon}_-^I\psi_{2+}^I
\tag{A.20}
$$

Let me end this appendix (and the lectures) on a speculative note. The potentially most important reason for advocating a chiral reduction is that the $N = 16$ theory described here may give rise to a new dual model which would be related to $D = 11$ supergravity [56] in the same way as the ordinary closed superstring [39] is related to $N = 2$ supergravity in ten dimensions[13]. Of course, it is well known that the conventional superconformal theories defining the various superstrings stop at $N = 4$ [57]. On the other hand, $N = 16$ supergravity is not a conventional superconformal theory. Among other things, this is evident from the fact that the algebra (A.7) is "soft" (i.e. the structure constants are field dependent). Despite these differences, the similarities with string theory are intriguing. The usual (super-)Virasoro generators are now replaced by the requirement that (A.10) and the energy momentum tensor (alias the Wheeler-DeWitt operator)

$$T_{++} = \lambda^{-1}\partial_+\lambda\,\partial_+\rho - \tfrac{1}{2}\partial_+^2\rho - \tfrac{1}{2}\rho P_+^A P_+^A + \tfrac{i}{2}\rho\bar{\chi}_+^{\dot{A}}\gamma_+ D_+\chi_+^{\dot{A}} + \dots \qquad (A.21)$$

annihilate the physical states. As one can see, the role of the unphysical target space coordinates $X^\pm$ and their fermionic partners is now played by the fields ρ, λ, ψ_+^I and ψ_{2+}^I, which originate from the higher-dimensional vielbein and gravitino. It is encouraging that, with the assumption that the bosonic and fermionic creation operators are related to the Fourier modes of $P_+^A(x^+)$ and $\chi_+^{\dot{A}}(x^+)$ and that the groundstate of this "superstring" is just the massless multiplet of $D = 11$ supergravity (consisting of the **44** and **84** representations of $SO(9)$ in the bosonic sector, and the **128** representation in the fermionic sector), we can adapt the arguments given in [58] to show that the states of the first excited level can be assembled into multiplets of $SO(10)$! The full $SO(1,10)$ Lorentz algebra would then have to be contained in the algebra $\mathbf{H}^{\infty\infty}$ in the same way as its transverse $SO(9)$ subalgebra is contained in $\mathbf{H}^\infty$.

References

1 *Solutions of Einstein's Equations: Techniques and Results*, ed. by C. Hoenselaers and W. Dietz (Springer, Berlin 1984)

2 T. Kaluza: Sitzungsberichte der preussischen Akademie der Wissenschaften **1921**, 966; O. Klein: Z.Phys. **37**, 895 (1926)

3 J. Ehlers: Dissertation, Hamburg University (1957);
G. Neugebauer and D. Kramer: Annalen der Physik **24**, 62 (1969);
R. Geroch: J. Math. Phys. **12**, 918 (1971);
D. Maison: Gen.Rel.Grav. **10**, 717 (1979)

4 R. Geroch: J.Math.Phys. **13**, 394 (1972)

5 W. Kinnersley and D. Chitre: J.Math. Phys. **18**,1538 (1977); **19**, 1926 and 2037 (1978)

---

13 A very similar proposal was already made in [19], where it was suggested to interpret the $d = 2, N = 16$ theory as a superstring. Apart from the technical difficulties of realizing this idea in practice, there is a problem with the number of physical degrees of freedom. The $d = 2$ theory still has both the left and right moving degrees of freedom (corresponding to pure supergravitational radiation as opposed to the solitonic excitations of the type encountered in section 3.4), and would therefore give rise to a kind of *closed* superstring. This would imply the factorizability of the states into products of right and left moving states. However, the massless multiplet of $D = 11$ supergravity does not admit such a factorization unlike the corresponding multiplet of $D = 10, N = 2$ supergravity. Clearly, this problem does not arise for the unidexterous reduction discussed here.

6 B.K.Harrison: Phys.Rev.Lett. **41**, 1197 (1978);
G. Neugebauer: J.Phys. **A12**, L67 (1979)

7 V.A. Belinskii and V.E. Sakharov: Zh.Eksp.Teor.Fiz. **75**,1955 (1978); **77**, 3 (1979)

8 D. Maison: Phys.Rev.Lett. **41**, 521 (1978)

9 F. Ernst and I. Hauser: Phys.Rev. **D20**, 362 (1979)
K. Ueno and Y. Nakamura: Phys.Lett. **117B**, 208 (1982)

10 P. Breitenlohner and D. Maison: in [1] and Ann. Inst. Poincare **46**, 215 (1987)

11 K. Khan and R. Penrose: Nature **229**, 286 (1970);
P. Szekeres: J.Math.Phys. **13**, 286 (1972);
M. Nutku and M. Halil: Phys.Rev.Lett. **39**,1379 (1977);
V. Ferrari: in *General Relativity and Gravitation 1989*, ed. by N. Ashby, D.F.Bartlett and W. Wyss (Cambridge University Press 1990)

12 V. Ferrari and J. Ibanez: Gen.Rel.Grav. **19**, 405 (1987)

13 F. Ernst and I. Hauser: J.Math.Phys. **28**, 2155 (1987); **28**, 2951 (1987); **29**, 581 (1988)

14 *Unification of The Fundamental Particle Interactions*, ed. by S. Ferrara, J. Ellis and P. van Nieuwenhuizen (Plenum Press, New York, 1980);
Unification of The Fundamental Particle Interactions II, ed. by S. Ferrara and J. Ellis (Plenum Press, New York, 1983);
Supersymmetry and Supergravity '82, ed. by S. Ferrara, J.G. Taylor and P. van Nieuwenhuizen (World Scientific, Singapore, 1983)

15 E. Cremmer and B. Julia: Nucl.Phys. **159B**,141 (1979)

16 B. Julia: in *Superspace and Supergravity*, ed. by S.W. Hawking and M. Rocek (Cambridge University Press, 1980)

17 B. Julia: in *Johns Hopkins Workshop on Current Problems in Particla Physics: Unified Field Theories and Beyond*, Johns Hopkins University, Baltimore (1981)

18 D. Maison: preprint MPI-PAE/PTh 80/88 (1988)

19 H. Nicolai: Phys. Lett. **194**, 402 (1987)

20 H. Nicolai and N.P. Warner: Comm.Math.Phys. **125**, 384 (1989)

21 N. Marcus and J.H. Schwarz: Nucl.Phys. **B228**, 145 (1983)

22 V.G. Kac: Adv. in Math. **26**, 8 (1977)

23 V.G. Kac: *Infinite Dimensional Lie Algebras*, (Birkhäuser, Boston, 1983)

24 B Julia: in *Vertex Operators in Mathematics and Physics*, ed. by J. Lepowsky, S. Mandelstam and I. Singer (Springer Verlag, Heidelberg, 1984)

25 E. Del Giudice, P. DiVecchia and S. Fubini: Ann.Phys. **70**, 378 (1972);
J. Scherk: Rev.Mod.Phys. **47**,123 (1975)

26 P. Breitenlohner, D. Maison and G.W. Gibbons: Comm.Math.Phys. **120**, 295 (1988)

27 D.Z. Freedman, S. Ferrara and P. van Nieuwenhuizen: Phys.Rev. **B13**, 3214 (1976);
S. Deser and B. Zumino: Phys.Lett. **62B**, 335 (1976)

28 P. van Nieuwenhuizen: Phys. Rep **68**,189 (1981);
D.Z. Freedman and B. de Wit: in *Supersymmetry*, ed. by K. Dietz, R. Flume, G. v. Gehlen and V. Rittenberg (Plenum Press, New York, 1985)

29 B. de Wit and H, Nicolai: Nucl.Phys. **274**, 363 (1986)

30 S. Coleman, J. Wess and B. Zumino: Phys.Rev. **177**, 2239 (1969)

31 H. Eichenherr and M. Forger: Nucl.Phys. **B155**, 381 (1979)

32 C. Misner, K. Thorne and J.A. Wheeler: *Gravitation* (Freeman, San Francisco 1973);
C. Isham: in *General Relativity and Gravitation* , Cambridge University Press (1987)

33 H. Nicolai: Nucl.Phys. **B353**, 493 (1991)

34 P.A.M. Dirac: *Lectures on Quantum Mechanics*, Academic Press, London (1967)

35 J.A. Wheeler: in *Relativity, Groups and Topology*, ed. by C. DeWitt and B. DeWitt (Gordon and Breach, 1964);
B. DeWitt: Phys. Rev. **160**,113 (1967); **162**,1195 (1967)

36 V. Husain and L. Smolin: Nucl.Phys. **B327**, 205 (1989)

37 A. Ashtekar: *New Perspectives in Canonical Gravity* (Bibliopolis, Naples, 1988)

38 H. Nicolai: Phys.Lett. **235B**,195 (1990)

39 J.H. Schwarz: Phys.Rep. **89**, 223 (1982)

40 M. Lüscher and K. Pohlmeyer: Nucl.Phys. **B137**,46 (1978)

41 M.C. Davies, P.J. Houston, J.M. Leinaas and A.J. Macfarlane: Phys.Lett. **119B**,187 (1982);
H.J. de Vega, H. Eichenherr and J.M. Maillet: Comm.Math.Phys. **92**, 507 (1984)

42 F. Ernst: Phys.Rev. **167**, 1175 (1968)

43 F. Ernst and I. Hauser: J.Math.Phys. **30**, 872 (1989)
44 K. Pohlmeyer: Comm.Math.Phys. **46**, 207 (1976)
45 W. Zakrzewski: Durham preprint DTP-87/3 (1987)
46 D. Maison and P. Breitenlohner: MPI report, unpublished
47 H. Eichenherr and M. Forger: Comm. Math.Phys. **82**, 227 (1981)
48 P.C. Aichelburg and R. Güven: Phys.Rev.Lett. **51**, 1613 (1983)
49 Bo-Yu Hou and Wei Li: Lett.Math.Phys. **13**, 1 (1987)
50 F. Ernst and I. Hauser: J.Math.Phys. **22**,1 051 (1981)
51 P. Goddard and D.I. Olive: in *Vertex Operators in Mathematics and Physics*, ed. by J. Lepowsky, S. Mandelstam and I. Singer (Springer Verlag, Heidelberg, 1984); Int.J.Mod.Phys. **A1**, 303 (1986)
52 B. Zumino: in *Anomalies, Cocycles and Schwinger Terms* ed. by W.A. Bardeen and A.R. White (World Scientific, Singapore, 1986)
53 C. Saclioglu: preprint CERN-TH 4854 (1987)
54 I.B. Frenkel and V.G. Kac: Inv.Math. **62**, 23 (1980), G. Segal: Comm.Math.Phys. **80**, 301 (1981)
55 A.J. Feingold and I.B. Frenkel: Math.Ann. **263**, 87 (1983)
56 E. Cremmer, B. Julia and J. Scherk: Phys.Lett. **76B**, 409 (1978)
57 M. Ademollo et al.: Phys.Lett. **B62**,105 (1976); L. Brink and J.H. Schwarz: Nucl.Phys. **B12**, 285 (1977); P. Ramond and J.H. Schwarz: Phys.Lett. **B64**, 75 (1976)
58 I. Bars: Nucl.Phys. **308**, 462 (1988)

Recent developments in relativistic thermal field theories

M. Le Bellac

Institut Non Linéaire de Nice*,
Université de Nice Parc Valrose, 06034 Nice Cedex, France
Unité Mixte de Recherche 129 du CNRS

1. Introduction

Field theory at non-zero temperature and/or non-zero chemical potential has been first developed in a non-relativistic context mainly during the years 1955-1970, in the framework of the so-called "many-body" problem. Many applications to condensed matter physics and to nuclear physics were worked out during these fifteen years; a rather complete account of the subject can be found in the book by Fetter and Walecka[1].

The first relativistic calculations were performed in the early sixties by Silin[2] and by Fradkin[3] in thermal QED (not surprisingly, this work was rediscovered ten years later by physicists from the West). Then, in the mid-seventies, some physicists became interested in the problem of symmetry restoration at high temperature[4]-[6]: just as high temperature restores rotational symmetry in a ferromagnet above the Curie point, so does high temperature restore the $SU(2) \times U(1)$ symmetry of the Salam-Weinberg model above $T_c \simeq 200$ GeV, because the effective potential is temperature-dependent. Such temperatures were of course available in the early Universe, and symmetry restoration has obvious applications to Cosmology. Another, more speculative, symmetry restoration may also occur in the case of grand unification.

However the main motivation for being interested today in finite temperature quantum field theory (FTQFT) is the strong belief that we might be able to produce a deconfined state of matter in laboratory experiments. At sufficiently high temperatures or densities, quarks and gluons are no more confined into hadrons, and constitute what has been called a "quark-gluon plasma". This quark-gluon plasma was present in the early Universe up to times $\sim 10^{-5}$s, but it is not easy to find observable consequences of this fact, while it may be possible (although not so easy!) to detect signals of a quark-gluon plasma created in the laboratory by heavy-ion collisions[7].

The main basis for our belief in this new phase of matter comes from lattice simulations. In the case of a pure $SU(3)$ gauge theory, experts now agree that they observe a first-order deconfinement phase transition at a temperature $T_c \simeq 230$ MeV[8]. The character (and even the existence) of the transition depends in a complicated way on the number of flavours and the masses of the quarks, and the situation does not seem to be fully settled in lattice simulations. In the realistic case of two light quarks and one heavy quark (u, d, and s flavours), results are consistent with a first-order transition, corresponding to chiral symmetry restoration and deconfinement, at a critical temperature $T_c \simeq 150$ MeV.

In order to express T_c in MeV, one has to make a comparison with another prediction. In general the comparison is made with the calculation of the ρ-mass, m_ρ, and the quoted values for T_c have been given with that choice. Unfortunately the m_π/m_ρ ratio has not yet reached its physical value in lattice calculations, so that it is reasonable to allow for a rather large error bar on T_c; we shall take as a conservative estimate T_c in the range 150-200 MeV.

Another estimate of the critical temperature comes from low energy effective Lagrangians. The quark condensate can be computed in perturbation theory and its low temperature expansion is given by[9]

$$< q\bar{q} >_T = < q\bar{q} >_{T=0} \left(1 - \frac{T^2}{8F^2} - \frac{T^4}{384F^4} - \frac{T^6}{288F^6} \ln \frac{\Lambda_q}{T} + O(T^8)\right) \tag{1.1}$$

For simplicity the formula has been written for massless quarks. The parameter F is closely related to the pion decay constant f_π and Λ_q to the $\pi - \pi$ scattering lengths; their numerical values are

$$F = (1.057 \pm 0.012) f_\pi = 88.3 \pm 11 \text{ MeV} \quad ; \quad \Lambda_q = 470 \pm 110 \text{ MeV}$$

The scale of the low temperature expansion is set by $\sqrt{8}F \simeq 250$ MeV and the expansion seems to be reliable up to $T \simeq 150$ MeV. A tempting (but bold) extrapolation to higher values of T leads to a vanishing condensate at $T_c \simeq 190$ MeV. Various corrections (massive quarks and massive hadrons) do not modify this estimate in an essential way.

At temperatures much higher than T_c, one expects the quark-gluon plasma to behave almost as an ideal gas of free quarks and gluons: indeed, because of asymptotic freedom, the QCD coupling constant $g(T)$ tends to zero when T goes to infinity. As T is the only scale at our disposal we must have

$$\alpha_S(T) = \frac{g^2(T)}{4\pi} = \frac{6\pi}{(33 - 2N_f)\ln(cT/\Lambda_{\overline{MS}})} = \frac{1}{8\pi\beta_0 \ln(cT/\Lambda_{\overline{MS}})} \tag{1.2}$$

the constant c being unknown at present (more details will be given in subsection 2.7); as usual β_0 denotes the first coefficient of the β-function and $\Lambda_{\overline{MS}}$ the QCD scale in the $\overline{MS}$ renormalization scheme. In the limit of infinite temperature, we thus get the Stefan-Boltzmann (SB) law for the energy density

$$\varepsilon_{SB} = \frac{\pi^2}{15}(N_c^2 - 1 + \frac{7}{4}N_c N_f)T^4 \tag{1.3}$$

where N_c is the number of colours and N_f the number of flavours. In order to get an order of magnitude estimate of the energy density around T_c, we simply plug in (1.3) the value of the critical temperature and find

$$\varepsilon \simeq 1.0 \text{ GeV}/(\text{fm})^3 \quad T_c = 150 \text{ MeV}$$

$$\varepsilon \simeq 3.3 \text{ GeV}/(\text{fm})^3 \quad T_c = 200 \text{ MeV}$$

The other extreme way of reaching the chirally symmetric phase is to increase the baryonic density, and thus the chemical potential μ at small values of T; for simplicity we take $T = 0$. Recall that the nuclear density $d_0 \simeq 0.15$ nucleon/(fm)^3, and that the chemical potential μ is given from d_0 by

$$d_0 = \frac{2}{3\pi^2}(\mu^2 - m^2) \tag{1.4}$$

where m is the nucleon mass. Unfortunately the region $T = 0$, μ increasing, is not accessible to present lattice calculations. In order to estimate the critical chemical potential μ_c, one has to rely on the bag model or on effective chiral Lagrangians[10]. It is likely that a transition occurs at a density $\sim 4 - 5$ times the nuclear density; taking into account (1.4) and the fact that the quark chemical potential $\mu_q = \mu/3$ at the transition, this corresponds to $\mu_q \simeq 300$

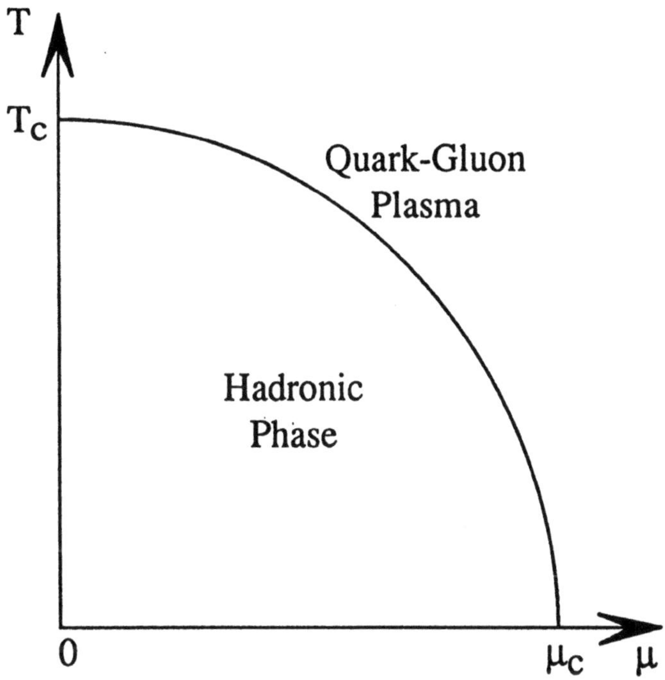

Fig. 1. The phase diagram in the $\mu - T$ plane.

MeV. Thus we can guess that the phase diagram is roughly that drawn in fig.1. However we must stress that the exact shape of the curve in fig.1, as well as the character of the transition for $\mu \neq 0$, are still very uncertain. Let us only mention that the situation with $T \simeq 0$ and $d \simeq 5d_0$ could be reached in the core of some neutron stars, which could thus contain a quark-gluon plasma.

Let us come back to the case $\mu = 0$ and to the results of lattice calculations; one of the best signals of a first-order phase transition is the jump of the energy density ε at T_c (fig.2). Above T_c, it is interesting to compare lattice results with those of perturbative calculations, since for $T \gg T_c$ we expect the SB-law to hold. We can even do a little better since the perturbative corrections to the SB-law have been evaluated; one finds for $N_c = 3$

$$\epsilon/T^4 = \frac{8\pi^2}{15} + \frac{7\pi^2}{20}N_f - (\frac{1}{2} + \frac{5}{24}N_f)g^2 + \frac{2g^3}{\pi}(1 + \frac{1}{6}N_f)^{3/2} + ... \tag{1.5}$$

Notice the term of order g^3, and not g^4, as could be expected from a perturbative expansion. This feature arises from the infrared behaviour of any (renormalizable) relativistic thermal field theory and is not typical of QCD; it happens for instance in the case of the φ^4 scalar field theory. Infrared divergent terms are resummed in the so-called "ring-diagrams" (see subsection 4.1). Terms of order $g^4 \ln g^2$ have also been computed, but they are not meaningful in the absence of a complete $O(g^4)$ calculation.

It is interesting to see how far down in T one can extrapolate the perturbatively corrected SB-law (1.5); however it is necessary to correct first (1.5) , which is a continuum formula, for lattice effects. In other words one has to make the comparison with a perturbative calculation on the lattice. Such a calculation has been performed[11] and one can see on fig.2 that there is a very good agreement between the lattice simulations and the perturbative calculation down to $T/T_c \simeq 1.2$. However the situation does not look so bright when one examines the pressure P. In an ideal ultrarelativistic gas one should have $\varepsilon = 3P$; the ratio $(\varepsilon - 3P)/T^4$

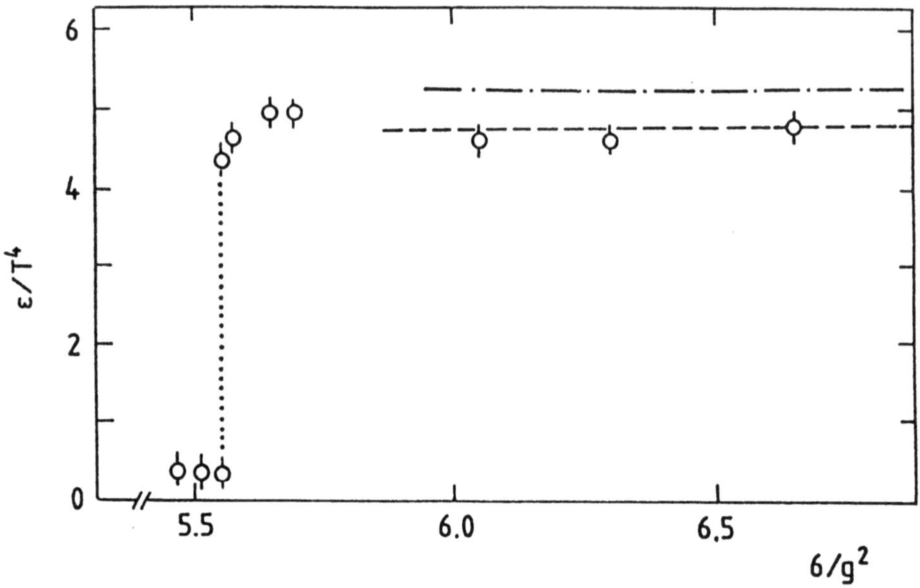

Fig. 2. The ratio ϵ/T^4 as a function of $6/g^2$ in pure $SU(3)$[11].

is shown in fig.3[12], and one can see that there are certainly strong non-perturbative effects up to $T/T_c \simeq 2$. In (continuum) perturbation theory the leading contribution to $(\varepsilon - 3P)/T^4$ is of order g^4

$$(\varepsilon - 3P)/T^4 = -\frac{2}{3}\beta_0 g^4(\frac{1}{2} + \frac{5}{24}N_f) + O(g^5) \tag{1.6}$$

and this has the wrong sign if compared to lattice results.

Another physically meaningful quantity is the Debye screening length r_D, or its inverse the electric mass m_{el} ($m_{el} = r_D^{-1}$), which governs the screening of the heavy quark potential above T_c. One expects this potential to be given by

$$\begin{aligned} V(r,T) &= \frac{\alpha_S(T)}{r} + r\sigma(T) \quad\quad T < T_c \\ V(r,T) &= \frac{T\alpha_S(T)}{(rT)^n} e^{-\mu(T)r} + c(T) \quad\quad T > T_c \end{aligned} \tag{1.7}$$

with $n \simeq 2$ and $\mu \simeq 2m_{el}$ for $T \gg T_c$. One finds on the lattice in pure $SU(3)$

$$\mu(T)/T \sim 2 - 3 \quad \text{for} \quad T_c \leq T \leq 2T_c \tag{1.8}$$

while perturbation theory (subsection 3.4) gives

$$m_{el}^2(T) \simeq (\frac{1}{3}N_c + \frac{1}{6}N_f)\; g^2(T)\; T^2 \tag{1.9}$$

With $g(T) \sim 1$, there is rough agreement between the lattice and perturbative calculations.

Thus our motivation for studying the perturbative aspects of FTQFT is that perturbation theory seems to be a good approximation for $T/T_c \geq 2$, and might even be a good qualitative guide for lower values of this ratio. In any case, comparison with lattice results, and matching

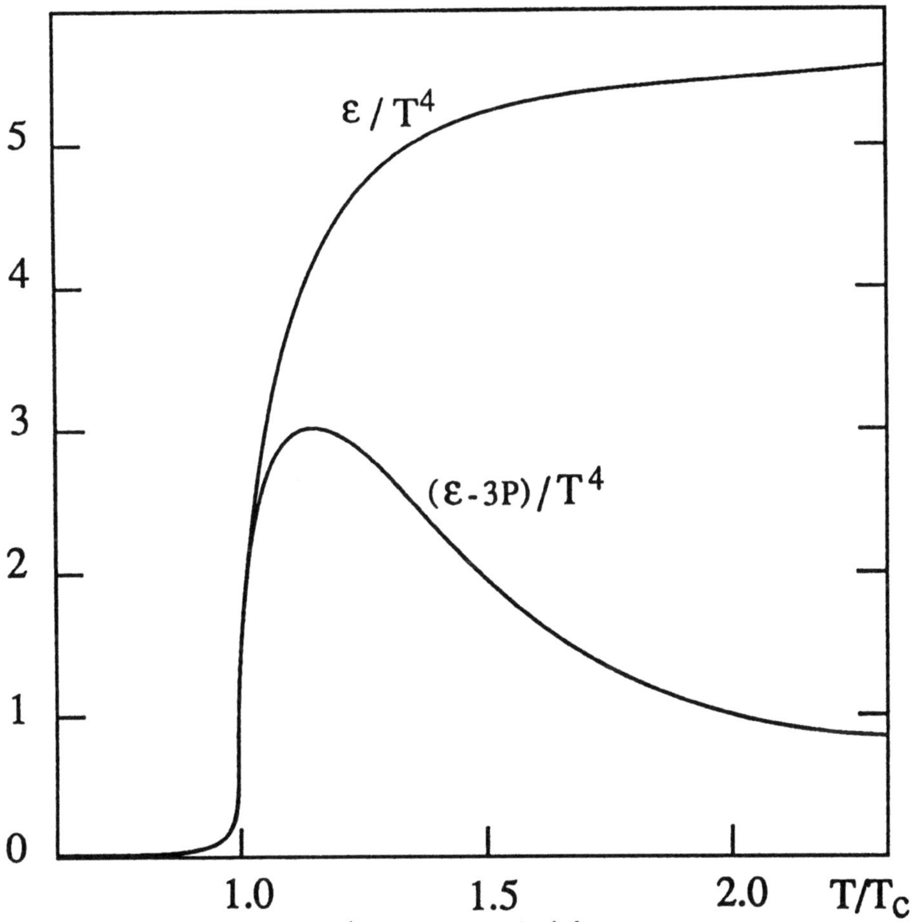

Fig. 3. The ratios ϵ/T^4 and $(\epsilon - 3P)/T^4$ as a function of T/T_c [12].

of these results with the perturbative ones around $T \simeq 2T_c$ may give useful indications on the behaviour of the quark-gluon plasma, and in particular may give useful hints on the non-perturbative effects which occur at lower values of T/T_c.

2. Feynman Rules at Finite Temperature

The formalism of FTQFT relies on a formal analogy between inverse temperature and imaginary time. Since only the time variable plays a role in this analogy, it is possible to understand the formalism by ignoring first the spatial degrees of freedom, and by concentrating on quantum mechanics at finite temperature or, equivalently, quantum statistical mechanics. Actually we shall take as our starting point the quantum statistical mechanics of a single particle, moving in one dimension in a potential $V(q)$, where q is the position variable. By taking $V(q)$ to be the harmonic oscillator potential $\frac{1}{2}\omega^2 q^2$, we shall of course have the analogue of a free field in zero space dimension. Since a free field is nothing but a superposition of independent frequencies, it is clear that we shall be able to derive for instance the expression of the free propagator at finite temperature.

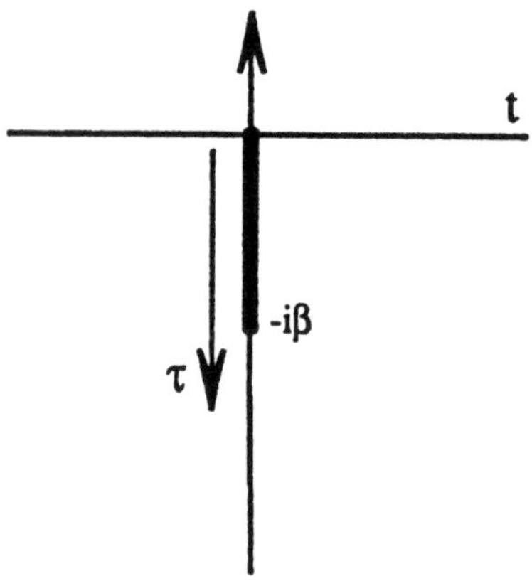

Fig. 4. The imaginary time axis.

2.1 The Imaginary Time Axis

One possible starting point for studying the motion of a particle in a potential $V(q)$ is the probability amplitude for finding the particle at point q' at time t', when one knows that it was located at point q at time t :

$$F(q',t';q,t) = < q'|e^{-iH(t'-t)}|q > \tag{2.1}$$

This amplitude can be analytically continued to imaginary time

$$t \to -i\tau \qquad t' \to -i\tau'$$

$$F(q',-i\tau';q,-i\tau) = < q'|e^{-H(\tau'-\tau)}|q > \tag{2.2}$$

and one can write a path integral for F, exactly as one writes it in real time. Actually the mathematically correct way of writing down the path-integral is to start in imaginary time, and then continue analytically to real time with an inverse Wick rotation[13]. The path-integral representation of (2.2) is

$$F = \int \mathcal{D}q \; e^{-\int_\tau^{\tau'} (\frac{1}{2}m\dot{q}^2 + V(q(\tau'')))d\tau''} \tag{2.3}$$

Rather than F, it will be convenient to consider the partition function $Z(\beta)$:

$$Z(\beta) = \mathrm{Tr}\; e^{-\beta H} = \int dq < q|e^{-\beta H}|q > \tag{2.4}$$

Its path-integral representation is deduced immediately from (2.3) and (2.4) :

$$Z(\beta) = \int \mathcal{D}q \; e^{-\int_0^\beta d\tau(\frac{1}{2}m\dot{q}^2 + V(q))} \tag{2.5}$$

where, because of the trace in (2.4), the paths are periodic functions in imaginary time : $q(0) = q(\beta)$; one sees on (2.5) the formal analogy between imaginary time and the inverse of the temperature. In the complex t-plane, the time-path must go from 0 to $-i\beta$, and the most obvious choice is of course the straight line along the imaginary axis from 0 to $-i\beta$ (fig.4). However other choices are possible, and some other possibilities will be explored later on. At this point we could pursue with the path-integral formalism, but it will be convenient, for later purposes, to revert to the operator formalism.

In statistical mechanics, the average value of an operator $\hat{A}$ (in all what follows, operators will be characterized by a caret) is given by :

$$< \hat{A} >= \frac{1}{Z}\mathrm{Tr}\,(\hat{A}\,\mathrm{e}^{-\beta H}) \tag{2.6}$$

We shall be interested by thermal averages of products of position operators $\hat{q}(t)$ in the Heisenberg picture, since they are the analogue in quantum mechanics of the products of field operators. We define the two-point functions

$$D_+(t,t') =< \hat{q}(t)\hat{q}(t') > \tag{2.7}$$

$$D_-(t,t') =< \hat{q}(t')\hat{q}(t) > \tag{2.8}$$

and the propagator

$$D(t,t') =< T(\hat{q}(t)\hat{q}(t')) >= \theta(t-t')D_+(t,t') + \theta(t'-t)D_-(t,t') \tag{2.9}$$

By inserting a complete set of eigenstates of the Hamiltonian (see(2.21)), one can show that that $D_+(t,t')$ is certainly defined for $-\beta \leq \mathrm{Im}(t-t') \leq 0$ and $D_-(t,t')$ for $0 \leq \mathrm{Im}(t-t') \leq \beta$. Now in the Heisenberg picture

$$\hat{q}(t) = \mathrm{e}^{iHt}\hat{q}(0)\mathrm{e}^{-iHt} \tag{2.10}$$

so that

$$\mathrm{e}^{-\beta H}\hat{q}(t)\mathrm{e}^{\beta H} = \hat{q}(t+i\beta) \tag{2.11}$$

Using the cyclicity of the trace, it is straightforward to derive

$$D_+(t,t') = D_-(t+i\beta,t') \tag{2.12}$$

This is the Kubo-Martin-Schwinger (KMS) condition, which is of course consistent with the domain of definition of D_+ and D_-; in a finite system, it is equivalent to the Boltzmann distribution $P_{eq} \propto \exp\,(-\beta H)$. However it is more general, since it also applies in the thermodynamic limit $V \to \infty$, where the Hamiltonian is not defined.

The KMS condition (2.12) implies that, on the imaginary axis, the propagator is a periodic function of time, with period β; let us indeed define :

$$\Delta(\tau) = D(-i\tau) =< T_\tau(\hat{q}(-i\tau)\hat{q}(0)) > \tag{2.13}$$

where the time-ordering T_τ is taken along the imaginary axis, from $+i\infty$ to $-i\infty$. We have from (2.12) in the interval $0 \leq \tau \leq \beta$

$$\Delta(\tau-\beta) = \Delta(\tau) \tag{2.14}$$

The property $\Delta(0) = \Delta(\beta)$ can also be easily derived from the path-integral representation of the propagator, taking into account the periodicity of the trajectories ($q(0) = q(\beta)$).

2.2 The Spectral Function $\rho(k_0)$

The two-point Green's function will appear in a variety of versions : imaginary time, real time, advanced, retarded. All these versions depend only on one single function, which we call the spectral function $\rho(k_0)$. Let us define the Fourier transforms of $D_+(t)$ and $D_-(t)$:

$$D_+(k_0) = \int_{-\infty}^{\infty} \mathrm{d}t\mathrm{e}^{ik_0t}D_+(t) \tag{2.15}$$

$$D_-(k_0) = \int_{-\infty}^{\infty} \mathrm{d}t\mathrm{e}^{ik_0t}D_-(t) = \int_{-\infty}^{\infty} \mathrm{d}t\mathrm{e}^{ik_0t}D_+(t-i\beta) \tag{2.16}$$

where the second equality follows from the KMS condition (2.12). Comparing (2.15) and (2.16) one deduces

$$D_-(k_0) = e^{-\beta k_0} D_+(k_0) \tag{2.17}$$

The spectral function $\rho(k_0)$ will be defined by

$$\rho(k_0) = D_+(k_0) - D_-(k_0) \tag{2.18}$$

and we can write from (2.17)

$$D_+(k_0) = (1 + f(k_0))\rho(k_0) \quad ; \quad D_-(k_0) = f(k_0)\rho(k_0) \tag{2.19}$$

with :

$$f(k_0) = (e^{\beta k_0} - 1)^{-1} \tag{2.20}$$

An explicit expression for $\rho(k_0)$ can be obtained by inserting a complete set of states of the Hamiltonian H in the expression (2.7) which defines $D_+(t)$

$$D_+(t) = \frac{1}{Z} \sum_{n,m} e^{-\beta E_n} e^{i(E_n - E_m)t} | < n|\hat{q}(0)|m > |^2 \tag{2.21}$$

In order to obtain $\rho(k_0)$ we compute $D_+(k_0) - D_-(k_0)$, making use of (2.17)

$$\rho(k_0) = \frac{2\pi}{Z} \sum_{n,m} e^{-\beta E_n} (1 - e^{-\beta k_0}) \delta(k_0 + E_n - E_m) | < n|\hat{q}(0)|m > |^2 \tag{2.22}$$

Equation (2.22) shows explicitly that

(*i*) $\rho(k_0)$ is an odd function of k_0: $\rho(k_0) = -\rho(-k_0)$

(*ii*) $\rho(k_0)$ obeys the positivity condition $\varepsilon(k_0)\rho(k_0) > 0$, where $\varepsilon(k_0)$ is the sign function.

Furthermore $\rho(k_0)$ obeys a sum rule which can be deduced from the equal-time commutation relation

$$[\hat{q}(t), \frac{d}{dt'}\hat{q}(t')|_{t'=0}] = i \tag{2.23}$$

Indeed we have

$$\int \frac{dk_0}{2\pi} \, k_0 e^{-ik_0 t} (D_+(k_0) - D_-(k_0)) = i\frac{d}{dt} (D_+(t) - D_-(t))$$

and from the very definition of D_+ and D_-

$$D_+(t) - D_-(t) = < [\hat{q}(t), \hat{q}(0)] > \tag{2.24}$$

Taking the $t \to 0$ limit of the above equations yields the sum rule

$$\int_{-\infty}^{+\infty} \frac{dk_0}{2\pi} k_0 \rho(k_0) = 1 \tag{2.25}$$

This sum rule can also be obtained from a simple manipulation of (2.22).

Finally it will be interesting to write down the analogue in quantum mechanics of the propagator; it is easily found from the expression of the position-operator in the case of the harmonic oscillator (with m=1)

$$\hat{q}(t) = \frac{1}{\sqrt{2\omega}} (a e^{-i\omega t} + a^\dagger e^{i\omega t}) \tag{2.26}$$

and equation (2.24) which leads to

$$D_+(t) - D_-(t) = \frac{1}{2\omega} < [a, a^\dagger]e^{-i\omega t} + [a^\dagger, a]e^{i\omega t} >$$

Since $[a, a^\dagger] = 1$, one finds immediately by Fourier transform the free (F) spectral function $\rho_F(k_0)$

$$\rho_F(k_0) = 2\pi\varepsilon(k_0)\delta(k_0^2 - \omega^2) \tag{2.27}$$

It is easily checked that this free spectral function obeys the positivity properties and the sum rule (2.25). It may also be worth noticing that ρ_F is temperature independent.

2.3 The Matsubara-(or Imaginary-Time) Propagator

We define the Fourier transform of the imaginary-time propagator $\Delta(\tau)$ (2.13) through :

$$\Delta(i\omega_n) = \int_0^\beta d\tau e^{i\omega_n \tau}\Delta(\tau) \tag{2.28}$$

Since $\Delta(\tau)$ is periodic ($\Delta(\tau + \beta) = \Delta(\tau)$), the frequencies ω_n take discrete values

$$\omega_n = \frac{2\pi n}{\beta} \tag{2.29}$$

which are called "Matsubara frequencies". If we choose τ in the interval $[0, \beta]$ we have $\Delta = D_+$ and from (2.15)

$$\Delta(\tau) = \int \frac{dk_0}{2\pi} e^{-k_0\tau} D_+(k_0) \tag{2.30}$$

Using the representation (2.19) of $D_+(k_0)$ in order to introduce the spectral function $\rho(k_0)$ leads to

$$\Delta(i\omega_n) = -\int_{-\infty}^{\infty} \frac{dk_0}{2\pi} \frac{\rho(k_0)}{i\omega_n - k_0} \tag{2.31}$$

In the free-field case, we use the expression (2.27) of $\rho(k_0)$ and we get :

$$\Delta_F(i\omega_n) = \frac{1}{\omega_n^2 + \omega^2} \tag{2.32}$$

The Matsubara (full) propagator (2.31) is defined for discrete values (2.29) of the frequencies only. Thus its analytic continuation to arbitrary values of the frequencies is not unique. However one can define from (2.31) a unique analytic continuation provided one requires that[14]

(i) $|\Delta(z)| \to 0$ if $|z| \to \infty$

(ii) $\Delta(z)$ is analytic outside the real axis.

Then the analytic continuation is provided by

$$\Delta(z) = -\int_{-\infty}^{\infty} \frac{dk_0}{2\pi} \frac{\rho(k_0)}{z - k_0} \tag{2.33}$$

The continuation to real values of z, or more exactly to $z = k_0 \pm i\epsilon$, with k_0 real, leads to the retarded and advanced functions $D_R(t)$ and $D_A(t)$:

$$D_R(t) = < \theta(t)[\hat{q}(t), \hat{q}(0)] > \tag{2.34}$$
$$D_A(t) = - < \theta(-t)[\hat{q}(t), \hat{q}(0)] > \tag{2.35}$$

From the usual representation of the θ-function

$$\theta(t) = i \int \frac{dk_0'}{2\pi} \frac{e^{-ik_0't}}{k_0' + i\epsilon} \tag{2.36}$$

one finds for example the Fourier transform $\overline{D}_R(k_0)$ of $D_R(t)$:

$$\overline{D}_R(k_0) = i \int \frac{dk_0'}{2\pi} \frac{\rho(k_0')}{k_0 - k_0' + i\epsilon} \tag{2.37}$$

so that :

$$\overline{D}_R(k_0) = i\Delta(k_0 + i\epsilon) \;\; ; \;\; \overline{D}_A(k_0) = -i\Delta(k_0 - i\epsilon) \tag{2.38}$$

We note that the free retarded and advanced Green's functions are temperature independent, because the free spectral function (2.27) is itself temperature independent. This is not surprising: as will be explained in the next section, the physical content of these Green's functions is that they give the energies and the lifetimes of the elementary excitations. In the free case we have simply $k_0 = \pm\omega$ and the excitations are of course stable.

Before going to the Feynman rules, we mention that the generalization to fermions is straightforward : instead of periodicity conditions on the trajectories we must have antiperiodicity conditions : $q(\beta) = -q(0)$, so that $\Delta(\tau)$ is now antiperiodic :

$$\Delta(\tau + \beta) = -\Delta(\tau) \tag{2.39}$$

and the Matsubara frequencies are now given by

$$\omega_n = \pi T(2n + 1) \tag{2.40}$$

Finally in the case of a non-zero chemical potential μ one has to make in the propagator the substitution

$$\omega_n \to \omega_n - i\mu \tag{2.41}$$

2.4 Feynman Rules in Imaginary Time

The expression (2.32) of the Matsubara propagator for the harmonic oscillator is easily generalized to field theory by taking into account the spatial degrees of freedom; it suffices to make in (2.32) the substitution $\omega^2 \to \omega_p^2 = p^2 + m^2$:

$$\Delta_F(p) = \frac{1}{\omega_n^2 + p^2 + m^2} \tag{2.42}$$

This expression is of course quite analogous to that of the propagator of an Euclidean $T = 0$ field theory, except that $\omega_n = p_4$ is restricted to discrete values. Because of this, in a loop integral $\int dp_4$ will be replaced by a sum over ω_n :

$$\int \frac{d^4p}{(2\pi)^4} \to T \sum_n \int \frac{d^3p}{(2\pi)^3} \tag{2.43}$$

The rules for the vertices are the same as those of the Euclidean $T = 0$ field theory[15]; they could be easily deduced from the generalization to field theory of the path-integral representation (2.5). For example in a $g^2\varphi^4$ scalar theory (the use of g^2 for the coupling constant, rather than g, allows a convenient comparison with other field theories), to each vertex will be associated a factor $-g^2$. For particles of spin 1/2 and for gauge particles, the spin structure of the propagator will be taken from the Euclidean $T = 0$ field theory.

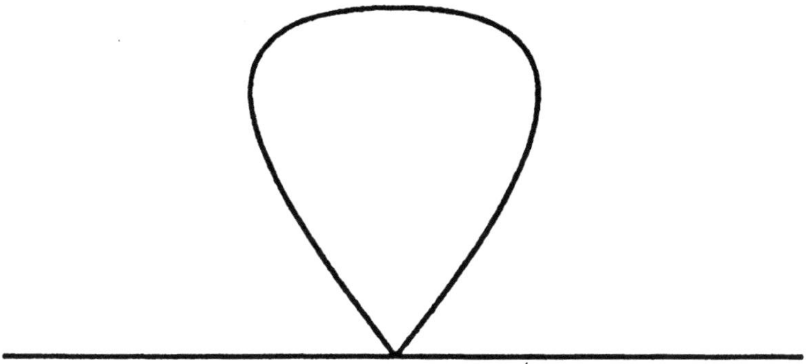

Fig. 5. The tadpole in $g\phi^4$.

In order to become familiar with these rules, let us work out a simple example : the tadpole in $g^2\varphi^4$ (fig.5)

$$\Pi = \frac{g^2 T}{2} \sum_n \int \frac{d^3 p}{(2\pi)^3} \frac{1}{\omega_n^2 + p^2 + m^2} \tag{2.44}$$

The standard trick, in order to perform the summation over n in (2.43) is to use a contour integral (fig.6)

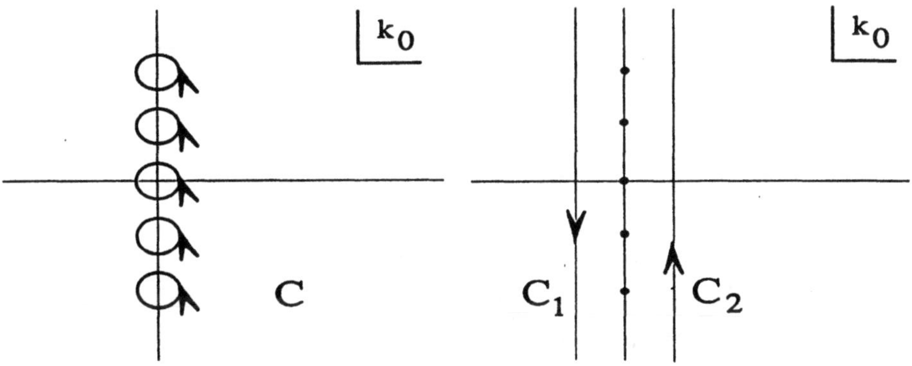

Fig. 6. The integration contour in the p_0 complex plane.

$$T \sum_{n=-\infty}^{\infty} f(p_0 = i\omega_n) = \frac{T}{2i\pi} \int_C dp_0 f(p_0) \frac{\beta}{2} \coth \frac{\beta p_0}{2} \tag{2.45}$$

which is valid provided $f(p_0)$ has no singularities on the imaginary axis. After some simple algebra one finds

$$\frac{1}{2i\pi} \int_{-i\infty}^{i\infty} dp_0 \frac{1}{2}(f(p_0) + f(-p_0)) + \frac{1}{2i\pi} \int_{C_2} dp_0 (f(p_0) + f(-p_0)) \frac{1}{e^{\beta p_0} - 1} \tag{2.46}$$

This allows to separate Π into a vacuum part, Π_{vac}, and a thermal part Π_β ; using the explicit value of $f(p_0)$:

$$f(p_0) = \frac{1}{\omega_p^2 - p_0^2} \qquad \omega_p^2 = p^2 + m^2$$

we get :

$$\Pi_{vac} = \frac{g^2}{2} \int \frac{d^4p}{(2\pi)^4} \frac{1}{p_4^2 + p^2 + m^2} \tag{2.47}$$

$$\Pi_\beta = \frac{g^2}{2} \int \frac{d^3p}{(2\pi)^3} \frac{1}{\omega_p(e^{\beta\omega_p} - 1)} \tag{2.48}$$

Π_{vac} is of course divergent, but this divergence will be taken care of by the usual $T = 0$ renormalization, while Π_β is easily evaluated for $T \gg m$:

$$\Pi_\beta = \frac{g^2T^2}{24} = \delta m_\beta^2 \tag{2.49}$$

Equation (2.49) shows that in a thermal bath, massless particles acquire a mass of order gT.

Instead of using the contour integration technique, it may be more elegant to start from the representation (2.30) of $\Delta(\tau)$. Taking into account the spatial degrees of freedom we get from (2.27) the spectral function of the free propagator :

$$\rho_F(p_0) = 2\pi\varepsilon(p_0)\delta(p^2 - m^2) \tag{2.50}$$

whence

$$\Delta_F(\tau, p) = \frac{1}{2\omega_p}[(1 + n_B(\omega_p))e^{-\omega_p\tau} + n_B(\omega_p)e^{\omega_p\tau}] \tag{2.51}$$

Then one uses the representation (2.28) of $\Delta(i\omega_n,\ p)$ which allows to perform easily the sum over n. For example for the graph of fig.7 we have:

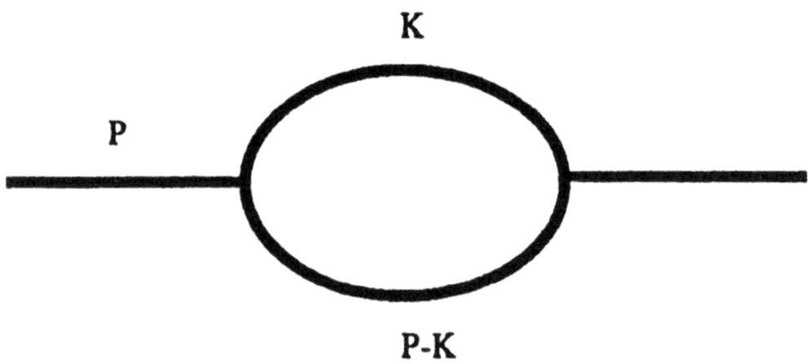

Fig. 7. A one-loop contribution to the self-energy.

$$\sum_n \int d\tau d\tau' e^{i\omega_n\tau} e^{i(\omega - \omega_n)\tau'} = \beta \int d\tau e^{i\omega\tau} \tag{2.52}$$

This formula gives automatically the correct continuation to real energies. However there is a tricky point in more complicated loop calculations : if an integration variable does not belong to the interval $[0, \beta]$, one has to extend (2.51) by enforcing the periodicity condition (2.14).

2.5 Real Time Formalism

We now wish to establish the expression of the thermal propagator in real-time. Let us revert for a while to the case of quantum statistical mechanics, as in subsections (2.1) and (2.2). The Fourier tranform $\overline{D}(p_0)$ of the propagator (2.9) reads

$$\overline{D}(p_0) = \int \mathrm{d}t \mathrm{e}^{ip_0 t}(\theta(t)D_+(t) + \theta(-t)D_-(t)) \tag{2.53}$$

From the representation (2.36) of the θ-function one derives

$$\overline{D}(p_0) = i\int \frac{\mathrm{d}p_0'}{2\pi}\,\frac{\rho(p_0')}{p_0 - p_0' + i\epsilon} + f(p_0)\rho(p_0) \tag{2.54}$$

From this expression it is clear that $\overline{D}(p_0)$ is not the analytic continuation to real frequencies of the Matsubara propagator. Indeed we have to define a continuation D' which depends on the sign of p_0

$$D'(p_0) = -i\Delta(p_0 + i\epsilon p_0) \tag{2.55}$$

and we have

$$\overline{D}(p_0) = (1 + n_B(p_0))D'(p_0) \; + \; n_B(p_0)D'^*(p_0) \tag{2.56}$$

where

$$n_B(p_0) = (\mathrm{e}^{\beta|p_0|} - 1)^{-1} \tag{2.57}$$

It is now easy to find the free propagator by plugging into (2.54) the expression (2.27) of the free spectral function

$$\overline{D}_F(p_0) = \frac{i}{p_0^2 - \omega^2 + i\epsilon} + 2\pi n_B(p_0)\delta(p_0^2 - \omega^2) \tag{2.58}$$

There are two important points to be noticed in the previous formula. First one sees that (2.58) allows to separate a zero-temperature part and a thermal part, which vanishes in the limit $T = 0$. Secondly one has to notice that it is the absolute value $|p_0|$ which is featured in (2.58), while p_0 appears in (2.20).

Generalizing (2.58) to field theory is straightforward: including spatial degrees of freedom yields the real-time, or Dolan-Jackiw propagator[4]

$$D_F(p) = \frac{i}{p^2 - m^2 + i\epsilon} + 2\pi n_B(p_0)\delta(p^2 - m^2) \tag{2.59}$$

At this point it would seem that we only need to take the $T = 0$ Feynman rules and use the real-time propagator (2.59) in order to compute in perturbation theory at $T \neq 0$. Unfortunately this is not the whole story: when one wants to write Feynman rules for real-time Green's functions, namely for Green's functions whose time-arguments are real, one runs into problems[16]. Let us assume that we want to compute

$$D(x,y) = \frac{1}{Z}\mathrm{Tr}(\mathrm{e}^{-\beta H}T(\varphi(x)\varphi(y))) \tag{2.60}$$

where the times x^0 and y^0 are real. The problem is that x^0 and y^0 belong to the real axis, and if we want to write a path-integral, we know that if the initial time is t_i, the final time must be $t_i - i\beta$. We have thus to choose a time-path starting from some time t_i, which goes through x^0 and y^0, and lands into $t_i - i\beta$. Taking into account the domain of definition of

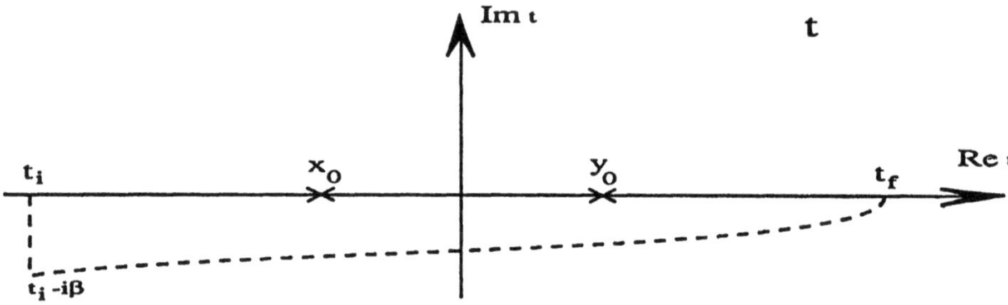

Fig. 8. A possible time-path in the real-time formalism.

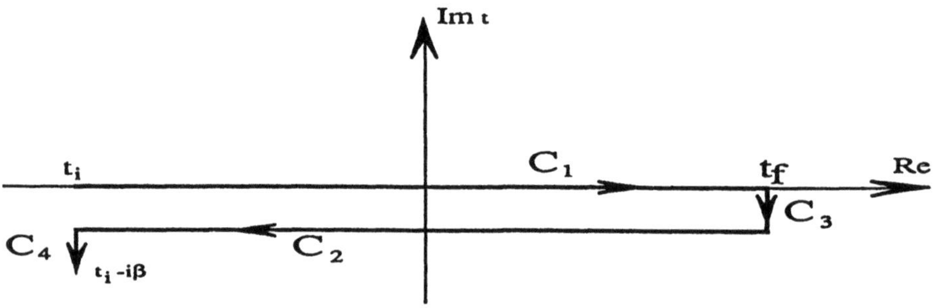

Fig. 9. The usual time-path in the real-time formalism.

D_+ and D_-, we see that the imaginary part of t must be non-increasing along the contour (fig.8).

The standard choice is that of fig.9

, with $t_i \to -\infty$ and $t_f \to +\infty$. This leads to a doubling of the number of degrees of freedom, because one has to introduce a ghost field φ_2 which lives on the contour C_2. The propagator takes a matrix form, the indices 1 and 2 corresponding to the normal (φ_1) and ghost (φ_2) field respectively :

$$D^F_{ab}(p) = U_{ac}(p) \begin{pmatrix} D^F_0(p) & 0 \\ 0 & D^{F*}_0(p) \end{pmatrix}_{cd} U_{db}(p) \tag{2.61}$$

where $D^F_0(p)$ is the $T = 0$ Feynman propagator

$$D^F_0(p) = \frac{i}{p^2 - m^2 + i\epsilon} \tag{2.62}$$

and the matrix U is given by

$$U(p) = \begin{pmatrix} \sqrt{1 + n_B(p)} & \sqrt{n_B(p)} \\ \sqrt{n_B(p)} & \sqrt{1 + n_B(p)} \end{pmatrix} \tag{2.63}$$

We have suppressed the bar over D^F for notational simplicity. The Dolan-Jackiw propagator (2.59) is nothing but the (11)-matrix element of D^F :

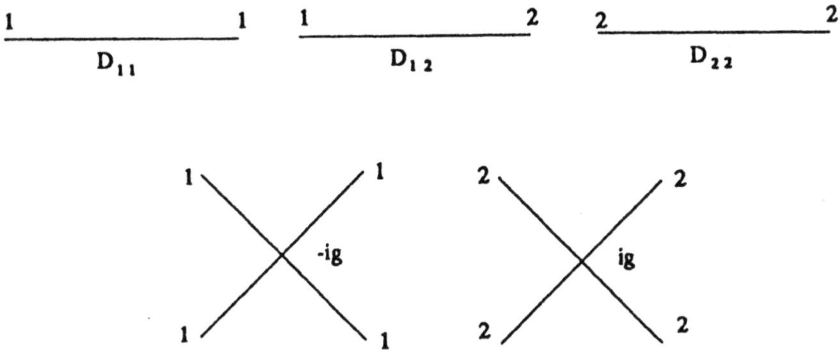

Fig. 10. Feynman rules in real-time.

$$D_{11}^F(p) = \frac{i}{p^2 - m^2 + i\epsilon} + 2\pi n_B(p_0)\delta(p^2 - m^2) \tag{2.64}$$

There are two types of vertices : in a scalar theory, to normal field vertices are associated factors $(-ig)$, while to ghost vertices are associated factors ig. Note that the fields are mixed through propagators, and not through vertices (fig.10)

We shall not attempt to give a derivation of these Feynman rules, since this has been worked out in detail in the literature[16], but we shall illustrate the necessity of the doubling of degrees of freedom on a simple example. Assume that we have a free Lagrangian describing particles of mass $m^2 + \mu^2$, but that we treat μ^2 as a perturbation :

$$\delta\mathcal{L} = -\frac{1}{2}\mu^2\varphi_1^2 - \frac{1}{2}\mu^2\varphi_2^2 \tag{2.65}$$

where φ_1 is the normal field and φ_2 the ghost field. To first order in perturbation theory we have (fig. 11)

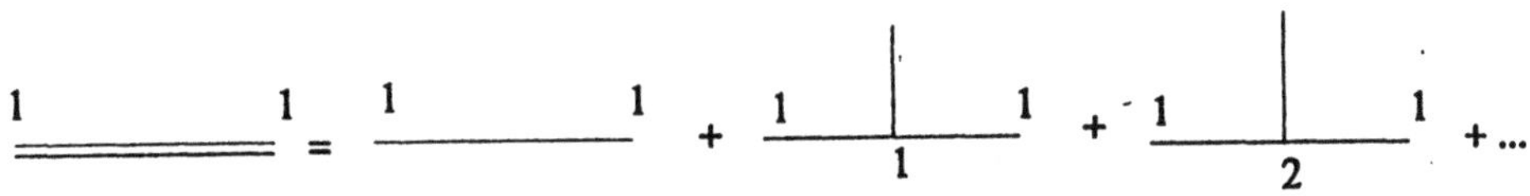

Fig. 11. The perturbative expansion of the propagator.

$$D_{11} = D_{11}^F(m^2) - i\mu^2[(D_{11}^F(m^2))^2 - (D_{12}^F(m^2))^2] + ... \tag{2.66}$$

where D_{12}^F is given from (2.61) :

$$D_{12}^F = 2\pi n_B(p_0)e^{\beta|p_0|/2}\delta(p^2 - m^2) \tag{2.67}$$

In order to interpret (2.66), we need a regularization of δ-functions ; because of (2.36) the correct regularization is

$$\delta(x) = \lim_{\epsilon\to 0}\frac{1}{\pi}\frac{\epsilon}{x^2 + \epsilon^2} \tag{2.68}$$

which gives :

$$\frac{1}{x + i\epsilon}\delta(x) = -\frac{1}{2}\delta'(x) \; - \; i\pi(\delta(x))^2 \tag{2.69}$$

This regularization is also useful at $T = 0$, and is not typical of finite temperature. One discovers that the unwanted δ^2 disappear in (2.66), which may be rewritten as

$$D_{11} = D_{11}^F(m^2) + \mu^2 \frac{\partial D_{11}^F}{\partial m^2} + ... = D_{11}^F(m^2 + \mu^2)$$

as it should. This formula is known as the mass-derivative formula, and has been used as a heuristic trick before the doubling of degrees of freedom was discovered.

2.6 The Self-Energy in the Real-Time Formalism

The above example was given in order to illustrate the doubling of the number of degrees of freedom. There is of course a much faster (and better) way of handling the calculation, which makes use of the diagonal form (2.61) of the propagator. We start from Dyson's equation in matrix form and write the full propagator $D_{ab}(p)$ by introducing the self-energy $\Sigma_{ab}(p)$

$$D_{ab}(p) = D_{ab}^F(p) + D_{ac}^F(p)(-i\Sigma_{cd}(p))D_{db}(p) \tag{2.70}$$

From the KMS condition (2.12), one can show that the same matrix $U(p)$ which diagonalizes the free propagator $D_{ab}^F(p)$ also diagonalizes the full propagator $D_{ab}(p)$[17]; then $U^{-1}(p)$ diagonalizes Σ_{ab}

$$-i\Sigma_{ab}(p) = U_{ac}^{-1}(p) \begin{pmatrix} -i\Sigma(p) & 0 \\ 0 & (-i\Sigma(p))^* \end{pmatrix}_{cd} U_{db}^{-1}(p) \tag{2.71}$$

The calculation in the end of the previous subsection could have been performed at once by noting that $\Sigma = \mu^2$ and by making in the diagonal form of the propagator the substitution

$$\frac{i}{p^2 - m^2 + i\epsilon} \rightarrow \frac{i}{p^2 - (m^2 + \mu^2) + i\epsilon}$$

Furthermore the matrix equation (2.71) leads to relations between various matrix elements

$$\begin{aligned} \mathrm{Re}\Sigma_{11}(p) &= \mathrm{Re}\Sigma(p) \\ \mathrm{Im}\Sigma_{11}(p) &= \coth(\tfrac{1}{2}\beta|p_0|)\mathrm{Im}\Sigma(p) \\ \Sigma_{12}(p) &= \frac{-i}{\sinh(\beta|p_0|/2)}\mathrm{Im}\Sigma(p) \end{aligned} \tag{2.72}$$

Finally using (2.55) and (2.56) one derives the relation between the real- and imaginary-time self-energies

$$\Sigma(p) = \overline{\Sigma}(p_0 + i\epsilon p_0, \boldsymbol{p}) \tag{2.73}$$

where the imaginary-time self-energy $\overline{\Sigma}(z, \boldsymbol{p})$ is related to the imaginary-time propagator through

$$\Delta^{-1}(z, \boldsymbol{p}) = \Delta_F^{-1}(z, \boldsymbol{p}) + \overline{\Sigma}(z, \boldsymbol{p}) \tag{2.74}$$

We recall that $\Delta(z, \boldsymbol{p})$ may be expressed in terms of the spectral function $\rho(p_0, \boldsymbol{p})$ (cf.(2.33))

$$\Delta(z, \boldsymbol{p}) = \int_{-\infty}^{+\infty} \frac{\mathrm{d}p_0}{2\pi} \frac{\rho(p_0, \boldsymbol{p})}{p_0 - z} \tag{2.75}$$

and that the function $\rho(p_0,\boldsymbol{p})$ obeys the positivity condition $\varepsilon(p_0)\rho(p_0,\boldsymbol{p}) \geq 0$ and the sum rule (cf.(2.25))

$$\int_{-\infty}^{+\infty} \frac{\mathrm{d}p_0}{2\pi} p_0 \rho(p_0,\boldsymbol{p}) = 1 \tag{2.76}$$

which follows from the equal-time commutation relations of the field $\hat{\varphi}(\boldsymbol{x},t)$ and its time derivative

$$[\hat{\varphi}(\boldsymbol{x},t), \frac{\mathrm{d}}{\mathrm{d}t'}\hat{\varphi}(\boldsymbol{x}',t')|_{t'=t}] = i\delta^{(3)}(\boldsymbol{x}-\boldsymbol{x}') \tag{2.77}$$

In a relativistic context sum rules like (2.76) are in general useless because of renormalization problems: $\Sigma(p)$ has to be subtracted in order to be made finite. In relativistic thermal field theories, such sum rules may hold in the limit of high temperatures, because in many important cases (see the next two sections) the leading term of Σ is proportional to T^2 and the $T=0$ terms are negligible. This leading term is finite in perturbation theory, at least to one-loop order, and the sum rule can then be applied.

Finally we may note that the relation between real and imaginary time Green's functions has been investigated recently by Kobes and relations generalizing (2.73) have been established for 3- and 4-point Green's functions[18].

2.7 Renormalization at Non-Zero Temperature

In computing loop diagrams at finite temperature, one encounters divergences which must be taken care of. Fortunately it is easy to see that renormalization at zero temperature suffices to make the theory finite at non-zero T. The reason is not difficult to understand: temperature does not modify the theory at distances $\ll 1/T$, and thus the UV (short-distance) singularities are the same as at $T=0$[19]. A more formal argument is as follows: let us start from the expression of the $T=0$ free propagator in Euclidean space

$$\Delta_F(\tau,\boldsymbol{x};T=0) = \int \frac{\mathrm{d}^4p}{(2\pi)^4} \frac{\mathrm{e}^{-i(p_4\tau+\boldsymbol{p}.\boldsymbol{x})}}{p_4^2+\omega_p^2} \tag{2.78}$$

with $\omega_p^2 = \boldsymbol{p}^2+m^2$, and compute the sum

$$\sum_{n=-\infty}^{+\infty} \Delta_F(\tau+n\beta,\boldsymbol{x};T=0) = \int \frac{\mathrm{d}^3p}{(2\pi)^3}\mathrm{e}^{-i\boldsymbol{p}.\boldsymbol{x}} \sum_{n=-\infty}^{+\infty} \Delta_F(\tau+n\beta,\boldsymbol{p};T=0) \tag{2.79}$$

A straightforward calculation gives

$$\sum_{n=-\infty}^{+\infty} \Delta_F(\tau+n\beta,\boldsymbol{p};T=0) = \sum_{n=-\infty}^{+\infty} \frac{1}{2\omega_p}\mathrm{e}^{-\omega_p|\tau+n\beta|} =$$

$$= \frac{1}{2\omega_p}\left((1+n_B(\omega_p))\mathrm{e}^{-\omega_p\tau} + n_B(\omega_p)\mathrm{e}^{\omega_p\tau}\right) = \Delta_F(\tau,\boldsymbol{p};T) \tag{2.80}$$

where we have used (2.51) in order to identify the thermal propagator in the second line of (2.80). We thus discover that

$$\Delta_F(\tau,\boldsymbol{x};T) = \sum_{n=-\infty}^{+\infty} \Delta_F(\tau+n\beta,\boldsymbol{x};T=0) \tag{2.81}$$

This expression can also be obtained by appealing to the summation formula of Poisson, or by noticing that $\Delta_F(\tau, \boldsymbol{x}; T)$ is the only solution of the equation

$$(\partial_\tau^2 + \partial^2 + m^2)\Delta_F(\tau, \boldsymbol{x}; T) = \delta(\tau)\delta^{(3)}(\boldsymbol{x}) \tag{2.82}$$

with the correct periodicity property (2.14). Since the UV singularities of the Euclidean propagator in x-space occur for $\tau^2 + \boldsymbol{x}^2 \to 0$, one sees from (2.81) that the short distance singularities of the thermal propagator $\Delta_F(\tau + n\beta, \boldsymbol{x}; T)$ come from $n = 0$ only, and that they are identical to those of the $T = 0$ propagator.

In p-space, one can start from the real-time expression of the free propagator D^F (2.61-64): this propagator can be split into a $T = 0$ part and a T-dependent part, which contains a factor decreasing at least as

$$\mathrm{e}^{-\beta|p_0|/2}\delta(p^2 - m^2) \tag{2.83}$$

Writing $D^F = D^F_{T=0} + D^F_\beta$, we see that divergences can only arise from loop integrations $\int \mathrm{d}^4p$ which do not contain D^F_β, since the δ-function in (2.83) puts the p-line on mass-shell, and the d^3p-integration is then damped exponentially. Of course divergences could come from integrations over other loop momenta which do not contain D^F_β factors, but the corresponding divergences are taken care of by $T = 0$ counter-terms. In conclusion, we see that the $T \neq 0$ theory is renormalized by temperature-independent counter-terms.

Let us now come to the use of the renormalization group (RG). In problems where there is no other scale than the temperature or the chemical potential, RG-arguments are very simple to apply. In order to be specific, let us look in QED at the expression for the pressure P in the case $T = 0, \mu \neq 0$: this example is chosen because it is the simplest one in which the calculation has been pushed far enough. The fine structure constant $\alpha(M)$ is defined by subtracting the photon propagator at the Euclidean point $k^2 = -M^2$. One find for a zero electron mass

$$\begin{aligned} P = \frac{\mu^4}{12\pi^2}(1 - \frac{3}{2}\frac{\alpha(M)}{\pi} - \frac{3}{2}\frac{(\alpha(M))^2}{\pi^2}\ln(\frac{\alpha(M)}{\pi}) - \\ - (\frac{\alpha(M)}{\pi})^2 \ln(\frac{\mu}{M}) + (2.11 \pm .12)(\frac{\alpha(M)}{\pi})^2) \end{aligned} \tag{2.84}$$

As usual RG improved perturbation theory implies that one should choose $M \simeq \mu$ in order to minimize the logarithmic term in (2.84). For $T \neq 0$, one expects in QED and QCD terms of the form $\alpha^2(M)\ln(T/M)$ which are summed by the RG, but they have not been computed yet.

When the problem at finite-T includes an intrinsic scale M, one can try to use T-dependent counter-terms, which are of course related to the $T = 0$ ones by a finite renormalization. For example one could impose the following condition for the renormalization of the propagator

$$\frac{\mathrm{d}}{\mathrm{d}k_0^2}\mathrm{Re}(\Sigma(k_0^2, \boldsymbol{k}^2))|_{k^2=-M^2} = 0 \tag{2.85}$$

where $\Sigma(k_0^2, \boldsymbol{k}^2)$ is the self-energy defined in (2.71), and similar condition on the vertices. This leads to a T-dependent coupling constant $g(T, M)$ which obeys a RG-type equation. This equation is strongly dependent on the details of the renormalization conditions. For example one finds in QCD (without quarks), using the imaginary time vertices[20],

$$T\frac{\mathrm{d}\alpha_S(T, M)}{\mathrm{d}T} \simeq N_c\left(\frac{31\pi^2}{18}(\frac{T}{M}) - \frac{11}{6}\right)\alpha_S^2(T, M) \tag{2.86}$$

in the limit M fixed, $T \to \infty$, and asymptotic freedom is lost in this limit! The physical significance of this result is not entirely clear, at least to the author.

A closely related subject is that of "dimensional reduction"[21]-[23]. From the expressions (2.29) and (2.32) of the Matsubara frequencies and of the imaginary-time propagator, one sees that the "non-static" modes with $n \neq 0$ depend on the factor $(2\pi nT)^2$ which acts like a mass term; thus the non-static modes ought to decouple in the limit $T \to \infty$, by appealing to the Appelquist-Carazzone decoupling theorem. Then the theory is equivalent to that of static modes ($n = 0$) only, namely an effective field theory in dimension three with coupling constant g^2T. However the decoupling theorem holds up to terms of order $(p/T)^2$ and $(m/T)^2$, and since the propagation in a heat bath induces mass terms of order g^pT, it may happen that decoupling holds only up to terms of order g^{2p}. A complete discussion is rather involved; it depends on the use of a T-dependent renormalization scheme: see e.g.[23]. The same line of arguments shows that dimensional reduction may also occur in the infrared limit, since for $p \to 0$, the $n = 0$ mode is clearly dominant, being the most infrared singular.

3. Collective Excitations in a Quark-Gluon Plasma

In this section and in the following one, we shall always assume that the temperature is much larger than the masses of the particles which we consider : $T \gg m$. This is of course true for gluons, and also for u- and d-quarks. In what follows, we shall simply take massless quarks and gluons, in order to avoid unnecessary notational complications. It turns out that one must distinguish three scales for the momenta of the external particles

- momenta $\sim g^2T$: magnetic mass scale
- momenta $\sim gT$: soft momenta
- momenta $\sim T$: hard momenta.

In what follows, we shall adopt the following convention: four-momenta will be denoted by upper case letters, while lower case letters will denote the corresponding energies and three-momenta: $P_\mu = (p_0, \mathbf{p})$. When we write $P \sim gT$, we mean that p_0 and $p \sim gT$. The quasi-particle spectrum will be especially interesting for soft external lines, since, as we already saw in the $g^2\varphi^4$-theory, thermal masses are of order gT.

3.1 Fermionic Excitations (QED or QCD)

The quasi-particle spectrum is found from the fermion self-energy $\Sigma(P)$[24]-[25] ; we shall be interested by the leading term in T only, which has the remarkable property of being gauge-invariant. This is a particular case of a general statement which will be discussed in the next section, in the framework of the effective expansion. We perform in QCD the one-loop calculation of Σ in the real-time formalism ; we only need the (11)-component of the fermion and gluon propagators

$$S_F(P) = \not{P} \left(\frac{i}{P^2 + i\epsilon} - 2\pi n_F(p_0)\, \delta(P^2) \right) \tag{3.1}$$

$$D_F^{\mu\nu}(K) = -g^{\mu\nu} \left(\frac{i}{K^2 + i\epsilon} + 2\pi n_B(k_0)\, \delta(K^2) \right) \tag{3.2}$$

The Fermi-Dirac and Bose-Einstein factors are given by

$$n_F(p_0) = \frac{1}{e^{\beta|p_0|}+1} \quad ; \quad n_B(k_0) = \frac{1}{e^{\beta|k_0|}-1}$$

The gluon propagator has been taken in the Feynman gauge ; one has to be careful when using other gauges[26]. From (3.1) and (3.2) we immediately get for the real part of the thermal self-energy Σ_β (fig.12)

Fig. 12. The fermion self-energy to one-loop order (full lines: quarks; wavy lines: gluons).

$$\mathrm{Re}\Sigma_\beta = 2g^2 C_F \mathrm{P} \int \frac{d^4K}{(2\pi)^3} \frac{1}{(P+K)^2} ((\not{P}+\not{K}) n_B(k_0) + \not{K} n_F(k_0))\, \delta(K^2) = C\not{P} + \not{D} \quad (3.3)$$

where P denotes a principal value. Since Lorentz invariance has been lost, we can build Lorentz scalars other than $\not{P}$; indeed we have at our disposal $u_\mu = (1,0)$, which defines the rest-frame of the plasma. It is very important to remark that chiral invariance is maintained since obviously

$$\{\gamma_5, \Sigma_\beta\} = 0 \quad (3.4)$$

With massive fermions, we would get in addition a scalar term (B1) and a term of the form $E\not{P}\not{u}$, which of course do not anticommute with γ_5 and break chiral invariance. Equation (3.3) leads to a <u>chiral invariant mass</u> ; indeed the full propagator $S(P)$ is :

$$-iS(P) = \frac{1}{\not{P}(1-C)-\not{D}} = \frac{(1-C)\not{P}-\not{D}}{P^2(1-C)^2 - 2P.D(1-C) + D^2} \quad (3.5)$$

and to first order in perturbation theory we discover that the thermal mass is $2m_f^2 = 2P.D$. Let us evaluate m_f^2 from (3.3) ; it is easy to find that

$$m_f^2 = \frac{g^2 C_F}{2\pi^2} \int_0^\infty \omega d\omega \, (n_B(\omega) + n_F(\omega)) \; + \; O(g^2 P^2)$$

namely

$$m_f^2 \simeq \frac{g^2 C_F T^2}{8} \quad (3.6)$$

In order to complete the evaluation of Σ_β, we need the integral

$$\int_{-1}^{1} \frac{d(\cos\theta)}{P^2 + 2\eta p_0 k + 2pk\cos\theta} = \frac{1}{2pk} \ln \frac{E_\eta(E_{-\eta}+k)}{E_{-\eta}(E_\eta+k)} \quad (3.7)$$

where

$$\eta = \pm 1 \;\; , \;\; E_\eta = \frac{1}{2}(p_0 + \eta p) \quad (3.8)$$

One then notices that a sum of terms of the form $[\eta = +1]+[\eta = -1]$ leads to a logarithmic divergence in the k-integration, while a difference $[\eta = +1] - [\eta = -1]$ leads to a quadratic divergence, resulting in a $T^2 \ln E_+/E_-$ factor in the final result. Such a quadratic divergence will appear in the calculation of D_0, since k_0 can be either positive or negative. We find from (3.3)

$$D_0 = \frac{g^2 C_F}{4\pi^2 p} \ln\frac{E_+}{E_-} \int_0^\infty d\omega\, \omega(n_B(\omega) + n_F(\omega)) = \frac{m_f^2}{p} Q_0(\frac{p_0}{p}) \tag{3.9}$$

where Q_0 is a Legendre function of the second kind :

$$Q_0(x) = \frac{1}{2} \ln\frac{x+1}{x-1} \tag{3.10}$$

We shall also need $Q_1(x)$

$$Q_1(x) = xQ_0(x) - 1 \tag{3.11}$$

Knowing D_0 and $P.D$, it is straightforward to compute the space components D_i, using rotational invariance :

$$|D| = \frac{1}{p}(p_0 D_0 - P.D) = \frac{m_f^2}{p} Q_1(\frac{p_0}{p}) \tag{3.12}$$

On the other hand it is obvious that the terms proportional to $\not P$ are non-leading, and we may write the final result, to leading order in T ($\hat{p} = p/p$) :

$$iS^{-1}(P) = \not P - \Sigma = A_0\gamma_0 - A_S \gamma.\hat{p} \tag{3.13}$$

$$A_0 = p_0 - \frac{m_f^2}{p} Q_0(p_0/p) \tag{3.14}$$

$$A_S = p - \frac{m_f^2}{p} Q_1(p_0/p) \tag{3.15}$$

In order to find the quasi-particle spectrum, we must locate the poles of $S(P)$. Now we have to be a bit careful. Indeed if we write

$$-iS(P) \simeq \frac{1}{\not P - \not D} = \frac{\not P - \not D}{P^2 - 2P.D + D^2} \tag{3.16}$$

naive perturbation theory allows to neglect D^2, which is of order g^4, and gives a pole at $p^2 = 2m_f^2$, in agreement with the previous discussion. However this is no longer the case if the external momentum is of order gT ; because of the m_f^2/p factor in (3.9) and (3.12), all terms in the denominator of (3.16) are now of order g^2T^2, and all of them are to be taken into account. A convenient way of rewriting (3.13) is

$$-iS(P) = \frac{1}{2(A_0 + A_S)}(\gamma_0 + \gamma.\hat{p}) + \frac{1}{2(A_0 - A_S)}(\gamma_0 - \gamma.\hat{p}) \tag{3.17}$$

There are two dispersion laws[25],[27]: the first one corresponds to the solutions of $A_0 + A_S = 0$. The corresponding quasi-particles have a negative helicity/chirality ratio ($\chi = -1$), while those corresponding to $A_0 = A_S$ have $\chi = +1$. We denote the dispersion laws by $\omega^\pm(p)$

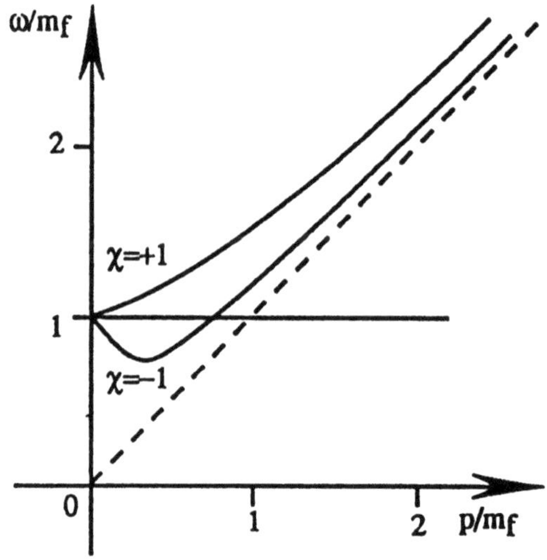

Fig. 13. The dispersion laws for the fermionic excitations.

; the two functions are displayed in fig.13. It is interesting to give their behaviour for $p \to 0$ and for $p \to \infty$, as well as that of the residue $Z_2(p)$ at the pole ; in the case $\chi = +1$ we have

$$p \to 0 : \omega(p) \simeq m_f + \frac{1}{3}p \quad ; \quad Z_2 \simeq \frac{1}{2} + \frac{p}{6m_f} \tag{3.18a}$$

$$p \to \infty : \omega(p) \simeq p + m_f^2/p \quad ; \quad Z_2 \simeq 1 + \frac{m_f^2}{2p^2} \tag{3.18b}$$

while for $\chi = -1$ one finds :

$$p \to 0 : \omega(p) \simeq m_f - \frac{1}{3}p \quad ; \quad Z_2 \simeq \frac{1}{2} - \frac{p}{6m_f} \tag{3.19a}$$

$$p \to \infty \quad Z_2 \simeq \exp(-2p^2/m_f^2) \tag{3.19b}$$

One sees that the mode with $\chi = -1$ decouples at large p. The physical interpretation is clear : $\chi = +1$ corresponds to ordinary $T = 0$ Dirac particles ; at large values of p the propagation in the plasma induces only weak effects, in the form of a thermal mass $\sqrt{2}m_f$, while the $\chi = -1$ quasi-particles decouple. On the contrary, when $p \to 0$, helicity is no more defined and one cannot distinguish between $\chi = +1$ and $\chi = -1$ excitations: they both have the same mass m_f. Both $\chi = +1$ and $\chi = -1$ quasiparticles are collective excitations, but the latter is more collective than the former!

Let us conclude this subsection with three remarks

(*i*) We have not computed the imaginary part of Σ: it can be shown that this imaginary part is non-zero only below the light-cone, i.e. for $p_0^2 < p^2$, and it is just given by the imaginary part of the Q_l functions in (3.10) and (3.11)

$$\ln \frac{x+1}{x-1} = \ln |\frac{x+1}{x-1}| - i\pi\theta(1-|x|) \tag{3.20}$$

We shall see in section 4 that this is a general feature of the leading term in T which is related to the so-called "Landau damping" mechanism.

(*ii*) It has been argued by Lebedev and Smilga[28] that one cannot observe in fact the behaviour (3.19b): for $p \geq gT$ the one-loop calculation is unstable and the $\chi = -1$ excitation disappears below the light-cone.

(*iii*) In general one does not expect $Z_2(\chi = +1) + Z_2(\chi = -1) = 1$, as could be concluded from (3.18)-(3.19). Indeed one can derive a sum rule analogous to (2.76), which tells us that $Z_2(\chi = +1) + Z_2(\chi = -1) \leq 1$. The continuum contribution is negligible for $p \to 0$ or $p \to \infty$, but in the general case one finds numerically[27]

$$0.8 \leq Z_2(\chi = +1) + Z_2(\chi = -1) \leq 1 \tag{3.21}$$

3.2 Linear Response Theory

We are now going to study the gluonic collective excitations. In order to put the discussion in a more general framework, let us recall some elements of linear response theory[29]. Let H be a time independent Hamiltonian and $V(t)$ be an external perturbation. The state vector $|\Psi_S(t)>$ in the Schroedinger picture is given at time t by

$$|\Psi_S(t)>= e^{-iH(t-t_0)}U(t,t_0)|\Psi_S(t_0)> \tag{3.22}$$

where the operator $U(t,t_0)$ obeys the following differential equation and boundary condition

$$i\frac{dU(t,t_0)}{dt} = V_H(t,t_0)\, U(t,t_0) \quad ; \quad U(t_0,t_0) = 1 \tag{3.23}$$

$$V_H(t,t_0) = e^{iH(t-t_0)}\, V(t)\, e^{-iH(t-t_0)} \tag{3.24}$$

Of course $V_H(t,t_0)$ is nothing but the operator $V(t)$ in the Heisenberg picture. Usually H is the free Hamiltonian H_0, and then (3.24) defines the perturbation in the interaction picture. Expanding U in powers of V one finds as usual

$$U(t,t_0) = 1 - i\int_{t_0}^{t} dt'\; V_H(t',t_0) \;+\; ... \tag{3.25}$$

whence

$$|\Psi_S(t)>= e^{-iH(t-t_0)}\;|\Psi_S(t_0)> -ie^{-iH(t-t_0)}\int_{t_0}^{t} dt' V_H(t',t_0)|\Psi_S(t_0)> \;+ ... \tag{3.26}$$

Let $\hat{O}(t)$ be any operator in the Schroedinger picture and $<\hat{O}(t)>$ its expectation value

$$<\hat{O}(t)> \;=\; <\Psi_S(t)|\hat{O}(t)|\Psi_S(t)> \tag{3.27}$$

If we denote by $\delta <\hat{O}(t)>$ the variation with respect to the expectation value in the absence of $V(t)$, we have from (3.26), to first order in V

$$\delta <\hat{O}(t)>= i\int_{t_0}^{t} dt' \;\; <\Psi_H|[V_H(t'),\hat{O}_H(t)]|\Psi_H> \tag{3.28}$$

where$|\Psi_H>= e^{iH(t-t_0)}U(t,t_0)|\Psi_S(t)>$ is the state vector in the Heisenberg picture. Equation (3.28) features the retarded commutator

$$\theta(t-t')\;[V_H(t')\;,\;\hat{O}_H(t)] \tag{3.29}$$

An important example corresponds to the following choice of $V(t')$ and $\hat{O}(t)$ (in the Schroedinger picture)

$$V(t') = \int \mathrm{d}x' \; j(\boldsymbol{x}', t')\varphi_S(\boldsymbol{x}')$$
$$\hat{O}(t) = \varphi_S(\boldsymbol{x})$$

where $j(x')$ is some external (c-number) source. We get from (3.28) and (3.29)

$$\delta < \varphi(\boldsymbol{x}, t) > = -i \int \mathrm{d}x' \; j(x') D_R(x - x') \tag{3.30}$$

$$D_R(x - x') = \theta(t - t') < [\varphi(x), \varphi(x')] > \tag{3.31}$$

Equation (3.28) holds obviously for thermal averages

$$\frac{1}{Z} \sum_n \mathrm{e}^{-\beta E_n} < n|....|n >$$

and thus the expectation value in (3.31) will correspond to a thermal average in finite-T field theory (at $T = 0$, it would be a vacuum expectation value). In Fourier space (3.30) is translated into

$$\delta < \varphi(k_0, \boldsymbol{k}) > = -ij(k_0, \boldsymbol{k}) \; D_R(k_0, \boldsymbol{k}) \tag{3.32}$$

In the case of a time-independent external perturbation $j(\boldsymbol{x})$, (3.32) reduces to

$$\delta < \varphi(\boldsymbol{k}) > = -ij(\boldsymbol{k}) \; D_R(k_0 = 0, \boldsymbol{k}) \tag{3.33}$$

Let us instead apply an impulsive perturbation

$$j(\boldsymbol{x}, t) = j(\boldsymbol{k})\mathrm{e}^{i\boldsymbol{k}.\boldsymbol{x}}\delta(t) \tag{3.34}$$

Then we learn from(3.30) that

$$\delta < \varphi(\boldsymbol{x}, t) > = -ij(\boldsymbol{k})\mathrm{e}^{i\boldsymbol{k}.\boldsymbol{x}} \int_{-\infty}^{\infty} \frac{\mathrm{d}k_0}{2\pi} \mathrm{e}^{-ik_0 t} D_R(k_0, \boldsymbol{k}) \tag{3.35}$$

Assume now that $D_R(k_0, \boldsymbol{k})$ can be approximated by the following formula

$$D_R(k_0, \boldsymbol{k}) = \frac{iR(k_0, \boldsymbol{k})}{k_0 - \omega(\boldsymbol{k}) + i\gamma(k_0, \boldsymbol{k})} \tag{3.36}$$

The response to the impulsive perturbation (3.34) is then a damped travelling wave with dispersion law $k_0 = \omega(\boldsymbol{k})$ and damping rate $\gamma(\omega(\boldsymbol{k}), \boldsymbol{k})) = \gamma$

$$\delta < \varphi(\boldsymbol{x}, t) > = -ij(\boldsymbol{k}) \; R(\omega(\boldsymbol{k}), \boldsymbol{k})) \exp\left(i(\boldsymbol{k}.\boldsymbol{x} - \omega(\boldsymbol{k})t - \gamma t\right) \tag{3.37}$$

For a free field we would have $\omega(\boldsymbol{k}) = (\boldsymbol{k}^2 + m^2)^{1/2}$ and $\gamma = 0$. The amplitude of the travelling wave is proportional to the residue at the pole of the retarded Green's function. We now have at our disposal all the necessary tools in order to study the response to an external perturbation of a QED or QCD plasma. Let us examine first the former case.

3.3 The Photon Propagator in a QED Plasma

We write the photon propagator $D^{\mu\nu}$ in a covariant gauge, exhibiting the transverse, longitudinal and gauge-fixing parts[30]

$$iD^{\mu\nu} = \frac{1}{G - K^2} P_T^{\mu\nu} \; + \; \frac{1}{F - K^2} P_L^{\mu\nu} \; + \; \frac{\rho}{K^2} \frac{K^\mu K^\nu}{K^2} \tag{3.38}$$

with

$$P_T^{00} = P_T^{0i} = 0 \quad P_T^{ij} = \delta^{ij} - \frac{k^i k^j}{k^2}$$

$$P_L^{\mu\nu} = \frac{K^\mu K^\nu}{K^2} - g^{\mu\nu} - P_T^{\mu\nu} \tag{3.39}$$

and ρ is the gauge parameter; in fact one can build a fourth tensor structure which cannot play any role in QED and seems also unimportant in QCD; thus we shall neglect this structure for simplicity. The polarization operator $\Pi^{\mu\nu}$ is defined as the difference between the full and free inverse propagators

$$\Pi_{\mu\nu} = -i(D^{-1}_{\mu\nu} \; - \; D^{-1}_{F,\mu\nu}) = FP^L_{\mu\nu} + GP^T_{\mu\nu} \tag{3.40}$$

Since Lorentz invariance is lost at finite T, the advantage of working in a covariant gauge are much less evident than at $T = 0$: popular choices of non-covariant gauges are the Coulomb and time-axial (TAG) gauges.

The results of a one-loop calculation have a rather simple expression in the high-temperature limit ($T \gg m$); keeping only terms proportional to T^2 one finds ($k = |\boldsymbol{k}|$)[2],[31]

$$F(k_0,k) = \frac{e^2T^2}{3}\left[1 - \frac{k_0^2}{k^2}\right]\left[1 - \frac{k_0}{2k}(\ln\left(\frac{k_0+k}{k_0-k}\right) - i\pi\theta(k^2-k_0^2))\right] \tag{3.41a}$$

$$G(k_0,k) = \frac{e^2T^2}{6}\left[\frac{k_0^2}{k^2} + (1 - \frac{k_0^2}{k^2})\frac{k_0}{2k}(\ln\left(\frac{k_0+k}{k_0-k}\right) - i\pi\theta(k^2-k_0^2))\right] \tag{3.41b}$$

Let us apply to the QED plasma a (weak) external classical electric field $\boldsymbol{E}_{cl}$; the perturbation is then

$$V = \int \mathrm{d}^3x \; \boldsymbol{E}.\boldsymbol{E}_{cl} \tag{3.42}$$

and from (3.30) the response of the plasma is a modification of the average external field $\boldsymbol{E}$, which of course vanishes in the absence of $\boldsymbol{E}_{cl}$. This average is given by the retarded commutator of electric fields

$$\delta < E_i(\boldsymbol{x},t) >= -i\int \mathrm{d}^4x' \; E_{j,cl}(\boldsymbol{x}',t')\theta(t-t') < [E_i(x), E_j(x')] > \tag{3.43}$$

The retarded commutator is to be computed from the corresponding retarded commutator of the A_μ, namely

$$D^R_{\mu\nu} = \theta(t-t') \; < [A_\mu(x), A_\nu(x')] > \tag{3.44}$$

A straightforward calculation gives

$$E_i^{tot} = E_{i,cl} + \delta < E_i >= -i\int \mathrm{d}^4x' \; E_{j,cl}(x')\left(-\partial_i\partial'_j D^R_{00} + \partial_i\partial'_0 D^R_{0j} + \partial_0\partial'_j D^R_{i0} - \partial_0\partial'_0 D^R_{ij}\right) \tag{3.45}$$

For a <u>static</u> electric field, this formula is easily translated into

$$E_i^{tot}(\boldsymbol{k}) = \frac{k_i k_j E_{j,cl}(\boldsymbol{k})}{k^2 + F(k_0 = 0, \boldsymbol{k})} \tag{3.46}$$

In the $k_0 = 0, k \to 0$ limit, we get from (3.41a) the electric mass, which controls the screening of the Coulomb potential of a static charge Q

$$m_{el}^2 = F(k_0 = 0, k \to 0) = -\Pi_{00}(k_0 = 0, k \to 0) \simeq \frac{1}{3}e^2T^2 \tag{3.47}$$

Let us now turn to the plasma excitations, which are controlled by the transverse (G) and longitudinal (F) components of the polarization operator. We have for the transverse modes

$$\omega_T^2 = k^2 + \mathrm{Re}G(\omega_T, k)$$
$$\gamma_T = -\frac{1}{2\omega}\mathrm{Im}G(\omega_T, k) \tag{3.48}$$

From the explicit expression (3.41b) of G one derives the following results

$$\omega, k \gg eT: \quad \omega_T^2 \simeq k^2 + m_P^2 \quad m_P^2 = \frac{1}{6}e^2T^2 \tag{3.49}$$

$$\omega, k \sim eT: \quad \omega_T^2 \simeq \omega_P^2 + \frac{6}{5}k^2 \quad \omega_P^2 = \frac{1}{9}e^2T^2 \tag{3.50}$$

where ω_P is the plasma frequency, and for $k \to 0$[32]

$$\gamma_T = \frac{e^2}{24\pi}\omega_P n_F(\frac{\omega_P}{2}) \simeq \frac{e^2}{24\pi}\omega_P \tag{3.51}$$

This last result is easy to understand on dimensional grounds, since γ_T is proportional to the available phase space, which is nothing but ω_P. Furthermore since $\omega_P \ll T$, the Fermi-Dirac distribution can be approximated by one; the linear dependence of (3.51) on n_F will be explained in the next section.

The dispersion law for longitudinal modes is given by

$$\omega_L^2 = k^2 + \mathrm{Re}F(\omega_L, k)$$

We find that they do not propagate when $\omega, k \gg eT$, as the residue of the pole tends to zero exponentially; this is quite analogous to what happens for $\chi = -1$ fermions (see (3.21)). For $\omega, k \sim eT$ one finds

$$\omega_L^2 \simeq \omega_P^2 + \frac{3}{5}k^2 \qquad \gamma_L = \gamma_T \tag{3.52}$$

The physical interpretation of the above results is quite analogous to that found for fermions: at $T = 0$, the physical modes are the transverse modes, and these are the only modes which propagate at $T \neq 0$ for large values of k; however the photons acquire a (gauge invariant) mass$\sim eT$: this behaviour is quite reminiscent of that found for $\chi = 1$ fermions. On the other hand the longitudinal modes, which are unphysical at $T = 0$, do not propagate at $T \neq 0$ for large values of the momenta. On the contrary, for $k = 0$, we have $\omega_T = \omega_L$ since one cannot distinguish between transverse and longitudinal modes, and for the same reason we must have $\gamma_T = \gamma_L$. We can see that we have three different masses, all of order eT, for dimensional reasons. However it must be understood that they have a quite different physical interpretation as they are obtained by taking different limits:

$$(i) \text{ the electric mass}: \; m_{el}^2 = \frac{1}{3}e^2T^2 \;\; (k_0 = 0, k \to 0)$$

$$(ii) \text{ the T}-\text{dependent mass}: \; m_P^2 = \frac{1}{6}e^2T^2 \;\; (k_0, k \gg eT)$$

$$(iii) \text{the plasma frequency}: \; \omega_P^2 = \frac{1}{9}e^2T^2 \;\; (k_0 \sim eT, k = 0)$$

3.4 The Gluon Propagator in the QCD Plasma

At first sight the situation in QCD looks very similar to that in QED, except for colour factors and the substitution $e \to g$, where g is the strong coupling constant. At one-loop order one finds in the high T-limit that the gluon polarization tensor is given by (3.41) with the substitution

$$e^2 \to g^2(N_c + \frac{1}{2}N_f)$$

where N_c is the number of colours and N_f the number of flavours. Thus one finds

(*i*) $m_{el}^2 = \frac{1}{3}g^2(N_c + N_f/2)T^2$. This electric mass gives the screening of the heavy quark potential, which may play a crucial role in the J/ψ suppression, if this suppression comes from the formation of a plasma. As already discussed, it can also be calculated on the lattice, with results in rough agreement with those of the perturbative calculation.

(*ii*) $m_P^2 = \frac{1}{6}g^2(N_c+N_f/2)T^2$, which gives the temperature-dependent mass of high energy, transversely polarized gluons in the plasma.

(*iii*) $\omega_P^2 = \frac{1}{9}g^2(N_c + N_f/2)T^2$, which gives the frequency of long wave-length transverse or longitudinal collective excitations, or plasma oscillations.

The crucial difference between QED and QCD is that the functions F and G in (3.38) are gauge dependent; fortunately the results for m_{el}, m_P and ω_P turn out to be gauge independent, at least to one-loop order and to leading order in T. However the plasmon damping rate γ does not seem to be gauge independent: this problem has been known as the "plasmon puzzle". Furthermore the linear response theory is complicated by the fact that the chromoelectric field is not linear in A

$$E_i^a = \partial_i A_0^a - \partial_0 A_i^a - g f_{abc} A_i^b A_0^c \tag{3.53}$$

where f_{abc} represents the Lie algebra structure constants. Thus the retarded commutator $[E, E]$ will involve not only terms bilinear in A, but also A^3 and A^4 terms. This problem is overcome in the TAG-gauge where

$$E_i^a = -\partial_0 A_i^a \tag{3.54}$$

but the TAG-gauge has problems of its own. The plasmon decay rate was first computed in this gauge and found to be[32]

$$\gamma = \frac{g^2 N_c T}{24\pi} \tag{3.55}$$

With respect to (3.51), the absence of a factor ω_P can be understood as follows : the main decay mechanism is the decay into two gluons, which implies a factor $n_B(\omega_P) \sim T/\omega_P$, because $\omega_P \sim gT \ll T$. On the other hand the contribution of $q\overline{q}$ pairs is negligible, and the result (3.55) is independent of N_f.

The calculation has been repeated in the Coulomb gauge [33], with the same result as in (3.55), and in covariant or background gauges, where it is found that (3.55) should be multiplied by a factor ~ -5 to -10. This would lead to plasma instability (as γ would be negative!), and some authors have argued that perturbation theory is unstable, because the expansion is performed around the wrong vacuum. However we shall see in the next section that the one-loop calculation is incomplete, as the final state gluons are bare transverse gluons, while they should be gluons dressed by their interactions in the plasma. Actually the decay of the plasmon into two plasmons (dressed gluons) is kinematically forbidden !

At this point it may be useful to add some comments on the gauge independence of perturbative calculations at finite temperature. At $T = 0$ it is well-known that physical quantities like S-matrix elements can be proven to be gauge independent order by order in

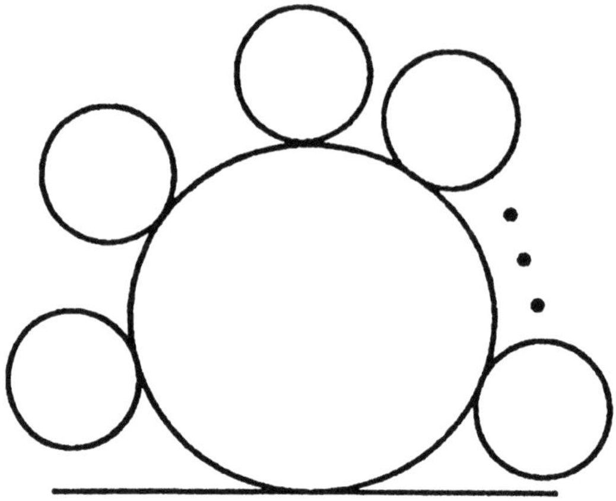

Fig. 14. An infrared divergent contribution to the propagator.

the loop expansion, at least if one is dealing with a broken gauge theory in order to avoid infrared divergences. However in the calculation of a dispersion law the loop expansion breaks down, as will be explained in detail in the next section. Nevertheless Kobes et al.[34] have been able to prove that dispersion laws <u>are</u> gauge independent: the locations of the physical poles in the gluon propagator are gauge independent in spite of the gauge dependence of the propagator itself. It follows that in any <u>consistent</u> approximation scheme, the dispersion laws should be found to be gauge independent. The loop expansion is certainly not such a scheme, since contributions from higher order loops are of the same order in g as the lower order terms. As will be shown in the next section, the effective expansion seems to provide such a consistent scheme: at least the results for the plasmon damping rate have been found to be identical in a general covariant gauge and in the Coulomb gauge. Athough this result is certainly very encouraging, it cannot be considered as a proof that the approximation scheme does give all the leading order terms!

4. Resummation Methods in Perturbation Theory

In this section we shall examine resummation methods which are necessary in order to supplement perturbation theory, because of the singular infrared behaviour.

4.1 The φ^4-model revisited

Let us come back to the massless $g^2\varphi^4$- model which was already studied in subsection 2.4 ; it was shown there that the first order correction to the propagator is of order g^2T^2. This result has the following consequence: assume that we look at the propagator of a particle with a "soft momentum", which means that its energy p_0 and its momentum p are $\sim gT$ (we recall our convention that 4-momenta are denoted by upper case letters, energies and 3-momenta by the corresponding lower case letters: $P^\mu = (p_0, \boldsymbol{p})$). For such a "soft particle", the inverse propagator is of order g^2T^2, but the first order correction Π_1 to Π_β (see (2.49)) is also of order g^2T^2. This means that one cannot limit oneself to a naïve perturbative expansion, but that one must use some kind of resummation. The solution in this simple case is obvious : one has to use the summation drawn in fig.14[35], or even better one can use a self-consistent equation[36]:

$$\Pi = \frac{g^2}{2} T \sum_n \int \frac{d^3p}{(2\pi)^3} \frac{1}{\omega_n^2 + p^2 + \Pi} \tag{4.1}$$

Having performed the summation over n and the $T = 0$ mass renormalization, we are led for the T-dependent part to the equation

$$1 = \frac{g^2}{4\pi^2} \int_1^\infty \frac{\sqrt{x^2-1}\, dx}{\exp(\beta \Pi_\beta^{1/2} x) - 1} \tag{4.2}$$

The expansion in powers of g is obtained from that of the function $F(u)$:

$$F(u) = \int_1^\infty dx \frac{\sqrt{x^2-1}}{e^{ux}-1} = \frac{2\pi^2}{u^2}\left(\frac{1}{12} - \frac{u}{4\pi} + O(u^2 \ln u)\right) \tag{4.3}$$

The results are plotted in fig.15[36], where one can see that the deviation with respect to the first order calculation is rather small.

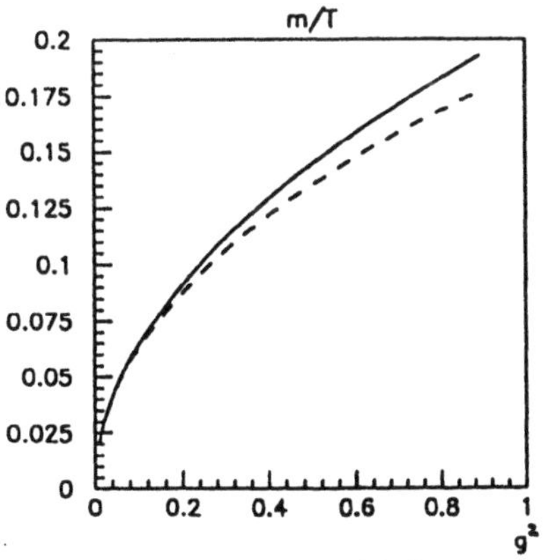

Fig. 15. The ratio m/T in $g\phi^4$ [36].

However the expansion in powers of g reveals an interesting feature :

$$\Pi_\beta = \frac{g^2 T^2}{24}\left[1 - 3\left(\frac{g^2}{24\pi^2}\right)^{1/2} + ...\right] \tag{4.4}$$

We see that the next term in Π_β is not in g^4, as one would expect from naïve perturbation theory, but in g^3. This is what could be called a "mild violation" of perturbation theory, to be contrasted with the strong violations which will be seen in the next section.

The same feature also appears in the perturbative expansion of the partition function. The same resummation method leads to the so-called "ring diagrams" (fig.16), which give the following contribution

$$\frac{1}{2}\beta VT \sum_n \int \frac{d^3p}{(2\pi)^3} \sum_{N=2}^\infty \frac{1}{N}(-\Pi_1(\omega_n,p)\Delta_F(\omega_n,p))^N$$

$$= -\frac{1}{2}\beta VT \sum_n \int \frac{d^3p}{(2\pi)^3} (\ln(1 + \Pi_1(\omega_n,p)\Delta_F(\omega_n,p)) - \Pi_1(\omega_n,p)\Delta_F(\omega_n,p)) \tag{4.5}$$

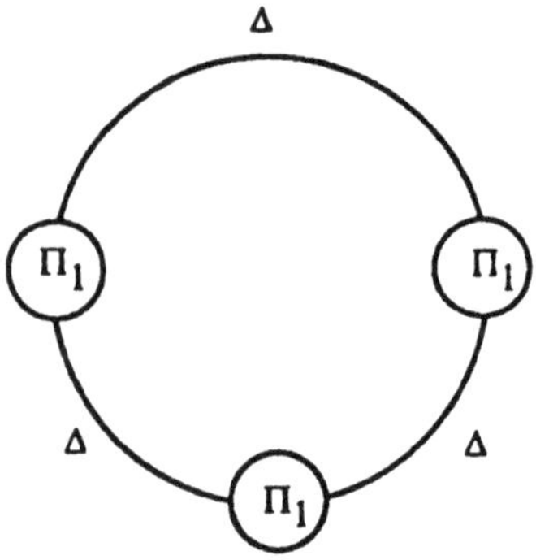

Fig. 16. An example of ring diagram contributing to order g^3 to the partition function.

The summation over N begins at $N = 2$, because the graph of fig.17 has already been taken into account at first order ; $\Pi_1 = g^2T^2/24$ is the first order temperature-dependent self-energy, and the explicit expression of (4.5) is, with $\lambda = g^2/24$

$$-\frac{1}{2}\beta VT\sum_n \int \frac{\mathrm{d}^3p}{(2\pi)^3}\left(\ln(1+\frac{\lambda T^2}{\omega_n^2+p^2}) - \frac{\lambda T^2}{\omega_n^2+p^2}\right) \tag{4.6}$$

The mode $\omega_n = 0$ is easily seen to give a contribution proportional to $\lambda^{3/2}$, corresponding again to a mild violation of perturbation theory. From the explicit evaluation of (4.6) and of the graph in fig.17, one obtains for the pressure[35]

Fig. 17. The graph contributing to order g^2 to the partition function.

$$P = \frac{\pi^2T^4}{90}[1 - \frac{15}{8}(\frac{\lambda}{\pi^2}) + \frac{15}{2}(\frac{\lambda}{\pi^2})^{3/2} + ...] \tag{4.7}$$

The first term corresponds to the free boson gas, the second one to the graph of fig.17, and the last one to the ring diagrams.

In φ^4-theory, this use of a dressed propagator is all we have to do in order to deal with the IR behaviour. Indeed the dominant correction at one-loop order is the tadpole graph of fig.5 which behaves as T^2. The one-loop correction to the coupling constant is only logarithmically divergent, and, as we shall see below, its thermal part is in fact proportional to $\ln T$.

4.2 Remarks on the One-Loop Correction to the Propagator

Before going to QCD, it is useful to examine in some detail a one-loop graph more complicated than a tadpole. Still in the case of a scalar theory, we consider a $\varphi\varphi_1\varphi_2$ coupling and the one-loop correction to the φ-propagator : see fig.7, which also defines the kinematics. From the Feynman rules and the techniques described in section 2, we get[37]

$$\Pi = g^2 \int \frac{d^3k}{(2\pi)^3} \frac{1}{4E_1E_2} \left(\frac{1+n_1+n_2}{p_0 - E_1 - E_2} + \frac{n_2 - n_1}{p_0 - E_1 + E_2} + 1 \leftrightarrow 2 - \frac{1+n_1+n_2}{p_0 + E_1 + E_2} \right) \quad (4.8)$$

where :

$$E_1 = E_k = k \ ; \ E_2 = E_{|p-k|} = |p-k| \quad (4.9)$$

while n_1 and n_2 are the corresponding Bose-Einstein factors. The physical interpretation of the various terms in Π is made clearer if one looks at the discontinuity of Π; since Π is the continuation to real-time of the imaginary-time self-energy, it is fact a retarded Green's function, and p_0 in (4.9) should be interpreted as $p_0 + i\epsilon$. Thus

$$\begin{aligned} \text{Disc } \Pi = -2i\pi g^2 \int \frac{d^3k}{(2\pi)^3} \frac{1}{4E_1E_2} &[(1+n_1+n_2)\delta(p_0 - E_1 - E_2) \\ &+ (n_1 - n_2)\delta(p_0 - E_1 + E_2) + 1 \leftrightarrow 2 - (1+n_1+n_2)\delta(p_0 + E_1 + E_2)] \end{aligned} \quad (4.10)$$

Let us look at the Bose-Einstein factor $1+n_1+n_2$ in the first term of (4.10). The discontinuity of Π is proportional to the difference (the sum for fermions)

$$\Gamma = \Gamma_d - \Gamma_c \quad (4.11)$$

between the decay and creation rates of the particle φ; the decay process, for $p_0 = E_1 + E_2$, is $\varphi \rightarrow \varphi_1 + \varphi_2$, and the creation process is the inverse process $\varphi_1 + \varphi_2 \rightarrow \varphi$. The Bose-Einstein factor is thus :

$$(1+n_1)(1+n_2) - n_1n_2 = 1 + n_1 + n_2 \quad (4.12)$$

At zero temperature, one has of course $\Gamma_c = 0$ and Γ_d is proportional to the factor of one in (4.12). Similarly the second term in (4.10), with $p_0 = E_1 - E_2$, corresponds to the reactions $\varphi_1 \rightarrow \varphi + \overline{\varphi}_2$ (creation) and $\overline{\varphi}_2\varphi \rightarrow \varphi_1$ (decay), the Bose-Einstein factor being

$$n_2(1+n_1) - n_1(1+n_2) = n_2 - n_1 \quad (4.13)$$

In all cases we have of course the detailed balance condition

$$\Gamma_d = e^{\beta p_0} \Gamma_c \quad (4.14)$$

In the case of a small deviation from thermal equilibrium, Γ should be interpreted as the inverse of the relaxation time which governs the approach to equilibrium[37].

The terms where $p_0 = E_1 - E_2$ or $p_0 = E_2 - E_1$ correspond to the so-called "Landau damping" mechanism: particles disappear or are created through scattering in the bath, and not via the processes which are avaible at zero temperature.

We are going to be interested in a situation where external particles are soft: p_0, $p \sim gT \ll T$. Momenta of order T are called hard momenta : since the T-dependent part of loop integrations is limited by the thermal Bose-Einstein or Fermi-Dirac factors, these are typical momenta running around a loop. We need kinematical approximations for loop momenta $k \sim T$ and external momental $p \sim gT \ll T$. Take as an example the expression (4.8) ; we have

$$\begin{aligned} E_1 = k \quad &; \quad E_2 = |p-k| \simeq k - p\cos\theta \\ p_0 - E_1 - E_2 \simeq -2k \quad &; \quad p_0 - E_1 + E_2 \simeq p_0 - p\cos\theta \\ n_1 \simeq n_2 &\simeq \frac{1}{e^{\beta k} - 1} \sim 1 \\ n_1 - n_2 \simeq p\cos\theta \, \frac{dn}{dk} &= -\beta p \cos\theta \, n(k)(1+n(k) \sim -\beta p\cos\theta \end{aligned} \quad (4.15)$$

The last approximations for n_1, n_2 and $n_1 - n_2$ are to be used for power counting only. Using these approximations, it is clear that the first and last term of (4.8) diverge as $\ln T$, while the second and third term behave as a constant.

4.3 Hard Thermal Loops in QCD and the Effective Expansion

In what follows we shall call hard thermal loops (HTL) one-loop integrals which diverge as T^2[38],[39]. The simplest example is of course the tadpole in $g^2\varphi^4$. From naïve power counting one could imagine that the only HTL correspond to loops which diverge quadratically when $T = 0$. Thus the gluon propagator for instance should have HTL. However this reasoning misses most of the HTL : even diagrams which are UV-convergent, such as the N-gluons Green's function with $N \geq 5$ do have HTL. Except in the case $N = 2$, HTL come uniquely from terms analogous to the 2nd and 3rd term in (4.8). In (4.8), one can see that one power of k is gained for soft external momenta since

$$p_0 - E_1 + E_2 \simeq p_0 - p\cos\theta$$

Thus the integral diverges linearly, and not logarithmically.

As we just mentioned, the case of the gluon propagator is special because one finds quadratic divergences at $T = 0$. The 4-gluon vertex will contribute via the tadpole graph, and there will be a further term coming from the combination $(n_1 + n_2)$ in (4.8). This term is tadpole like, in the sense that its imaginary part is zero at order T^2. On the other hand there will be terms corresponding to the combination $n_1 - n_2$: since

$$\int_{-1}^{1} \frac{\mathrm{d}(\cos\theta)}{p_0 - p\cos\theta} = \frac{1}{p}\ln\frac{p_0 - p}{p_0 + p} \tag{4.16}$$

these terms will have an imaginary part of order T^2 in the region $-p \leq p_0 \leq p$, i.e. below the light cone. This property is typical of Landau damping.

The quark propagator is linearly divergent at $T = 0$; thus the terms of order T^2 which we found in subsection 3.1 were typical of Landau damping. Let us now examine the one-loop correction Γ_1 to the 3-gluon vertex (fig.18).

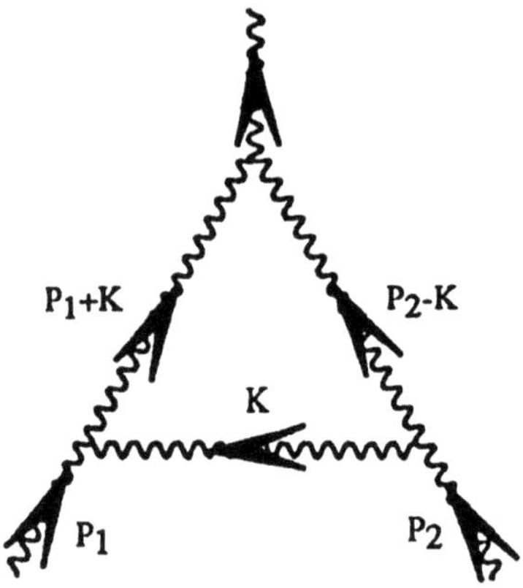

Fig. 18. The one-loop coorection to the three-gluon vertex (wavy lines: gluons).

Since the external momenta are supposed to be soft, the 3-gluon coupling can be simplified, by retaining only terms proportional to K^μ. After summation over n, the leading contribution to Γ_1 is given by a sum of terms of the form :

$$\Gamma_1 \sim g^3 \int \frac{\mathrm{d}^3k}{(2\pi)^3} \frac{K^{\mu_1}K^{\mu_2}K^{\mu_3}}{E_k E_{p_1+k} E_{p_2-k}} (n(E_k) - n(E_{p_1+k}))$$

$$\frac{1}{(p_{10} - E_k + E_{p_1+k})(p_{20} - E_k + E_{p_2-k})} \tag{4.17}$$

Power counting is very simple (see (4.15))

$$\Gamma_1 \sim g^3 T^3 \, \frac{T^3}{T^3} \, \frac{P}{T} \, \frac{1}{P^2} \sim g^2 T$$

Since $\Gamma_0 \sim g(gT) = g^2T$, we see that the HTL is of the same order of magnitude as the zero-loop approximation. This means for example that the 2-loop correction to the gluon propagator in fig.19 is of the same order as the one-loop correction, and the effective expansion will use an effective vertex in order to sum Γ_0 and Γ_1.

Fig. 19. An effective vertex.

Power counting for HTL has been established in[38]

- $\int d^3k \rightarrow T^3$
- 1st propagator $\rightarrow (T\sum_n) : T^{-1}$
- each additional propagator : $(PT)^{-1}$
- powers of K^μ : T
- Statistical factor : P/T

Hard thermal loops enjoy remarkable properties[38]-[41]

(i) HTL exist only for Green's functions with N external gluons lines or $(N-2)$ external gluon and 2 external quark lines. There are no HTL with external ghost lines.

(ii) HTL are totally symmetric with respect to external gluon indices; the colour factors are $(N_c + N_f/2)$ for HTL with external gluons only and C_F for HTL with two external quark lines.

(iii) HTL do not involve the 4-gluon vertex in the Feynman and Coulomb gauges, except in the case of the gluon propagator (contribution to the tadpole)

(iv) HTL are gauge independent: at least it has been shown in [38] and [39] that they are the same in general covariant, Coulomb and axial gauges; a general proof has been given in[34].

(v) HTL obey Ward identies analogous to those of QED; calling $\delta\Pi$ etc. the HTL correction and ignoring the colour factors for notational simplicity we have :

$$\text{propagator} \;: P^\mu \delta\Pi_{\mu\nu}(P) = 0 \tag{4.18}$$

$$3-\text{gluon vertex} \;: K^\mu \delta\Gamma_{\mu\nu\rho}(K,P,Q) = \delta\Pi_{\nu\rho}(P) - \delta\Pi_{\nu\rho}(Q) \tag{4.19}$$

$$\text{quark gluon vertex} \;: K^\mu \delta\tilde{\Gamma}_\mu(K,P,Q) = i\delta\Sigma(P) - i\delta\Sigma(Q) \tag{4.20}$$

where Σ is the quark self-energy, Γ the 3-gluon vertex and $\tilde{\Gamma}$ the quark-gluon vertex. In general, K^μ dotted into a N-point HTL gives a combination of $(N-1)$-point HTL.

From our study of HTL we can establish the rules for an effective expansion which gives consistently the lowest order term in the coupling constant for soft external momenta. The effective expansion is similar to the ordinary perturbation theory, except that for soft external momenta bare propagators and vertices are replaced by propagators and vertices which include the contribution of HTL. Topologically many diagrams are the same as in the bare expansion, but one may also find diagrams built out of effective vertices which do not have bare counterparts. The effective propagators resum all insertions of HTL: $\delta\Pi$ for gluons and $\delta\Sigma$ for quarks. The effective vertices are formed by adding HTL to the bare vertices: $\Gamma \to \Gamma+\delta\Gamma$.

4.4 The Plasmon Damping Rate

The gluon inverse propagator is written

$$iD^{-1}_{\mu\nu} = P^2 g_{\mu\nu} - P_\mu P_\nu + G.F.T - \delta\Pi_{\mu\nu} - \Pi^*_{\mu\nu} \tag{4.21}$$

where the first three terms correspond to the bare inverse propagator, $\delta\Pi_{\mu\nu}$ is the HTL correction and $\Pi^*_{\mu\nu}$ is given by the effective expansion (fig.20).

Fig. 20. Effective expansion graphs for the plasmon damping rate (dotted lines: ghosts; black blobs: effective vertices and propagators).

One notes that $\delta\Pi_{\mu\nu}$ does not contribute to the damping rate, because Im$\delta\Pi$ = 0 "on mass-shell", from energy-momentum conservation : a plasmon of mass ω_P cannot decay into two plasmons of mass ω_P. Thus the damping rate will be down by one power of g with respect to the HTL, namely

$$\Pi^* \sim g(gT)^2 \quad ; \quad \mathrm{Im}\Pi^* \neq 0 \tag{4.22}$$

To leading order the plasmon polarisation $e_\mu(P)$ obeys :

$$(D^{-1}_{F,\mu\nu} - \delta\Pi_{\mu\nu})e^\nu(P) = 0 \tag{4.23}$$

and

$$P^\mu e_\mu(P) = 0 \quad \text{(covariant gauge)}$$

$$p_i e_i(P) = 0 \quad \text{(Coulomb gauge)}$$

The gauge invariant decay rate is given by

$$\mathrm{Im} e^\mu(P)\Pi^*_{\mu\nu}e^\nu(P)$$

in a covariant gauge, and by a similar expression in the Coulomb gauge. The calculation is done by using the spectral representation (2.75) of propagators

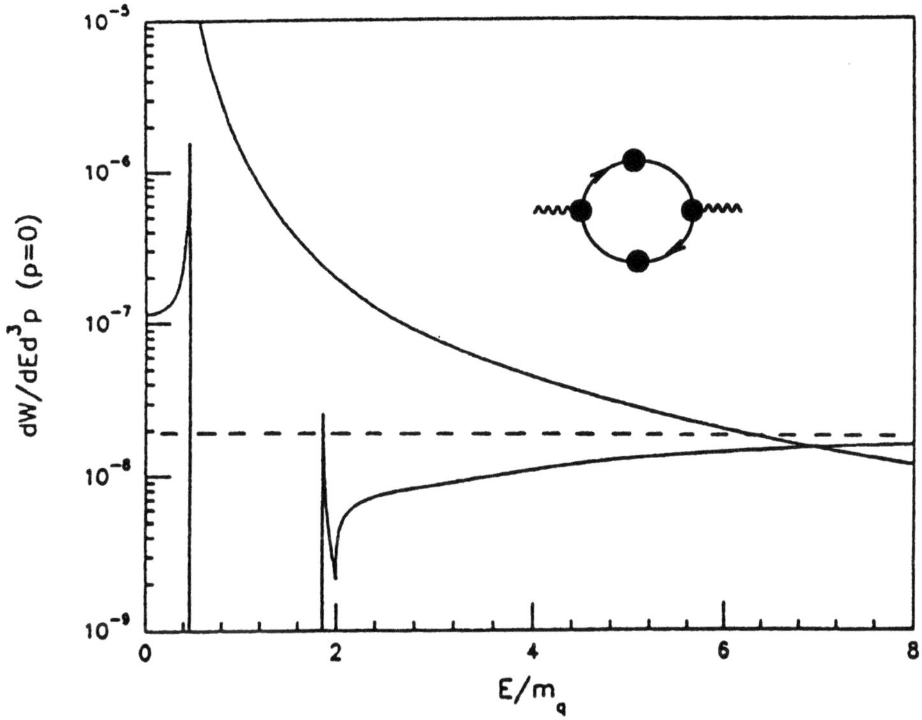

Fig. 21. The cross-section for lepton-pair production (wavy lines: external photons).

$$\rho(k_0, k) = Z(k)\left(\delta(k_0 - \omega(k)) - \delta(k_0 + \omega(k))\right) + \beta(k_0, k)\theta(k^2 - k_0^2) \tag{4.24}$$

and the result for the damping rate γ (cf. (3.55)) is[42]

$$\gamma = a\,\frac{g^2 N_c T}{24\pi} \quad ; \quad a \simeq 6.63 \tag{4.25}$$

the result being the same in a general covariant and in the Coulomb gauge. The factor of a in (4.25) is given by a rather long formula

$$\begin{aligned} a = 9 \int_0^{\infty} \mathrm{d}k \int_{-\infty}^{+\infty} \frac{\mathrm{d}k_0}{k_0} \int_{-\infty}^{+\infty} \frac{\mathrm{d}k_0'}{k_0'} \delta(\omega_P - k_0 - k_0') \\ (2(k^2 + k_0 k_0')^2 \rho_T(k_0, k)\rho_L(k_0', k) + k^4 \rho_L(k_0, k)\rho_L(k_0', k) - \\ -\frac{k^2}{\omega_P^2}(k^2 - k_0^2)^2 \rho_T(k_0, k)\rho_L(k_0', k) + \frac{k_0}{6k^3}(k^2 - k_0^2)^2 \theta(k^2 - k_0^2)\rho_T(k_0', k)) \end{aligned} \tag{4.26}$$

where ρ_T and ρ_L represent the spectral functions of the transverse and longitudinal gluons respectively.

4.5 Soft Photon Production in a Plasma

The same techniques have been used in order to compute the production rate of lepton pairs, via virtual photons in a plasma, when the virtual photons have a mass$\sim gT$. The result (fig.21) has a number of interesting features, due to the opening of various thresholds for $\chi = +1$ and $\chi = -1$ fermionic quasi-particles (see subsection 3.1). In the low-momentum region, the result is larger by several order of magnitudes than the naïve Drell-Yan expectation.

5. Some Other Problems Related to Infrared Behaviour

This last section will be devoted to three problems which are not a priori directly related; their common feature is to depend crucially on the IR behaviour of thermal field theories. We shall first examine a difficulty pointed out by Linde[44] more than ten years ago, and which is not yet well understood; then we shall turn to a recent estimate of transport cross-sections and finally to a generalization at $T \neq 0$ of the Kinoshita-Lee-Nauenberg (KLN) theorem[45]. Before going to these topics, let us discuss briefly the behaviour of non-leading terms in the gluon polarization tensor, in bare (not resummed) perturbation theory, in the limit $k_0 = 0,\ k \to 0$. The longitudinal part $F = -\Pi_{00}$ behaves in a pure $SU(N_c)$ gauge theory as[32]

$$F(k_0 = 0,\ k \to 0) = \frac{1}{3}g^2 N_c T^2 - \frac{1}{4}g^2 N_c T k + \dots \tag{5.1}$$

The first term is of course the leading term which was examined at length in section 3, and it gives rise to the Debye screening length $r_D = m_{el}^{-1}$. When one calculates the contribution of ring diagrams to the partition function Z, this term gives rise to terms of order g^3 in Z. It is thus certainly gauge independent. Similarly the second term in (5.1) gives rise to terms of order $g^2 \ln g^2$ and must also be gauge independent. Although higher order contributions to Z have not been computed yet, it should be noticed that one does not expect these higher order contributions to be independent of the gauge parameter and more generally of the gauge adopted in the calculation. Indeed the dependence with respect to the gauge parameter can be absorbed in the gauge dependence of the coupling constant. Calculations in various gauges confirm the gauge independence of the first two terms of the expansion in (5.1).

The real difficulties appear when one looks at the transverse part $G = \frac{1}{2}\Pi_{ii}$ of the gluon polarization tensor. In TAG one finds[32]

$$G(k_0 = 0, k \to 0) = -\frac{5}{16}g^2 N_c T k + \dots \tag{5.2}$$

while in a general covariant gauge with gauge parameter ρ the result is

$$G(k_0 = 0, k \to 0) = -\frac{1}{64}(8 + (1+\rho)^2) g^2 N_c T k + \dots \tag{5.2}$$

Clearly, even the first term in the expansion of G is not gauge independent! But whatever the gauge one does find that $G(k_0 = 0, k)$ vanishes in the limit $k = 0$, which means that <u>static</u> (chromo)magnetic fields are not screened: there is no magnetic mass, at least to first order of perturbation theory. It has been argued that the magnetic mass, if it is nonzero, cannot be computed in perturbation theory[46]; from dimensional arguments one may try to infer that m_{mag} should be related to the mass gap in the $D = 3$ theory, and should thus be of order $g^2 T$. Although the arguments sound plausible, this result cannot be considered as being firmly established.

5.1 Linde's Argument

Consider the $(l+1)$-loop graph of fig.22 giving a contribution Z_l to the partition function, and let us try to evaluate its leading infrared behaviour. We write each propagator, including the summation over Matsubara frequencies as

$$\sum_n \frac{1}{\omega_n^2 + k^2 + m^2} \quad ; \quad \omega_n = \frac{2\pi n}{\beta} \tag{5.4}$$

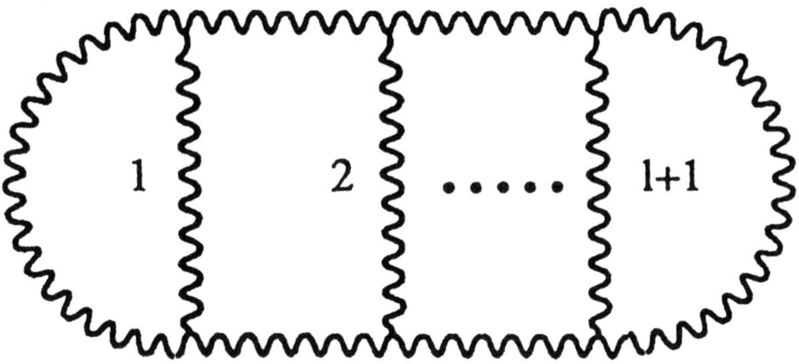

Fig. 22. The graph considered by Linde.

where m is some screening mass. The leading IR divergences occur for $n = 0$, since ω_n for $n \neq 0$ acts as an infrared cutoff. The leading IR behaviour can then be estimated from power counting

$$g^{2l}(T \int \mathrm{d}^3 k)^{l+1} k^{2l}(k^2 + m^2)^{-3l} \tag{5.5}$$

The various factors in (5.5) are obvious

- vertices lead to $g^{2l}k^{2l}$
- propagators to $(k^2 + m^2)^{-3l}$
- loop integrations to $(T \int \mathrm{d}^3 k)^{l+1}$

For $l < 3$, the behaviour is IR regular, while divergences occur for $l \geq 3$

$$l = 3 \qquad g^6 T^4 \ln \frac{T}{m} \tag{5.6a}$$

$$l > 3 \quad g^6 T^4 (\frac{g^2 T}{m})^{l-3} \tag{5.6b}$$

For longitudinal gluons, the screening mass is the electric mass $m_{el} \sim gT$, and thus for $l > 3$:

$$Z_l \sim g^{l+3} T^4 \tag{5.7}$$

Thus, instead of the expected g^{2l} behaviour of ordinary perturbation theory, one finds a g^{l+3} behaviour. This is reminiscent of what occurs in ring diagrams, and corresponds to what had been called in section 4 a mild breakdown of perturbation theory. However if we assume a magnetic mass $m_{mag} \sim g^2 T$, the situation is much worse for transverse gluons

$$Z_l \sim g^6 T^4 \tag{5.8}$$

independently of l. For $l > 3$, all graphs are of order g^6, independently of the order of perturbation theory[44]. This is of course a clear failure of perturbation theory!

One does not exactly know how to interpret this difficulty. It could be that power counting is misleading and that some cancellation occurs which invalidates the argument. It could be that the result is true, but it does not affect physics in an essential way since the phenomenon occurs at a rather high order of perturbation theory (remember that even the term of order g^4 has yet to be computed!). The most pessimistic view would be of course that perturbation theory is completely unreliable and should be abandoned. It is worth noticing that the absence of magnetic screening for static magnetic fields occurs also in QED; however no problem arises in that case since a photon can only give a $e^+ - e^-$ pair, and for fermions n can never vanish (see(2.41)): thus there is no infrared problem.

5.2 Screening and Transport Phenomena

Let us now turn to a completely different topic. G. Baym et al.[47] made recently an important observation on the computation of transport cross-sections in a quark-gluon plasma[48]: despite the absence of screening of static magnetic fields, Landau damping provides a cutoff for the exchange of transverse IR gluons, and this suffices to yield finite transport cross-sections. For simplicity the following discussion will be again held in the case of a pure $SU(3)$ gauge theory, and the reader is referred to the original article for the extension to the case where quarks are included. When computing the gluon-gluon differential cross-section one encounters at small angles the standard Rutherford-like behaviour

$$\frac{d\sigma}{d\Omega} \sim \frac{1}{\sin^4\theta} \tag{5.9}$$

The singular behaviour at $\theta = 0$ does not allow to compute the total cross-section σ which is needed for example to evaluate the mean-free path $\lambda = 1/(n\sigma)$. One usually takes the 90^0 cross-section $\sigma \sim (d\sigma/d\Omega)_{\theta=\pi/2}$ in order to get a rough estimate of λ[49], but this is clearly not entirely satisfactory. Fortunately when one wants to compute transport cross-sections, $d\sigma/d\Omega$ is weighted by a factor $(1-\cos\theta)$; this means physically that small angle cross-sections are very inefficient for exchanging momentum. Then the transport cross-section is only logarithmically divergent

$$\sigma_{tr} \sim \int d\Omega \frac{d\sigma}{d\Omega}(1-\cos\theta) \sim \int \frac{d\theta}{\theta} \tag{5.10}$$

The exchange of longitudinal gluons does not lead to any problem because of Debye screening; in this subsection we denote the Debye wave number by q_D ($q_D = m_{el}$), in order to follow the notations of [47]. The exchange of transverse gluons is usually neglected in non-relativistic plasmas, where magnetic effects are small. On the contrary magnetic effects are a priori as important as electric ones in an ultra-relativistic plasma.

Let us describe briefly the approach followed by the authors of [47]. They start from the Boltzmann equation which involves a $2 \to 2$ elastic reaction

$$p_1 + p_2 \to p_3 + p_4$$

The kinematics is defined in fig.23; one has

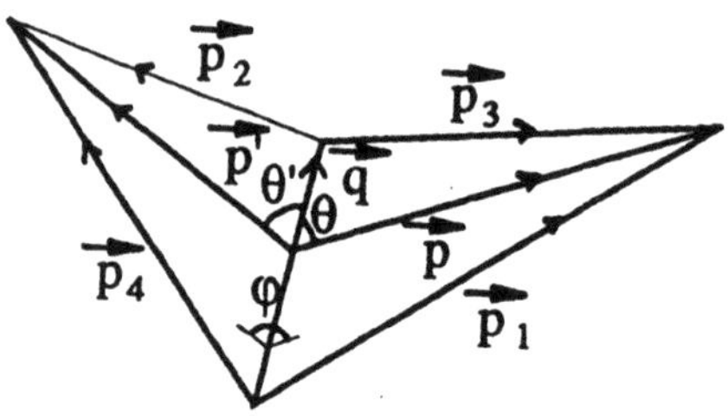

Fig. 23. Kinematics of the two-body collision.

$$\begin{aligned} p_1 &= p + q/2 \quad & p_3 &= p - q/2 \\ p_2 &= p' + q/2 \quad & p_4 &= p' - q/2 \end{aligned} \tag{5.11}$$

and φ denotes the angle between the planes (p_1, p_3) and (p_2, p_4). For small angle scattering the energy transfer is

$$\omega \simeq \hat{p}.q \simeq \hat{p}'.q \tag{5.12}$$

while the matrix element squared reads in bare perturbation theory

$$|M|^2 \sim \frac{1}{t^2} = \frac{1}{\omega^2 - q^2} \tag{5.13}$$

Setting $\mu = \omega/q$, we see that possible IR divergences arise from the region $\mu \to 0$, where magnetic fields are not screened. Separating out transverse and longitudinal exchanges, and including self-energy corrections in the propagator yields

$$|M|^2 \sim |\frac{1}{q^2 + q_D^2 \chi_L(\mu)} - \frac{(1-\mu^2)\cos\varphi}{q^2(1-\mu^2) + q_D^2 \chi_T(\mu)}|^2 \tag{5.14}$$

where to one-loop order we have from (3.41), recalling that $F = (K^2/k^2)\Pi_{00}$

$$\chi_L(\mu) = 1 - \frac{\mu}{2}\ln(\frac{\mu+1}{\mu-1}) \tag{5.15a}$$

$$\chi_T(\mu) = \frac{1}{2}\mu^2 + \frac{\mu(1-\mu^2)}{4}\ln(\frac{\mu+1}{\mu-1}) \tag{5.15b}$$

Note that $\chi_T(0) = 0$ corresponds of course to the absence of screening for static magnetic fields. Now, in the region $\mu \to 0$, $\chi_T(\mu)$ is non zero because of Landau damping; in fact it is pure imaginary and from (5.15b)

$$\chi_T \simeq -\frac{i\pi\mu}{4} \tag{5.16}$$

which leads to an ω-dependent screening length. The cross-section involved in the Boltzmann equation is of the form

$$\int \mathrm{d}\mu h(\mu) \int \mathrm{d}q\ q^3 |M|^2 \tag{5.17}$$

where $h(\mu)$ is regular as $\mu \to 0$. In the absence of screening, one would find a logarithmic divergence since $|M|^2 \sim q^{-4}$. Longitudinal exchange is of course screened, and one gets a factor

$$\ln\frac{q_{max}}{q_D} \sim \ln\frac{T}{q_D} \simeq \ln\frac{1}{g} \tag{5.18}$$

but one is still left with transverse exchange and the interference term. We shall look at the former case only, as the treatment of the interference term is quite similar. The integral to be computed is

$$\int \frac{q^3 \mathrm{d}q}{|q^2 - i\pi q_D^2 \mu/4|^2} \sim \int_{q_D\mu}^{q_{max}} \frac{\mathrm{d}q}{q} \sim \ln\frac{T}{q_D \mu} = \ln\frac{T}{q_D} - \ln\mu \tag{5.19}$$

Since the integral over μ in (5.17) is regular at $\mu = 0$, the factor of $\ln\mu$ does not lead to any singular behaviour when $\mu \to 0$, and one recovers a $\ln(T/q_D)$ screening factor. This factor is not affected by the possible existence of a magnetic mass. Thus, despite the absence of screening of static magnetic fields, transverse gluon exchange is effectively cut off by the Debye wave-number q_D. Variational estimates can be obtained for the shear viscosity η[47]

$$\eta \simeq 0.34 \frac{T^3}{\alpha_S^2 \ln 1/\alpha_S} \tag{5.20}$$

and for the relaxation time τ

$$\tau \simeq \frac{0.25}{T}\frac{1}{\alpha_S^2 \ln 1/\alpha_S} \tag{5.21}$$

Taking $T \simeq$200 MeV and $\alpha_S \simeq 0.3$ one finds $\tau \simeq$2.5 fm. In the realistic case of a quark-gluon plasma this time is reduced below 1 fm because other scattering processes give also important contributions.

5.3 The Kinoshita-Lee-Nauenberg Theorem at Non-Zero Temperature

Our last topic will be the $T \neq 0$ generalization of the Kinoshita-Lee-Nauenberg theorem[45]. Recall that at $T = 0$ the KLN theorem allows for example to state that the total decay rate of some massive particle remains finite order by order in perturbation theory, even if there are massless particles in the final state. A standard example is the total annihilation cross section $e^+ - e^- \rightarrow$hadrons (which can of course be viewed as the total decay rate of a heavy virtual photon into hadrons): in perturbative QCD, one computes the decay rate into quarks and gluons, and one finds a finite result despite the fact that gluons are massless. In fact one can even consider the case of massless quarks and still get a finite result. On the contrary the total cross-section for other processes, for instance the Drell-Yan process (lepton pair production in hadronic collisions) is not a priori finite in perturbative QCD: the singularities which occur in the perturbative calculation have to be absorbed via the factorization theorem into hadronic structure functions. One may wonder what happens at non-zero T if one tries for example to compute the total rate for muon-pair production in a quark-gluon plasma (one of the possible signals of this plasma). It is reasonable to assume that the weakly interacting muons do not have time to interact with the plasma once they have been produced, and they are not thermalized. Thus there is a second scale in the problem, in addition to the temperature, namely the invariant mass Q of the muon-pair. Of course this mass cannot be much larger than T: otherwise the production rate would be negligible, since muon pairs arise from the annihilation of quarks and antiquarks in the thermal bath. At finite T one always expects cross-sections to be finite because of screening. However if screening is responsible for the finiteness of cross-sections, one will find logarithms like $\ln^p(Q/gT)$ or even powers like $(Q/gT)^p$ or $(Q/g^2T)^p$. Then the cross-section is presumably not computable in perturbation theory. On the contrary if there are no singularities in (bare) perturbation theory, then such terms are absent and everything is computable in perturbation theory. A proof of the absence of singularities in the calculation of rates would constitute the required generalization of the KLN theorem.

Let us recall the origin of divergences: they arise from the degeneracy of a one-quark state with a one-quark + one-gluon state, in some special configurations: the gluon must be either collinear to the quark or must have a vanishingly small energy. The first case gives rise to a collinear divergence, the second case to an infrared divergence. One expects the $T \neq 0$ case to be more infrared singular than the $T = 0$ case, because of the Bose-Einstein factor which is featured in the real-time gluon propagator

$$\frac{1}{e^{\beta\omega} - 1} \simeq \frac{1}{\beta\omega} \tag{5.22}$$

Since we are now interested in computing rates, we need a $T \neq 0$ generalization of the $T = 0$ Landau-Cutkosky rules. Recall that at $T = 0$ one gets from the Feynman propagator the cut propagator

$$\frac{i}{p^2 - m^2 + i\epsilon} \rightarrow 2\pi\theta(p_0)\delta(p^2 - m^2) \tag{5.23}$$

One important feature of (5.23) is the θ-function, which implies that energy flows in one direction only. The $T \neq 0$ generalization of the cut propagator reads[17]

$$D_+(p) = 2\pi\left(\theta(p_0) + n_B(p_0)\right)\delta(p^2 - m^2) \tag{5.24}$$

For $p_0 > 0$ we get a factor $(1 + n_B(p_0))$, corresponding to a final particle in the bath, while for $p_0 < 0$ we get a factor $n_B(p_0)$, corresponding to an initial particle in the bath. Thus at $T \neq 0$ energy can flow in both directions.

Before going to the realistic case of QCD, let us try to understand the mechanism for the cancellation of singularities in a simpler case; a useful toy model is a massless scalar field theory with a cubic coupling in dimension $D = 6$[50], namely the dimension where this theory is renormalizable. On the other hand this theory, although asymptotically free, is very different from QCD on an important point: it displays collinear divergences only, since it is not a gauge theory. Because of this property, to lowest non-trivial order of perturbation theory, the only divergent graphs correspond to the topology of a self-energy insertion(fig.24). At $T = 0$ cut (a), corresponding to three particles in the final state, is singular in the configuration where the three particles have collinear momenta with values $(q/2, qz/2, q(1-z)/2)$ where z is a scaling variable $(0 \leq z \leq 1)$. If one uses a dimensional regularization $D = 6 + 2\varepsilon$ of the divergences, the singular contribution $\Gamma^{(a)}$ of the corresponding cut can be shown to be given by

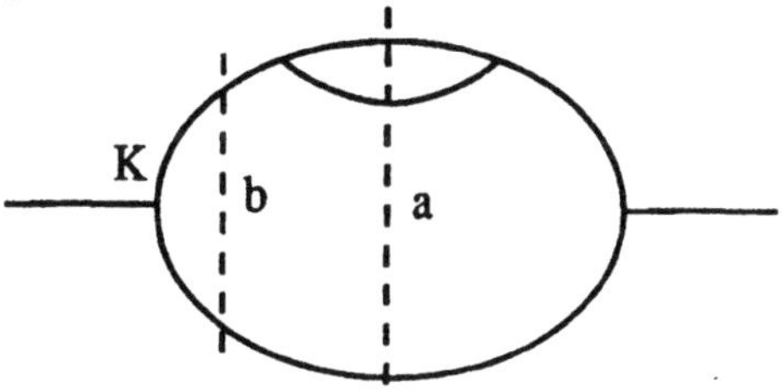

Fig. 24. Unitarity cuts in the proof of the KLN theorem.

$$\Gamma^{(a)}_{T=0} = \Gamma_B \frac{\alpha_S}{\varepsilon} \int_0^1 \mathrm{d}z\; z(1-z) \tag{5.25}$$

where Γ_B is the zeroth-loop order (Born approximation), $\alpha_S = g^2/(4\pi)^3$ and $z(1-z)$ is the Altarelli-Parisi kernel of the $g\varphi^3$ theory. It is easy to check that the singular contribution from cut (b) just cancels the singularity of (5.25) when $\varepsilon \to 0$.

At non-zero-T, one can write without any difficulty the generalization of (5.25) in the case of three final particles. It suffices to introduce Bose-Einstein factors for final particles: the singular contribution from emission (E) of three final particles is

$$\Gamma^{(a)}_E = \Gamma_B \frac{\alpha_S}{\varepsilon} \int_0^1 \mathrm{d}z\; z(1-z)(1 + n_B(\tfrac{q}{2}))(1 + n_B(\tfrac{qz}{2}))(1 + n_B(\tfrac{q(1-z)}{2})) \tag{5.26}$$

The product of the three Bose-Einstein factors in (5.26) can be rewritten as

$$(1 + n_B(\tfrac{q}{2}))^2(1 + n_B(\tfrac{qz}{2}) + n_B(\tfrac{q(1-z)}{2})) \tag{5.27}$$

One must not forget the absorption (A) of particles from the bath: the singular (collinear) configuration is then $(q(z-1)/2, -q/2, qz/2)$ where the first value corresponds to the initial particle and the last two values to final particles, while the scaling variable z now lies in the range $[1, \infty[$. The corresponding singular contribution reads

$$\Gamma_A^{(a)} = \Gamma_B \frac{2\alpha_S}{\varepsilon} \int_1^\infty dz\ z(z-1)(1+n_B(\frac{q}{2}))(1+n_B(\frac{qz}{2}))n_B(\frac{q(z-1)}{2}) \tag{5.28}$$

Again one can find for the three Bose-Einstein factors in (5.28) a rearrangement similar to that of (5.27). The total contribution from cut (a) is then $\Gamma^{(a)} = \Gamma_E^{(a)} + \Gamma_A^{(a)}$. The singular contribution from cut (b), which corresponds to a self-energy insertion on an external line, is

$$\Gamma^{(b)} = -\Gamma_B \frac{\alpha_S}{\varepsilon}(1+n_B(\frac{q}{2}))^2 \left(\frac{1}{6} + 4\int_0^\infty dz\ z n_B(\frac{qz}{2})\right) \tag{5.29}$$

A simple manipulation shows that the sum $\Gamma^{(a)} + \Gamma^{(b)}$ vanishes, thus establishing that the total rate is non-singular. In this simple case, one clearly sees that the mechanism for singularity cancellation is copied from $T = 0$: the only difference is that the Altarelli-Parisi kernel is now weighted by thermal factors.

Things are more involved in QCD, because of IR divergences[51]-[54]. Power-law singularities arise because of the Bose-Einstein factor; if one regulates the divergences by giving a mass λ to the gluon (although this regularization is not really correct, even at the one-loop level, the physics is clearer), one finds singularities of the kind

$$\frac{1}{\beta\lambda} \quad \text{and} \quad \frac{1}{\beta\lambda}\ln\frac{Q^2}{\lambda^2} \tag{5.30}$$

However these singularities cancel out when one sums over all possible cuts and the rate is finite, at least to this one-loop order. The rate can be written as

$$\Gamma = \Gamma_B\left(1 + \frac{\alpha_S C_F}{\pi}(\frac{3}{4} + F(\frac{Q}{T}))\right) \tag{5.31}$$

where $3/4$ is the $T = 0$ result (recall the $O(\alpha_S)$ correction to the ratio R in $e^+ - e^-$ annihilation), and the function $F(Q/T)$ has been explicitly computed[52],[54].

The verification of the KLN theorem has been pushed to the next order of perturbation theory, in the framework of the $g\varphi^3$ toy model described previously, by studying the most singular topology drawn in fig.25[55].

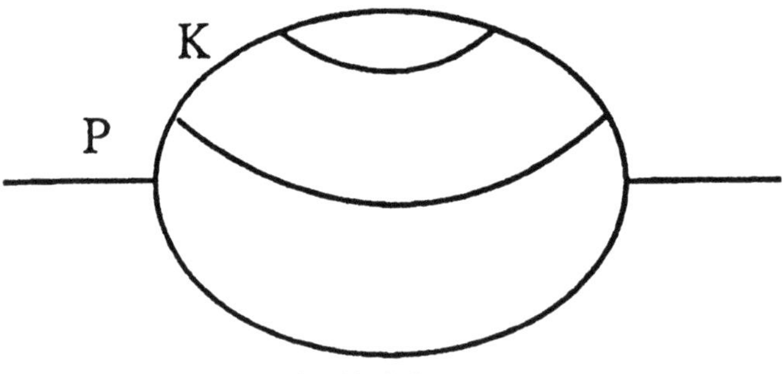

Fig. 25. The topology considered in [55].

New interesting features, which could not be seen at the one-loop order, arise at this order of the perturbative expansion. Indeed, when a self-energy graph is inserted in a line of four-momentum K (fig.25), one finds a T^2/k^2 behaviour when $k \to 0$. This behaviour, which is related to Landau damping, induces severe infrared divergences in a theory which at $T = 0$ displays only collinear divergences. In spite of these divergences, the KLN theorem still seems to hold true: using a $D = 6 + 2\varepsilon$ regularization scheme, one finds that the IR (power law) singularities cancel out, as well as the ε^{-3} and ε^{-2} singularities (note that at $T = 0$ the strongest singularities are in ε^{-2} for the topology studied in [55]). Due to the complexity of the calculation it has not been possible to check the cancellation of the ε^{-1} singularities (and to study other topologies) but there does not seem to be any reason against this cancellation. There are thus good indications for the general validity of the KLN theorem at finite temperature (see also[56]).

Conclusions and Outlook

We have seen that the quark-gluon plasma displays a very rich structure: one finds Debye screening, plasma oscillations, collective excitations with quark quantum numbers etc. Unfortunately there does not seem to exist a direct (or even indirect) way of detecting these structures in present experiments. Nevertheless it is clear that a good knowledge of these structures is essential if we want to make accurate predictions on possible signals of the plasma as well as on its evolution between formation and hadronization.

From a theoretical point of view, although important progress has been made recently in our understanding of the infrared structure of thermal QCD[38]-[43], much remains to be done before we can really claim to understand fully this IR behaviour. Also it is clear that the perturbative predictions can be at best qualitative since the region in T where it could be trusted ($T/T_c \sim 2$), the coupling constant g is of order one, and there is really no hierarchy of scales.

Finally, for lack of time and/or competence, a number of interesting topics had to be left out. Among these topics one could quote

- the problem of chiral anomalies at finite temperature[57]
- the perturbative calculation of interface tension[58]
- scalar particle emission (axions) from supernovae[59]
- the general theory of transport phenomena in the quark-gluon plasma[48]
- the role of instantons at finite temperature[46],[49].

In any case we hope to have convinced the reader that hot QCD is an exciting topic with a lot of open problems.

Acknowledgements

I am grateful to P. Mitter who gave me the opportunity to deliver these lectures and to F. Guerin, T. Grandou, H. Leutwyler, A. Rebhan and A. Smilga for discussions and comments on the manuscript. I wish also to thank A. Billoire for providing me with up to date reviews of lattice gauge theory and P. Coullet and J. Bettini for their help with T_EX and Illustrator.

References

1 A. Fetter and J.D. Walecka: "Quantum Theory of Many Particle Systems", Mc Graw Hill (1971)
2 V. Silin: Sov. Phys. JETP **11**, 1136 (1960)
3 E. Fradkin: Proc. Lebedev Inst. **29**, 6 (1965)
4 L. Dolan and R. Jackiw: Phys Rev **D9**, 3320 (1974)
5 S. Weinberg: Phys. Rev. **D9**, 3357 (1974)

6 D. Kirzhnits and A. Linde: Sov. Phys. JETP **40**, 628 (1975); A. Linde: Rep. Progr. Phys. **42**, 389 (1979)
7 For a recent review see e.g. A. Satz Plenary Talk at the ECFA Workshop on Large Hadron Colliders, Aachen, CERN preprint TH.5917/90 (1990).
8 For a recent review see e.g. A. Ukawa Plenary talk at the 25th International Conference on High Energy Physics, Singapore, preprint UTHE-P213 (1990)
9 P. Gerber and H. Leutwyler: Nucl. Phys. **B321**, 387 (1989); H. Leutwyler this volume
10 A. Barducci, R. Casalbuoni, S. De Curtis, R. Gatto and G. Pettini: Phys. Rev. **D41**, 1610 (1990)
11 U. Heller and F. Karsch: Nucl. Phys. **B251**, [FS13], 254 (1985); B258, 29 (1985); F. Karsch: CERN preprint TH.4851/87 (1987); F. Karsch: CERN preprint TH.5498/89 (1989) to appear in "Quark Gluon Plasma" R.C. Hwa ed. World Scientific
12 J. Engels et al.: Phys. Lett. **252**, 625 (1990); B. Petersson: invited talk at Quark Matter'90, Menton, France, Bielefeld preprint BI-TP 90/28 (1990)
13 See e.g. G. Parisi "Statistical Field Theory" Addison-Wesley (1988) chap.13
14 G. Baym and N.J. Mermin: Math. Phys. **2**, 232 (1961)
15 For a review of the imaginary time formalism see e.g.: J. Kapusta: "Finite temperature field theory", Cambridge University Press (1989)
16 For a review of the real time formalism, see e.g.: N. Landsman and Ch. van Weert: Phys. Rep. **145**, 142 (1987); this review contains a complete list of references prior to 1987, which have not been systematically quoted in the present article
17 R. Kobes and G. Semenoff: Nucl. Phys. **260**, 714 (1985); Nucl. Phys. **B272**, 329 (1986)
18 R. Kobes: Phys. Rev. **D42**, 562 (1990); University of Winnipeg preprint, to be published in Phys. Rev. (1990)
19 J.C. Collins and M. Perry: Phys Rev Lett. **34**, 1353 (1975); Collins JC. "Renormalization" Cambridge University Press (1984) chap.11.
20 R. Baier, B. Pire and D. Schiff: Phys. Lett. **238**, 367 (1990) ; Bielefeld preprint (1990)
21 T. Appelquist and R. Pisarski: Phys. Rev. **D23**, 2305 (1981)
22 S. Nadkarni: Phys. Rev. **D27**, 917 (1983); R. Alvarez-Estrada: Physica **A158**, 178 (1989)
23 N. Landsman: Nucl. Phys. **B322**, 498 (1989)
24 H.R. Weldon: Phys. Rev. **D26**, 2789 (1982)
25 V. Klimov: Sov. Journ. Nucl. Phys. **33**, 934 (1981); O. Kalashnikov: Forsch. Phys. **32**, 525 (1984)
26 K. Kowalski: Zeit. Phys. **C36**, 665 (1987)
27 H.R. Weldon: Phys. Rev. **D40**, 2410 (1990)
28 V. Lebedev and A. Smilga: Ann. Phys. (N-Y) **202**, 229 (1990); Bern preprint BUTP-90/38 (1990)
29 Fetter-Walecka, ref[1] chap5; Kapusta, ref.15 chap.6
30 Kapusta, ref.15 chap.5
31 H.A. Weldon: Phys. Rev. **D26**, 1394 (1982)
32 K. Kajantie and J. Kapusta: Ann. Phys. (N-Y) **160**, 477 (1985)
33 U. Heinz, K. Kajantie and T. Toimela: Ann. Phys. (N-Y) **176**, 218 (1987)
34 R. Kobes, G. Kunstatter and G. Rebhan: Phys. Rev. Lett. **64**, 2992 (1990); CERN preprint TH.5937/90 (1990)
35 Kapusta, ref.15. chap.4
36 T. Altherr: Phys. Lett. **B238**, 360(1990)
37 H.A. Weldon: Phys. Rev. **D28**, 2007 (1983)
38 R. Pisarski: Nucl. Phys. **B309**, 476 (1988); R. Pisarski: Phys. Rev. Lett. **63**, 1129 (1989); E. Braaten and E. Pisarski: Nucl. Phys. B337, 569 (1990)
39 J. Frenkel and J.C. Taylor: Nucl. Phys. **B334**, 199 (1990)
40 E. Braaten and R. Pisarski: Nucl. Phys. **B339**, 310 (1990)
41 J.C. Taylor and S.M. Wong: Nucl. Phys. **B346**, 115 (1990)
42 E. Braaten and R. Pisarski: Phys. Rev. **D42**, 2156 (1990)
43 E. Braaten, R. Pisarski and R. Yuan: preprint NUHEP-90-1(1990), to be published
44 A.D. Linde: Phys. Lett. **B96**, 289 (1980)
45 T.J. Kinoshita: Math. Phys. **3**, 650 (1962) ; T.D. Lee and M. Nauenberg: Phys. Rev. **B133**, 1549 (1964)
46 D. Gross, R. Pisarski and L. Yaffe: Rev. Mod. Phys. **53**, 43 (1981)
47 G. Baym, H. Monien, C. Pethick and D. Ravenhall: Phys. Rev. Lett. **64**, 1867 (1990)
48 For a review of transport phenomena in the quark-gluon plasma, see e.g. H.T. Elze and U. Heinz: Phys. Rep. **183**, 81 (1989)
49 E. Shuryak: "The QCD Vacuum, Hadrons and the Superdense Matter" World Scientific (1988) chap. 10
50 T. Grandou, M. Le Bellac and J.L. Meunier: Z. Phys. **C43**, 575 (1989)
51 R. Baier, B. Pire and D. Schiff: Phys. Rev. **D38**, 2814 (1988)
52 T. Altherr, P. Aurenche and T. Becherrawy: Nucl. Phys. **B315**, 436 (1989); T. Altherr and T. Becherrawy: Nucl. Phys. **B330**, 174 (1990); T. Altherr and P.Z. Aurenche: Phys. **C45**, 99 (1990)
53 J. Cleymans and I. Dadic: Zeit. Phys. C42, 133 (1989)
54 V. Gabellini, T. Grandou and D. Poizat: Ann. Phys. (N.Y.) **202**, 436 (1990)
55 T. Grandou, M. Le Bellac and D. Poizat: Phys. Lett. **B249**, 478 (1990); INLN preprint 1991/1 to be published in Nucl. Phys. B
56 T. Altherr: Annecy preprint LAPP-Th-325/91(1991)

57 A. Das and A. Karev: Phys. Rev. **D36**, 623 (1987); Y. Liu and G. Ni: Phys. Rev. **D38**, 3840 (1988); C. Contreras and M. Loewe: Zeit. Phys. **C40**, 253 (1988)
58 T. Bhattacharaya, A. Gocksch, C. Korthals-Altes and R. Pisarski: Phys. Rev. Lett. **66**, 998 (1991)
59 J. Ellis and P. Salati: Nucl. Phys. **B342**, 317 (1990); T. Altherr: Zeit. Phys. **C47**, 559 (1990) and LAPP preprint (1990), to be published in Ann. Phys. (N-Y).

Nonexistence of Scattering Theory at Finite Temperature

H. Narnhofer

Institut für Theoretische Physik,
Boltzmanngasse 5, Universität Wien

Abstract

We discuss the concept of scattering theory for finite temperature states. We demand that the Hamiltonian is unitarily equivalent to a free Hamiltonian with respect to quasiparticles. For translation invariant finite temperature states it is shown that if quasiparticles exist, they do not allow any freedom to distinguish between past and future particles and to find a scattering mechanism.

1. Introduction

When we claim that scattering theory does not exist at finite temperature we have first to explain what we demand from a functioning scattering theory. There the physical picture is the following: Initially, we have some particles, sufficiently separated from one another so that we can ignore their interaction and they move freely. In the course of time they approach and influence each other, separate again and move again freely. Evidently, such a mechanism cannot take place in the moment where we have particles densily distributed all over space, because no particle can escape the region of influence of all the others.

We have spoken in the language of classical particles, but quantum theory of finitely many particles does not change this picture – spreading of wave functions is usually a smaller effect than the movement of the particles. Also quantum field theory can be interpreted in this way, only care has to be taken in the description of a particle. We have spoken of the dressed physical particles that do not coincide with the naked particles, described by local fields. The task is to find the connection of the dressed particles with the fields, and this connection depends whether we take for the dressed particles the incoming or the outgoing particles. Their relation is expressed by the scattering matrix.

We could expect that something of this picture of dressed particles can be transfered to finite temperature (and finite density of naked particles). In fact, we are accustomed in solid state physics to talk of phonons, i.e. we consider excitations that have a particle-like behaviour and are called quasiparticles. Nonexistence of scattering theory now states that for homogeneous systems even for these excitations their asymptotic distribution does not change in the course of time. This is connected with the rigidness of our definition, what we consider to be a particle-like excitation. It does not mean that the calculation of Green functions will not give for finite times important insight into the structure of our theory, and the appearance of damping for $t \to \pm\infty$ corresponds to the instability of the particles.

2. Scattering Theory for Finitely Many Particles and in the Ground State of Quantum Field Theory

In nonrelativistic quantum mechanics where particles are stable, we can compare free time evolution given by

$$H_0 = \sum \frac{p_i^2}{2m}$$

with actual time evolution

$$H = \sum \frac{p_i^2}{2m} + \sum v(x_i - x_j) \tag{2.1}$$

by calculating

$$\text{st-}\lim e^{iHt} e^{-iH_0 t} = \Omega_\pm$$

resp.

$$\text{st-}\lim e^{iH_0 t} e^{-2iHt} e^{iH_0 t} = S_0 = \Omega_+^* \Omega_- .$$

Evidently, $f(H)\Omega_\pm = \Omega_\pm f(H_0)$ and therefore

$$[S_0, H_0] = 0. \tag{2.2}$$

We can consider

$$S_0|\varphi_{\text{in}}\rangle = |\varphi_{\text{out}}\rangle \tag{2.3}$$

where especially for $\varphi_{\text{in}}(|k|^2, \Omega) = g(|k|)f(\Omega)$ the corresponding φ_{out} has the same energy distribution $\varphi(k^2)$. Instead we can also work in the Heisenberg picture (we ignore for simplicity bound states) and calculate

$$\text{st-}\lim e^{iHt} e^{-i2H_0 t} e^{iHt} = S_H$$

for which

$$[S_H, H] = 0. \tag{2.4}$$

Further

$$\Omega_{\pm}^{*} H \Omega_{\pm} = H_0 \tag{2.5}$$

(again assuming that there are no bound states), so H is unitarily equivalent to H_0. What remains valid of this formalism in quantum field theory? Again we can write

$$\begin{aligned} H &= H_0 + H_I \\ H_0 &= \int d^3x \Phi \Delta \Phi \\ H_I &= \int d^3x \mathcal{L}_I^0(0,x) d^3x \end{aligned} \tag{2.6}$$

and we think of H_0 and H_I as expressed in terms of canonical fields with the commutation relations (we concentrate on a scalar boson field) [1,2]

$$\begin{aligned} [\Phi(0,x), \Phi(0,y)] &= 0 \\ [\Phi(0,x), \dot{\Phi}(0,y)] &= i\delta^3(x-y) \end{aligned} \tag{2.7}.$$

Everybody knows that for Poincaré invariant field theories such an ansatz fails. Explicit calculations demand for a renormalization procedure. Abstract reasoning gives the famous result of Haag [3].

H and H_0 expressed by the same fields over the same vacuum are only well defined operators if $H_I = 0$. Otherwise stated, if it is possible to compare a field with a free field by unitaries such that

$$V(t)\Phi(x,t)V(t)^{-1} = \Phi_{\text{free}}(x,t) \tag{2.8}$$

then $V \equiv 1$. The commutation relations that fix implicitly the time evolution can only be uniquely (up to unitary equivalence) realized.

This theorem only tells us that we have to be careful with mathematical expressions, but it does not exclude a reasonable scattering. We cannot calculate st-lim $e^{iHt}e^{-iH_0t}$, because e^{-iH_0t} is not defined in the same Hilbert space, but we can construct the analogue of (2.3) with the appropriate assumptions on the momentum energy spectrum. In the Haag-Ruelle Scattering Theory [4,5,6] we assume a discrete mass shell, Buchholz [7] succeeded to construct a scattering theory also for massless particles. Then we can construct subspaces $\mathcal{H}_{\text{in}}, \mathcal{H}_{\text{out}}$ so that every vector in $\mathcal{H}_{\text{in}}$ can be written as

$$|\varphi_{\text{in}}^n\rangle = \prod a^\dagger(f_1) \dots a^\dagger(f_n)|\Omega\rangle \tag{2.9}$$

and every vector in $\mathcal{H}_{\text{out}}$ as

$$|\varphi_{\text{out}}^n\rangle = \prod b^\dagger(f_1) \dots b^\dagger(f_n)|\Omega\rangle, \qquad f_i \in S(\mathbf{R}^3)$$

where $a(f)$, $b(f)$ are annihilation operators of dressed particles. They correspond to $|\varphi_{\text{in}}\rangle$, $|\varphi_{\text{out}}\rangle$ of the finite system but written in the spectral representation of H.

$|\varphi_{\text{in,out}}^n\rangle$ describes n incoming resp. outgoing particles, a, $a^\dagger$, and b, $b^\dagger$ satisfy the commutation relations of free particles and evolve in time like free particles of mass m determined by the mass shell in the spectrum of H. Again we can define an S-matrix in the Heisenberg picture, so $[S, H] = 0$ (this is the only choice, since H_0 does not exist)

$$S|\varphi^n_{\text{in}}\rangle = |\varphi^n_{\text{out}}\rangle \tag{2.10}$$

or

$$Sa^\dagger S^{-1} = b^\dagger.$$

If, in addition, we have $\mathcal{H}_{\text{in}} = \mathcal{H}_{\text{out}} = \mathcal{H}$ (asymptotic completeness), this is, in view of Haag's theorem equivalent to the statement that we can write

$$H = \int d^3k \sqrt{k^2 + m^2} a^\dagger_k a_k = \int d^3k \sqrt{k^2 + m^2} b^\dagger_k b_k.$$

We have two different possibilities to find creation and annihilation operators (in fact, if we were not to insist on a physical interpretation of incoming and outgoing particles even more possibilities cannot be excluded). a_k resp. b_k are not expressed linearly by the local fields, but with the appropriate assumptions on the spectral conditions they will vary in a fairly smooth way as functions of the fields (more precisely they are quasilocal operators).

Finally we have to mention that there is not only the abstract reasoning, why we are allowed to speak of particles and scattering, but we have also the explicit formula of the Lehmann-Symanzik-Zimmermann (LSZ) formalism how to calculate the S-matrix [8]

$$\begin{aligned}\langle\varphi^k_{\text{out}}|\varphi^\ell_{\text{in}}\rangle = (-1)^{k+\ell} \int f^*(\xi_1)\dots f(\eta_\ell) K_{\xi_1}\dots K_{\xi_k} K_{\eta_1}\dots K_{\eta_k} \\ \langle\Omega|T\Phi(\xi_1)\dots\Phi(\eta_\ell)|\Omega\rangle d\xi_1\dots d\eta_\ell.\end{aligned} \tag{2.11}$$

Here K_{ξ_1} is the differential operator $\mathbf{2}_m$ corresponding to the mass of the mass shell (so in calculations the renormalized mass), $f(\xi)$ satisfies $K_\xi f(\xi) = 0$, it corresponds to an asymptotically free particle with wave function f, T is the time ordered product, and $\langle\Omega|\dots|\Omega\rangle$ are the Wightman functionals and have to be calculated for the interacting theory. In practice this can only be done by perturbative methods.

3. Temperature States

For a system of N particles in the finite box Λ we know how to define the temperature state:

$$\omega_{\beta,\Lambda}(A) = Tre^{-\beta H_\Lambda}(A)/Tre^{-\beta H_\Lambda} \tag{3.1}$$

where A is some polynomial in smeared fields and we use second quantization. This works because H_Λ has discrete spectrum so that Tr $e^{-\beta H_\Lambda}$ exists.

For infinite extended systems Tr $e^{-\beta H_\Lambda}$ diverges, but nevertheless hfill $\lim_{\Lambda\to\infty}\omega_{\beta,\Lambda}(A)$ exists (or, at least, has weak limit points). The state we obtain in this way satisfies the famous KMS condition [9,10].

Theorem: Assume that the time evolution is well defined, i.e.

$$\lim_{\Lambda\to\infty} e^{iH_\Lambda t}Ae^{-iH_\Lambda t} = \lim_{\Lambda\to\infty} \tau_{t,\Lambda}A = \tau_t A \tag{3.2}$$

exists, then $\lim_{\Lambda\to\infty}\omega_{\beta,\Lambda}(A\tau_{t,\Lambda}B) = \omega_\beta(A\tau_t B)$ is analytic in t in the strip $\{t; 0 < \mathrm{Im}t < \beta\}$ with the boundary condition

$$\omega_\beta(A\tau_t B) = \omega_\beta(B\tau_{i\beta-t}A). \tag{3.3}$$

Remark: We have not specified in which sense the limit in (3.2) has to exist. In abstract reasoning it is preferred to assume that it exists in norm. The existence in the strong operator topology in the Fock space is not sufficient. This is not only a mathematical delicacy but has an important physical background. In fact, assume we are considering nonrelativistic local interaction that is strictly attractive, otherwise the potential is smooth. This guarantees that for every number of particles N the time evolution is well defined and this suffices to guarantee the existence of (3.2) in the Fock space. But increasing the number of particles the potential energy will decrease like $-N^2$ and accelerate the particles so that the kinetic energy $K \sim N^2$ or the velocity of the individual particle $|v| \sim \sqrt{N}$ [11]. This behaviour was in fact substantiated by computer calculations [12]. What can be done is to regularize the potential to make the interaction stable and show the existence of (3.2) in the norm [13,14]. It is conjectured that stability of the potential ($H_N \geq -AN$, the ground state energy increases proportional with the number of particles) is sufficient for a sound definition of temperature states (with finite particle density) but a proof is so far missing.

It seems tempting to ignore the time dependence and just calculate $\lim\omega_{\beta,\Lambda}(A)$. But doing this, analyticity properties and therefore the KMS condition might get lost [15]. In Poincaré invariant quantum field theory, where we need renormalization procedure already in the ground state, the problem is evidently even more delicate. We must be openminded that the necessary renormalization might be temperature dependent, and we can only hope for the best that the analyticity properties and the KMS condition survive.

Here it seems worthwhile to mention that the KMS condition is not just an æsthetic and mathematically powerful condition that emerges from the ansatz of the Gibbs state. On the contrary, physically plausible conditions on the equilibrium state like adiabatically invariance or the passivity of the state (which is equivalent to the assumption that some aspect of the second law of thermodynamics is satisfied) [16,17,18] lead to the KMS condition in the thermodynamical limit and therefore justify the Gibbs ansatz.

Let us therefore assume that the limit state can be obtained and satisfies the KMS condition. This holds e.g. in relativistic theories for massive free particles. We can find a representation in a Hilbert space though with remarkable differences from the vacuum representation.

Theorem: Let $\omega_\beta(A)$ be a KMS state with respect to the time evolution τ_t, i.e. for A, B smeared polynomials in the fields

$$\omega_\beta(\tau_t AB) = \omega_\beta(B\tau_{i\beta+t}A). \tag{3.3}$$

Then there exists a representation such that

$$\begin{aligned} \omega_\beta(A) &= \langle \Omega|\pi(A)|\Omega\rangle \\ \omega_\beta(A\tau_t B) &= \langle \Omega|\pi(A)e^{iHt}\pi(B)|\Omega\rangle \\ e^{iHt}|\Omega\rangle &= |\Omega\rangle \\ \pi(\tau_t B) &= e^{iHt}Be^{-iHt} \\ \overline{\bigcup \pi(A)|\Omega\rangle} &= \mathcal{H} \end{aligned} \tag{3.4}$$

In contrast to the properties in Fock space we have

$$\pi(A)|\Omega\rangle = \pi(B)|\Omega\rangle \nLeftrightarrow A = B. \tag{3.5}$$

Further

$$e^{-\beta H/2}\pi(A)|\Omega\rangle = J\pi(A)^\dagger|\Omega\rangle \tag{3.6}$$

defines an antiunitary operator J. This operator satisfies

$$[J\pi(A)J, \pi(B)] = 0 \qquad \forall A, B.$$

Evidently $J\pi(A)J \neq \pi(\bar{A})$ for any $\bar{A} \in \mathcal{M}$, so cannot be expressed in terms of field polynomials. It belongs to the commutant of $\mathcal{M} = \{\pi(A)\}''$ (the weak closure of bounded smeared polynomials), $\mathcal{M}' = \{B \in \mathcal{B}(H), [B, \pi(A)] = 0 \,\forall\, \pi(A) \in \mathcal{M}\}$ and J implements an antiisomorphism from $\mathcal{M}''$ into $\mathcal{M}'$. In [19] it is abbreviated by $J\pi(A)J = \pi(A)^\sim$.

H itself cannot be expressed in terms of field polynomials. In some sense it can be expressed in the form

$$H = \lim_{\Lambda\to\infty}(H_\Lambda - JH_\Lambda J). \tag{3.7}$$

If we think of its action on A, $[H, A] \simeq [H_\Lambda, A]$, if A is sufficiently local. Subtracting the corresponding part of the commutant corresponds to the fact that we can only measure energy differences whereas the energy of the whole system is infinite. However, we can only hope

$$\text{w-}\lim(H_\Lambda - JH_\Lambda J)|\Omega\rangle = 0.$$

There is no strong convergence because $\lim \|(1 - e^{-\beta H})H_\Lambda|\Omega\rangle\| \neq 0$ (there is always a boundary term). In fact this implies

$$\text{w-}\lim e^{i(H_\Lambda - JH_\Lambda J)t} \neq e^{iHt}.$$

If it were equal, we had strong convergence and therefore strong resolvent convergence of $H_\Lambda - JH_\Lambda J$.

Nevertheless we can often think of $e^{iHt} \sim e^{i(H_\Lambda - JH_\Lambda J)t}$, only exact reasonings ask for care, since the product of weakly convergent sequences need not converge even weakly.

The theory of the KMS states was first developed in [20] and independently from the mathematical view point in [21,22]. For free fields this construction can be carried through rather explicitly [19,23]. We start with creation and annihilation operators, a_k and $a_k^\dagger$, add $Ja_kJ \equiv b_k$ and $Ja_k^\dagger J \equiv b_k^\dagger$ and perform a Bogoliubov transformation γ, so that we get new creation and annihilation operators $A_{k,1}$, $A_{k,2}$, for which Ω is the vacuum and $\mathcal{H}$ the Fock state. Then (formally)

$$\int d^3k\varepsilon(k)(a_k^\dagger a_k - b_k^\dagger b_k) = \int d^3k\varepsilon(k)(A_{k,1}^\dagger A_{k,1} - A_{k,2}^\dagger A_{k,2})$$

where the separate terms on the left side are not defined but give ∞ expectation value, whereas the right side is well defined.

The fact that Ω is cyclic and separating (3.5) can most easily be understood in the example of a finite system. In the ground state H has eigenvectors $|j\rangle$ with eigenvalues h_j. Then

$$\begin{aligned}\frac{Tre^{-H}A}{Tre^{-H}} &= \frac{\sum e^{-h_j}\langle j|A|j\rangle}{\sum e^{-h_j}} \equiv \langle\Omega|A\otimes\mathbf{1}|\Omega\rangle = \\ &= \frac{\left\langle \sum_j e^{-h_j/2} j\otimes\tilde{j}|A\otimes 1|\sum_k e^{-h_k/2}k\otimes\tilde{k}\right\rangle}{\sum e^{-h_j}}\end{aligned} \tag{3.8}$$

with $\Omega = (\sum e^{-h_j})^{-1/2}|\sum_k e^{-h_k/2} k\otimes\tilde{k}\rangle$ in the Hilbert space $\mathcal{H}\otimes\tilde{\mathcal{H}}$. J acts as $J|k\otimes\tilde{\ell}\rangle = |\ell\otimes\tilde{k}\rangle$.

With $A = \sum a_{k\ell}|k\rangle\langle\ell|$

$$A\otimes 1|\Omega\rangle = \sum_{k\ell} a_{k\ell}|k\rangle\langle\ell|\otimes 1|\sum_j e^{-h_j/2} j\otimes\tilde{j}\rangle = |\sum_{k\ell} a_{k\ell}e^{-h_\ell/2}k\otimes\tilde{\ell}\rangle.$$

The same vector can be obtained by $1\otimes\sum a_{k\ell}e^{-(h_\ell-h_k)/2}|\ell\rangle\langle k| = hfill = 1\otimes e^{H/2}A^t e^{-H/2}$. Therefore A is uniquely determined by $A|\Omega\rangle$ and we have also found $JA\otimes 1J = 1\otimes A^\dagger$. H now is given by $\sum h_j(|j\rangle\langle j|\otimes 1 - 1\otimes|\tilde{j}\rangle\langle\tilde{j}|)$.

In the equilibrium state we can perform perturbation theory with respect to the interaction as one is used from the ground state, first considering local perturbations and then removing the local cut-off. In fact, this stability under perturbations can be used as a criterium of an equilibrium state and leads under some assumptions to the KMS condition. We assume that the unperturbed time evolution is implemented by e^{iHt}, where $H\Omega = 0$ ($H \notin \mathcal{M}''$). We add a local perturbation $V_\Lambda \in \mathcal{M}$ and calculate

$$\omega_{\beta,\Lambda} = \frac{\langle\Omega|e^{H/2}e^{-(H+V_\Lambda)/2}Ae^{-(H+V_\Lambda)/2}e^{H/2}|\Omega\rangle}{\langle\Omega|e^{H/2}e^{-(H+V_\Lambda)}e^{H/2}|\Omega\rangle}.$$

Though $e^{H/2}$ is an unbounded operator, $e^{H/2}e^{-(H+V_\Lambda)/2}$ can be expanded as a power series in V_Λ, and if V_Λ is bounded, the KMS condition guarantees that the power series converges and the limit belongs to $\mathcal{M}$. Using the KMS condition, we can also calculate

$$\langle \Omega | e^{H/2} e^{-(H+V_\Lambda)/2} A e^{-(H+V_\Lambda)/2} e^{H/2} | \Omega \rangle = \langle \Omega | A e^{-(H+V_\Lambda)} e^{H} | \Omega \rangle .$$

If we perform now a limit $\Lambda \to \infty$ or remove additional cut-offs, then the denominator will go to ∞ or 0. This means we are leaving the space of the representation: we have no operator convergence, neither in the weak nor in the strong sense, we can only rely on the expectation value and after the performance of the limit construct with them a new representation and check whether the new state satisfies the KMS condition with respect to the new time evolution: It will, again under assumptions on the convergence of the time evolution.

4. Nonexistence of Scattering Theory in Temperature States

We can ask two questions in this context: Does there exist an analogue to the LSZ formalism? Do there exist creation and annihilation operators so that we can write $H = \int d^3k \varepsilon(k) a_k^\dagger a_k$ and if so, do there exist other $b_k^\dagger$, b_k with $H = \int d^3k \varepsilon(k) b_k^\dagger b_k$ such that a_k and b_k are connected by a scattering mechanism?

We examine the LSZ formula under the standing assumption that the system is translation invariant. Evidently K_ξ should be replaced by a differential operator representing $\varepsilon(k)$ (remember that Poincaré invariance is broken as Galilei invariance is broken for nonrelativistic temperature states, so we have more freedom for $\varepsilon(k)$). Again $K_\xi f(\xi) = 0$ gives a restriction on the chosen f. $\langle \Omega | T\Phi(\xi_1) \ldots | \Omega \rangle$ has to be replaced by the temperature Green function that according to our last remarks should be calculated by perturbational methods. In order to get something $\neq 0$ it must be forbidden to transfer $K_\xi \langle \Omega | T\Phi(\xi_1) \ldots | \Omega \rangle$ by partial integration on $f(\xi)$. This means that $\langle \Omega | \Phi(0) e^{i(Ht-Px)} \Phi(0) \ldots | \Omega \rangle$ must contain a singularity in the joint spectrum of H and P.

Let us examine the spectral properties of H and P: Since the state is space and time translation invariant, H and P are well defined by

$$e^{i(Ht-Px)} \pi(A) | \Omega \rangle = \tau_t \sigma_x \pi(A) | \Omega \rangle . \tag{4.1}$$

$|\Omega\rangle$ is an eigenvector with eigenvalue 0 for both H and P. Apart from this eigenvector the spectrum of H and P is continuous and equals $\mathbf{R}$ resp. $\mathbf{R}^3$. (Notice that H in (3.7) is not bounded from below any more which corresponds to the physical fact that our system represents an infinite energy reservoir.) H and P commute, so they have a joined spectral resolution and we have the chance of a spectral singularity

$$d\sigma_{\rm sing}(\omega, k) = \delta(\varepsilon(k) - \omega) d\omega dk. \tag{4.2}$$

Apart from this singular part which we expect to correspond to quasiparticles we will also have a continuous part, and it covers in the temperature state $\mathbf{R} \times \mathbf{R}^3$. The projection operator on the singular part of the Hilbert space P_+ is orthogonal to $|\Omega\rangle\langle\Omega|$ and by assumption $\neq 0$. $\pi(A)|\Omega\rangle$ spans the Hilbert space, so there exists some A (we suppress in the following the π) such that

$$P_+ A|\Omega\rangle \neq 0.$$

We can consider $A(x,t) = \sigma_x \tau_t A$ and perform its Fourier transform denoted by $\tilde{A}(k,\omega)$. Now we smear with $f \in S$

$$\int f(x,t)A(x,t)dxdt = \int \tilde{f}(k,\omega)\tilde{A}(k,\omega)dkd\omega$$

such that $A(k,\omega)dkd\omega$ becomes an operator valued measure.

Theorem: [24] Take some $\tilde{h} \in \mathcal{D}(\mathbf{R}^1)$ a smooth function vanishing at infinity with $\tilde{h}(0) = 1$. Define

$$A_{Z,\pm}(f) = \int \tilde{h}(Z(\omega \mp \varepsilon(k))\tilde{f}(k)\tilde{A}(k,\omega)d\omega dk. \tag{4.3}$$

Then

$$n-\lim_{Z\to+\infty} A_{Z,\pm}(f)|\Omega\rangle$$

exists and is independent of the special $\tilde{h}$.

Remark: $\tilde{A}(k,\omega)\Omega$ is a vector valued distribution that by assumption is not orthogonal to P_+. The function $\tilde{h}$ concentrates more and more on $\omega = \pm\varepsilon(k)$. $\pm$ corresponds to the fact that due to the KMS condition a singularity at $\omega = \varepsilon(k)$ has its counterpart at $\omega = -\varepsilon(k)$.

So far we have copied the idea of Buchholz [7] how to show the existence of scattering theory for massless particles where the spectrum corresponding to the singular particle state is imbedded into a continuous spectrum and has to be filtered out. For this single particle state there is no distinction between past and future, a single particle is unscattered. The next step is to pass from the single particle state to the corresponding creation operator that enables us to construct states with several particles. Here our analysis has to depart from [7].For temperature states the essential point is that due to (3.5) the convergence to a single particle state is equivalent to

Theorem: There exist densily defined operators $a_\pm(f)$ affiliated to $\mathcal{M}''$ such that

$$\text{st}-\lim_{Z\to\infty} A_{Z,\pm}(f) = a_\pm(f)$$

and these operators are unique. $a_\pm(f)$ depends linearly on f and further

$$e^{iHt}a_\pm(f)e^{-iHt} = a_\pm(e^{\mp i\varepsilon(-i\nabla)t}f).$$

Proof: Vectors of the form $JBJ\Omega$, $B \in \mathcal{M}''$ are dense in $\mathcal{H}$, and we define

$$a_\pm(f)JBJ|\Omega\rangle = \lim_{Z\to\infty} A_{Z,\pm}(f)JBJ|\Omega\rangle = \lim_{Z\to\infty} JBJA_{Z,\pm}(f)|\Omega\rangle.$$

The statement that $a_\pm(f)$ is affiliated to $\mathcal{M}''$ means that $[a_\pm(f), JBJ] = 0$ $\forall\ B \in \mathcal{M}$. Further the operator is closable, because its adjoint is also densily defined. We do not know whether it is bounded (and then belongs to $\mathcal{M}''$) and we cannot expect a general proof, because this would forbid bosonic creation and annihilation operators.

We recall once more that there is no time dependence and the δ-type singularity only allows one type of fields due to (3.5). This analysis does not tell us how to find the appropriate $\varepsilon(k)$. Starting from $\varepsilon(k) = k^2$ perturbation theory will lead to a decay of the quasiparticle, but a new $\varepsilon(k)$ corresponding to a time dependent quasiparticle, a phonon, might emerge.

At this point there remains the question whether such δ-type energy excitations really have a particle-like structure, i.e. can be added in some sense. This cannot be done without further assumptions.

First possibility: We are inspired by the LSZ formalism and demand that for the n-point Green functions the singularities in the joined $(H - P)$ spectrum are determined by the two-point singularities. There are some plausibility arguments in support of this assumption and in some sense it only copies the Wightman axioms [24]. Under these assumptions it follows that the singularity in the spectrum really corresponds to a particle structure and it was proven in [24] that the corresponding truncated Green functions have to be trivial. Thus the quasiparticles behave like free particles.

Second possibility: Here our arguments are closer to Haag-Ruelle scattering theory: First we demand that the A from which we start has strictly compact positive (resp. negative) support in the energy. Then we are allowed to construct the creation and annihilation part out of $a_\pm$ in a way that is inspired from the Bogoliubov transformation used in [23]. This construction is fairly technical and does not improve the understanding of the mechanism, so we do not reproduce it here and we refer to [24].

More transparent is the next assumption that is needed.

$$D_{A,B}(x,t) = [A(x,t), B]_\pm||^{1/2}$$

should be a function integrable in x for fixed t, and its Fourier transform (in both variables) should have at most a δ-type singularity at $\varepsilon(k) = \omega$ (and in fact will have one for the appropriate A). Then we can prove that for the quantities in (4.3)

$$\lim_{Z\to\infty} [A[B_Z, C_Z]_\pm] = 0.$$

Thus the commutator (or anticommutator) for the limiting operators will be a c-number like for creation and annihilation operators. Furthermore one can show that

$$\lim_{Z\to\infty} \omega_4^t(B_{1Z}B_{2Z}B_{3Z}B_{4Z}) = 0$$

where ω_4^t are the truncated functions. The B_{jZ} are chosen such that they converge to some $a(f_j)$. Therefore the state is a quasifree state over the quasi-particles.

This is of course also true for the dressed past or future particles in the ground state. The difference in our analysis is that our construction does not refer to a time direction and therefore gives no freedom to find a transition of past particles to future particles, i.e. our scattering matrix is trivial. The origin of this triviality is (3.5) since one particle does not scatter, $a_{\text{in}}|\Omega \rightarrow\rangle = a_{\text{out}}|\Omega \rightarrow\rangle$ and therefore $a_{\text{in}} = a_{\text{out}}$. Physically this means that if a quasiparticle goes through the thermal background without any damping then it is also not affected by another quasiparticle.

It should be noted that there exist attempts to reproduce ideas of scattering theory also for temperature states. We refer e.g. to [25]. If there is no singular part in the joint spectrum of H and P, we can nevertheless adjust to it an operator that will decay in the course of time according to Riemann-Lebesgue, but by a proper choice of this operator this decaying can be very slow. If we do not insist that this operator belongs to $\mathcal{M}$, we are even able to find an operator in $\mathcal{B}(\mathcal{H})$ that satisfies creation and annihilation commutation relation. Ω is not separating for $\mathcal{B}(\mathcal{H})$ and the above argument does not tell $A_{\text{in}} = A_{\text{out}}$ for these operators. But then we lose the possibility to give a physical interpretation to these excitations, and this was in fact our purpose.

As a final remark we want to emphasize that our investigations were not inspired so much by the desire to construct and discuss a scattering theory for a relativistic quantum field theory at finite temperature but by the remarkable fact that the ansatz of particle-like excitations of the energy-momentum spectrum, i.e. to consider phonons and forget electrons and nucleons, is so extremely successful in solid state physics. So far we reduced this fact to a spectral condition (up to technical assumptions). To justify that such a spectral condition follows from deeper principles (like ergodic behaviour of space and time translation) is one of the challenging tasks for future research, since we have seen that this must be a nonperturbative phenomenon.

Acknowledgement: It is a pleasure to thank W. Thirring for critical discussions on the manuscript and for presenting this talk on my behalf at the Schlading Winter School.

References

1 R. Haag: Local Quantum Physics, Springer, to be published in 1992
2 C. Itzykson, J.B. Zuber: Quantum Field Theory Mc Graw Hill, New York (1980)
3 R. Haag: Dan Mat. Fys. Medd. **29/12** (1955)
4 D. Ruelle: Helv. Phys. Acta **35**, 147 (1982)
5 R. Haag: Phys. Rev. **112**, 669 (1958)
6 M. Reed, B. Simon: Methods of Modern Mathematical Physics III, Academic Press, New York (1979)

7 D. Buchholz: Commun. Math. Phys. **42**, 269 (1975); Commun. Math. Phys. **52**, 147 (1977)
8 H. Lehmann, K. Symmanzik, W. Zimmermann: Il Nuovo Cim. **1**, 206 (1955)
9 R. Kubo: J. Phys. Soc. Jap. **12**, 570 (1957) D.C. Martin, J. Schwinger: Phys. Rev. **115**, 1342 (1959)
10 L.P. Kadanoff, G. Baym: Quantum Statistical Mechanics, Benjamin, New York (1962)
11 F.J. Dyson: Phys. Rev. **85**, 631 (1952)
12 H. Posch, H. Narnhofer, W. Thirring: Phys. Rev.A **42**, 1880 (1990)
13 R.F. Streater: Commun. Math. Phys. **7**, 93 (1968) R.F. Streater, I.F. Wilde: Commun. Math. Phys.**17**, 21 (1970)
14 H. Narnhofer, W. Thirring: Phys. Rev. Lett. **64**, 1863 (1990)
15 H. Narnhofer: Acta Phys. Austr. **49**, 207 (1978)
16 H. Narnhofer, W. Thirring: Phys. Rev. A **26/6**, 3646 (1982)
17 R. Haag, D. Kastler, E. Trych-Pohlmeier: Commun. Math. Phys. **38**, 173 1974)
18 W. Pusz, S.L. Woronowicz: Commun. Math. Phys. **58**, 273 (1978)
19 H. Umezawa, H. Matsumoto, M. Tachiki: Thermo Field Dynamics and Condensed States, North Holland, Amsterdam, (1982)
20 R. Haag, N.M. Hugenholtz, M. Winnink: Commun. Math. Phys. **5**, 215 (1967)
21 M. Tomita: Quasi Standard von Neumann Algebras, mimeographed lectures (1967)
22 M. Takesaki: Tomita's Theory of Modular Hilbert Algebras and its Applications, Springer, Berlin (1970)
23 H. Narnhofer: Rep. Math. Phys. **16**, 1 (1979)
24 H. Narnhofer: M. Requardt, W. Thirring, Commun. Math. Phys. **92**, 247 (1983)
25 N.P. Landsmann: Ann. of Physics **186**, 141 (1988)

Lecture Notes in Physics

For information about Vols. 1–365
please contact your bookseller or Springer-Verlag

Vol. 366: M.-J. Goupil, J.-P Zahn (Eds.), Rotation and Mixing in Stellar Interiors. Proceedings, 1989. XIII, 183 pages. 1990.

Vol. 367: Y Osaki, H. Shibahashi (Eds.), Progress of Seismology of the Sun and Stars. Proceedings, 1989. XIII, 467 pages. 1990.

Vol. 368: L. Garrido (Ed.), Statistical Mechanics of Neural Networks. Proceedings. VI, 477 pages. 1990.

Vol. 369: A. Cassatella, R. Viotti (Eds.), Physics of Classical Novae. Proceedings, 1989. XII, 462 pages. 1990.

Vol. 370: H.-D. Doebner, J.-D. Hennig (Eds.), Quantum Groups. Proceedings, 1989. X, 434 pages. 1990.

Vol. 371: K.W. Morton (Ed.), Twelfth International Conference on Numerical Methods in Fluid Dynamics. Proceedings, 1990. XIV, 562 pages. 1990.

Vol. 372: F. Dobran, Theory of Structured Multiphase Mixtures. IX, 223 pages. 1991.

Vol. 373: C. B. de Loore (Ed.), Late Stages of Stellar Evolution. Computational Methods in Astrophysical Hydrodynamics. Proceedings, 1989. VIII, 390 pages. 1991.

Vol. 374: L. Ting, R. Klein, Viscous Vortical Flows. V, 222 pages. 1991.

Vol. 375: C. Bartocci, U. Bruzzo, R. Cianci (Eds.), Differential Geometric Methods in Theoretical Physics. Proceedings, 1990. XIX, 401 pages. 1991.

Vol. 376: D. Berényi, G. Hock (Eds.), High-Energy Ion-Atom Collisions. Proceedings, 1990. IX, 364 pages. 1991.

Vol. 377: W. J. Duschl, S.J. Wagner, M. Camenzind (Eds.), Variability of Active Galaxies. Proceedings, 1990. XII, 312 pages. 1991.

Vol. 378: C. Bendjaballah, O. Hirota, S. Reynaud (Eds.), Quantum Aspects of Optical Communications. Proceedings 1990. VII, 389 pages. 1991.

Vol. 379: J. D. Hennig, W. Lücke, J. Tolar (Eds.), Differential Geometry, Group Representations, and Quantization. XI, 280 pages. 1991.

Vol. 380: I. Tuominen, D. Moss, G. Rüdiger (Eds.), The Sun and Cool Stars: activity, magnetism, dynamos. Proceedings, 1990. X, 530 pages. 1991.

Vol. 381: J. Casas-Vazquez, D. Jou (Eds.), Rheological Modelling: Thermodynamical and Statistical Approaches. Proceedings, 1990. VII, 378 pages. 1991.

Vol. 382: V.V. Dodonov, V. I. Man'ko (Eds.), Group Theoretical Methods in Physics. Proceedings, 1990. XVII, 601 pages. 1991.

Vol. 384: M. D. Smooke (Ed.), Reduced Kinetic Mechanisms and Asymptotic Approximations for Methane-Air Flames. V, 245 pages. 1991.

Vol. 385: A. Treves, G. C. Perola, L. Stella (Eds.), Iron Line Diagnostics in X-Ray Sources. Proceedings, Como, Italy 1990. IX, 312 pages. 1991.

Vol. 386: G. Pétré, A. Sanfeld (Eds.), Capillarity Today. Proceedings, Belgium 1990. XI, 384 pages. 1991.

Vol. 387: Y. Uchida, R. C. Canfield, T. Watanabe, E. Hiei (Eds.), Flare Physics in Solar Activity Maximum 22. Proceedings, 1990. X, 360 pages. 1991.

Vol. 388: D. Gough, J. Toomre (Eds.), Challenges to Theories of the Structure of Moderate-Mass Stars. Proceedings, 1990. VII, 414 pages. 1991.

Vol. 389: J. C. Miller, R. F. Haglund (Eds.), Laser Ablation-Mechanisms and Applications. Proceedings. IX, 362 pages, 1991.

Vol. 390: J. Heidmann, M. J. Klein (Eds.), Bioastronomy - The Search for Extraterrestrial Life. Proceedings, 1990. XVII, 413 pages. 1991.

Vol. 391: A. Zdziarski, M. Sikora (Eds.), Ralativistic Hadrons in Cosmic Compact Objects. Proceedings, 1990. XII, 182 pages. 1991.

Vol. 392: J.-D. Fournier, P.-L. Sulem (Eds.), Large-Scale Structures in Nonlinear Physics. Proceedings. VIII, 353 pages. 1991.

Vol. 393: M. Remoissenet, M.Peyrard (Eds.), Nonlinear Coherent Structures in Physics and Biology. Proceedings. XII, 398 pages. 1991.

Vol. 394: M. R. J. Hoch, R. H. Lemmer (Eds.), Low Temperature Physics. Proceedings. XXX,XXX pages. 1991.

Vol. 395: H. E. Trease, M. J. Fritts, W. P. Crowley (Eds.), Advances in the Free-Lagrange Method. Proceedings, 1990. XI, 327 pages. 1991.

Vol. 396: H. Mitter, H. Gausterer (Eds.), Recent Aspects of Quantum Fields. Proceedings. XIII, 332 pages. 1991.

New Series m: Monographs

Vol. m 1: H. Hora, Plasmas at High Temperature and Density. VIII, 442 pages. 1991.

Vol. m 2: P. Busch, P. J. Lahti, P. Mittelstaedt, The Quantum Theory of Measurement. XIII, 165 pages. 1991.

Vol. m 3: A. Heck, J. M. Perdang (Eds.), Applying Fractals in Astronomy. IX, 210 pages. 1991.

Vol. m 4: R. K. Zeytounian, Mécanique des fluides fondamentale. XV, 615 pages, 1991.

Vol. m 5: R. K. Zeytounian, Meteorological Fluid Dynamics. XI, 346 pages. 1991.

MIX
Papier aus verantwortungsvollen Quellen
Paper from responsible sources
FSC® C105338

If you have any concerns about our products,
you can contact us on
ProductSafety@springernature.com

In case Publisher is established outside the EU,
the EU authorized representative is:
Springer Nature Customer Service Center GmbH
Europaplatz 3, 69115 Heidelberg, Germany

Printed by Libri Plureos GmbH
in Hamburg, Germany